Science and Technology Studies
Mario Bunge, Series Editor

Philosophy of Science, Volume I : From Problem to Theory,
Mario Bunge

Philosophy of Science, Volume 2: From Explanation to Justijication,
Mario Bunge

Critical Approaches to Science and Philosophy,
edited by Mario Bunge

Microfoundations, Method, and Causation: On the Philosophy of the Social Sciences, Daniel Little

Complexity: A Philosophical Overview,
Nicholas Rescher

A Dictionary of Biology,
M. Abercrombie, C.J. Hickman, and M.L. Johnson

The Prevention of Cancer: Pointers from Epidemiology,
Richard Doll

Biotechnology: Between Commerce and Civil Society,
Nico Stehr, editor

Philosophy of Science: The Historical Background,
Joseph J. Kockelmans, editor

REVISED EDITION

VOLUME TWO

Philosophy of SCIENCE

FROM EXPLANATION TO JUSTIFICATION

MARIO BUNGE

LONDON AND NEW YORK

Original edition published in 1967 by Springer-Verlag.

Published 1998 by Transaction Publishers

Published 2017 by Routledge
2 Park Square, Milton Park, Abingdon, Oxon OX14 4RN
605 Third Avenue, New York, NY 10017

Routledge is an imprint of the Taylor & Francis Group, an informa business

Library of Congress Catalog Number: 97-22359

Library of Congress Cataloging-in-Publication Data

Bunge, Mario Augusto.
Philosophy of science / Mario Augusto Bunge.—Rev. ed.
p. cm.—(Science and Technology Studies)
Rev. ed. of: Scientific research. Berlin, New York: Springer-Verlag, 1967.
Includes bibliographical references and index.
Contents: v. 1.From problem to theory—v. 2. From explanation to justification.
ISBN 0-7658-0415-8 (set : pbk.: alk. paper)—ISBN 0-7658-0413-1 (vol. 1 : pbk.: alk. paper)—ISBN 0-7658-0414-X (vol. 2 : pbk.: alk. paper)
1. Research—Methodology. 2. Science—Philosophy. II. Title. III. Series.

Q180.55.M4B64 1998
001.4'.—DC21 97-22359
CIP

ISBN 13: 978-0-7658-0414-3 (vol. 2) (pbk)
ISBN 13: 978-0-7658-0415-0 (set)

Preface

This volume is a logical sequel of Volume 1, *From Problem to Theory*; indeed, it concerns the ways theoretical systems are put to work and subjected to test. Yet it can be read independently by anyone familiar with some factual theories, referring back to Volume 1 when necessary.

Special Symbols

$A \subseteq B$	the set *A* is included in the set *B*
$A \cup B$	the union of the sets *A* and *B*
$A \cap B$	the common part of the sets *A* and *B*
$a \in A$	the individual *a* as in (or belongs to) the set *A*
Card (*A*)	cardinality (numerosity) of the set *A*
$A \times B$	Cartesian product of the sets *A* and *B*
$Cn(A)$	consequence(s) of the set *A* of assumptions
$=_{df}$	equals by definition
Df.	definition
$(\exists x)$	some *x* (or there is at least one *x* such that)
e	empirical datum
e^*	translation of *e* into a semiempirical, semitheoretical language
h	hypothesis
$m(\dot{r})$	measured value of the degree $\dot{r}$
$\bar{m}(\dot{r})$	average (or mean) value of a set of measured values of $\dot{r}$
$P \dashv T$	*T* presupposes *P*
p, q	arbitrary (unspecified) propositions (statements)
$P(x)$	*x* has the property *P* (or *x* is a *P*)
$\{x \mid P(x)\}$	set of the *x* such that every *x* is a *P*
$p \vee q$	*p* and/or *q* (inclusive disjunction)
$p \,\&\, q$	*p* and *q* (conjunction)
$p \rightarrow q$	if *p*, then *q* (conditional or implication)
$p \leftrightarrow q$	*p* if and only if *q* (biconditional or equivalence)
Σ_i	sum over *i*
t	theorem, testable consequence
t^*	translation of *t* into a semiempirical, semitheoretical language
T	theory
$A \vdash t$	*A*, therefore *t* (or *A* entails *t*, or *t* follows logically from *A*)
$\emptyset$	the empty set
U	the universal set
x	arbitrary (unspecified) individual
(x)	for every *x*
$\langle x, y \rangle$	ordered pair of the elements *x* and *y*

Contents

Part III—Applying Scientific Ideas

Part IV—Testing Scientific Ideas

Part III

Applying Scientific Ideas

In Part II we have analyzed the process that starts with questioning, proceeds with conjecturing, and ends up by building syntheses of hypotheses, i.e. theories. We studied, in particular, both the form and the content or reference of factual theories, i.e. of systems of hypotheses supposed to mirror — symbolically and sketchily — pieces of reality. But so far we have hardly inquired into the use of theories. This will be done presently. On the other hand Part IV will be concerned with the problem of the adequacy of hypotheses and theories.

Factual theories are built chiefly in order to explain, predict, plan, or act. Explanation incorporates initially isolated items into a unified body of knowledge and clams an intellectual itch — alas, only temporarily. Prediction and retrodiction relate the present to the future and the past and thereby serve to test theories. Combined, explanation and prediction take part in the rational planning and execution of actions — in particular, those occurring in scientific experience.

Theories can then be applied to cognitive as well as to practical aims. The cognitive applications of theories — i.e., explanations, predictions, and the planning of experiments — precede their practical applications. Indeed, before we do something in a rational way with a practical aim we must know what it is (description), understand why it should be so (explanation), find out whether it is nearly as our theoretical model says (testing), and reckon how it is likely to behave (prediction). Accordingly rational action, whether in industry, government, science, or private life, is action based on scientific knowledge both theoretical and experimental. Unlike spontaneous action, rational action is just the consummation of a choice made in the light of tested theories, or the execution of a plan built with their help.

Let us proceed to study three chief applications of scientific theory: explanation, prediction, and the action envisaged by the various technologies. A fourth application of theories, namely to the design and interpretation of empirical procedures, will be taken up in Part IV.

9

Explanation

The main rationale for the invention and test of hypotheses, laws, and theories, is the solution to why-problems, i.e. the explanation of facts and their patterns. We do not rest content with finding facts, but wish to know why they should occur rather than not; and we do not even rest content with establishing explanatory constant conjunctions of facts, but attempt to discover the mechanism explaining such correlations.

9.1. Answering Whys

Explanations are answers to why-questions. A why-question is a question of the form 'Why *q*?', where '*q*' stands for whatever is to be explained (see Sec. 4.2). This, the problem generator and the object of explanation, is called the *explanandum*: that which is to be explained. The explanandum may be a fact (or, rather, a proposition expressing a fact), a law, a rule, a precept, or even a whole theory. It may but need not be something unusual or conflicting with prior belief: in science, at least, what calls for an explanation is the *isolated* item, which may well be a familiar phenomenon or a common sense generalization. Explanation is, indeed, a kind of systematization.

A *rational,* cogent or grounded answer to "Why *q*?" will be of the form "*q* because *p*", where *p*, a complex formula, is the reason for *q*. Since "*q* because *p*" is the same as "*p*, therefore *q*", the reason, ground or explainer of *q* is a formula that entails it, i.e., such that "$p \rightarrow q$" is formally true. The reason or ground, usually a set of formulas, is called the *explanans*: that which explains. A *nonrational* answer to

'Why q?' will, by contrast, provide no reason for q other than q itself: it will often consist either in an assertion logically irrelevant to the explanandum q, as in 'q because I say so', or in a sentence such as 'q because q'. We shall accept only rational answers to well-formulated why-questions; any such answer will logically entail the generator of the question, i.e. it will be such that *Explanans*$\vdash$*Explanandum*. An argument providing a rational answer to a well-formulated why-question will be called a rational explanation. (For $\vdash$ see Sec. 7.3.)

For example, an answer to the question 'Why did *c* burn that draft?' might be '*c* burnt that draft because that draft was wrong'. The explanandum and problem generator is here "*c* burnt that draft". But the proposition "That draft was wrong'', given as a reason, (does not entail by itself the explanadum and cannot accordingly be the full explanans. In short, pointing to the additional information "That draft was wrong" is no sufficient ground for "*c* burnt that draft". A satisfactory explanation will require an additional premise such as, e.g., the generalization "*c* burns all wrong drafts". A rational explanation of the fact that c burnt that draft (i.e., a complete answer to the question 'Why did *c* burn that draft?') will then be the following:

c burns all wrong drafts (*Generalization*) That draft was wrong (*Circumstance*)	} *Explanans*
c burnt that draft (*Fact being explained*)	*Explanandum*

Note that the explanans has been split into a general and a singular proposition and that the explanandum is a logical consequence of them. (In general, the explanans may contain a number of generalizations and a number of data.) In other words, a rational explanation of a fact involves the *subsumption* of the explanandum under one or more *generalizations* via information concerning the *circumstances* accompanying the fact that is being explained. (In the above example the explanans contained an empirical generalization. The logical structure of the explanation of a fact would be the same if a law statement proper occurred in the explanans instead of an empirical generalization.) Consequently a pile of data, no matter how numerous and precise, will have no explanatory power: only logically strong formulas can provide explanations if conjoined with data.

The logical relation between the full explanans or ground and the explanandum or object is

$\{Generalization(\text{s}), Circumstance(\text{s})\} \vdash Explanandum$ [9.1]

or $\{G, C\} \vdash E$ for short. We may read this formula '*G* and *C*, therefore *E*', or '$\{G, C\}$ is an explanans for *E*'. Quite often the deduction will consist in a mere specification, i.e. in the replacement of a variable by a constant. Ex. 1: {All *A* are *B*, *c* is an *A* }$\vdash$*c* is a *B*. Ex. 2: $\{pV= 1$ litre-atmosphere, $p=2$ atmospheres$\} \vdash V=\frac{1}{2}$ litre.

*The situation is slightly different, yet substantially the same, when a rule or a precept rather than a law is invoked to explain something. In fact, suppose we tried to justify the burning of the draft in question by invoking the norm or precept: "All wrong drafts must be [or ought to be] burnt". The first step would be this:

All wrong drafts must be burnt (*Rule*) That draft was wrong (*Circumstance*)	} *Explanans*
That draft had to be burnt	*Mediating explanandum*

We are still half-way from the given explanandum, "*c* burnt that draft". In order to derive it we must supply an additional premise referring to *c*'s habits or principles, i.e. some generalization concerning *c*. Something like "Whatever has to be done is done by *c*" will do. In fact this new premise, jointly with the above consequence, yields the desired conclusion:

Whatever has to be done is done by *c* (*Generalization*) That draft had to be burnt (*Mediating explanandum*)	} *Mediating Explanans*
That draft was burnt by *c*	*Explanandum*

When the explanation consists in the subsumption under a set of rules, the pattern is somewhat more complex but still one of deducibility, namely, the chain

$\{Rule(s), Circumstance\ (s)\} \vdash Mediating\ explanand\ um$
$\{Generalization(s), Mediating\ explanandum\} \vdash Explanandum.$ [9.2]

*Explanations with the help of rules are typical though not exclusive of technology (applied science), whereas explanations in terms of laws are typical of pure science. What is typical and exclusive of technology is the explanation in terms of grounded rules, i.e. of rules based on laws rather than rules of thumb (see Sec. 11.2). But in either case we are given an explanandum (the question generator) and we search for a set of explanans premises that will entail the given

explanandum. Accordingly, a fuller statement of the question *Why q?* is *What is the (set of premises) p entailing q?* We shall see in a while, though, that there is no single set of explainer premises.*

There are different kinds of explanation according to the nature of the explanans premises. Thus, e.g., an explanation in terms of causal laws will not be the same as an explanation in terms of stochastic laws. But such differences in content are beyond the scope of logic: all rational explanations, even the wrong ones, have the same logical structure. In particular, an explanation in terms of stochastic or probabilistic laws (such as Maxwell's velocity distribution law) is not probable itself, just as reasonings in probability theory are, when valid at all, strictly deductive hence conclusive and not just likely. Recall that formal logic is not concerned with the content of the premises and conclusions of an argument but only with their formal relations: for the explanandum to follow logically from the explanans it is necessary and sufficient that the conditional "Explanans→Explanandum" be a logical truth.

The above discussion will be summed up in the following *Definition 1: A rational explanation* of a formula q is an answer to the question 'Why q?', consisting in an argument showing that q follows logically from a set of generalizations and/or rules, and data, not containing q. Notice, in the first place, that this definition refers to formulas (propositions and propositional formulas). It does not refer directly to facts because explanation is a conceptual operation and facts are not affected by just thinking about them—unless the facts concerned happen to be thoughts. The phrase 'explanation of facts' is elliptical and must not be taken literally: indeed, to explain a fact is nothing but to explain a proposition describing a fact. We shall continue writing about explaining facts but shall construe the expression in the sense just indicated. Secondly, our definition leaves room for more than one explanation of each formula: it speaks of *an* answer rather than of *the* answer to a why-question. There are several reasons for this: (i) one and the same fact may be described in a number of ways, hence a whole set of propositions may refer to one and the same fact and every one of them may deserve an explanation; (ii) even if there were a one-to-one correspondence between facts and factual propositions, explanations would be multiple because any given formula is deducible from infinitely many sets of premises; (iii) we may be interested in different aspects of one and the same question or in different levels of explanation, corresponding to a different level of analysis each; (iv) explanations may vary alongside the body of knowledge in which they take place. To sum up, in factual subject matters there is no unique and final explanation.

Thirdly, our definition rules that the explanans shall not contain the explanandum. This is intended to eliminate answers of the form 'q because q' and 'q because p and q', which are circular, hence unenlightening. Unlike the relation of deducibility, the explanans-explanandum one is nonreflexive. We shall come back to this point in the next section. Fourthly, according to our definition an explanation is not a formula—such as "$G\&C \rightarrow E$", which is implied by "$\{G, C\} \vdash E$"—but a sequence of formulas making up a valid argument. This does not entail the converse, i.e. that every deductive argument is explanatory. An explanation requires, in addition to deduction, (i) a previous why-question (pragmatic ingredient) and (ii) an explanans made up of at least one general formula (e.g., a law or a rule) and a singular proposition (a datum). In short, if p explains q then p entails q—but not conversely. A purely syntactical characterization of explanation would therefore be incomplete; moreover, it would render the term 'explanation' redundant, since 'deduction' would suffice.

So far we have mentioned the following traits of explanation: (i) its consisting in answering why-questions (*pragmatic* aspect); (ii) its reference to formulas, which may or may not refer in turn to facts and patterns (*semantical* aspect); and (iii) its consisting in a logical argument involving generals and particulars (*syntactical* aspect). A fourth interesting feature of explanation is of an *ontological* kind: from this viewpoint we may say that to explain a fact expressed by an explanandum is to fit the said fact into a nomic pattern expressed by the law(s) or the rule(s) involved in the explanans—i.e. to locate the fact in a system of lawfully interrelated items. A fifth aspect of explanation is *epistemological:* from this point of view explanation proceeds inversely to deduction. In fact, what is given at the outset in explanation is just the explanandum, whereas the members of the explanans must be found out:

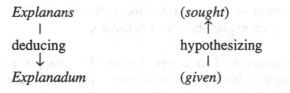

This last trait of explanation accounts for its seminal power, i.e. for its power to generate hypotheses and systems of hypotheses. Since the explanans must be logically stronger than the explanandum, proper questioning will lead to the building of higher and higher-level formulas, i.e. formulas both more general and containing more abstract concepts. By asking why facts happen as they do we are led to conjecture their laws; by asking why low-level laws are as they are we are led to build higher-level laws that entail the former—and so on without any other limitation than our present knowledge. This is why problems, and particularly why-problems, are the spring of science (see Sec. 4.1); and this is why the demand to stop explaining and concentrating on description or remain content with what has been explained leads to killing science. No why-questions, no explanation; no general, corroborated, and systematic hypotheses concerning objective patterns (i.e., law statements), no scientific answer to why-questions, i.e. no scientific explanation. In short, explanation is essential to science (see Sec. 9.7).

A sixth aspect of explanation that will concern us is the *psychological* one: i.e., explanation as supplying *understanding* (see Sec. 9.4). But before we can hope to understand this aspect of explanation we must delve deeper in the logic, the epistemology and the ontology of explanation. First of all, we should discover what distinguishes scientific from nonscientific explanation.

Problems

9.1.1. Etymologically, 'explaining' means unfolding or developing. Does this meaning match the sense specified by the definition proposed in the text?

9.1.2. Contrast the sense of 'explanation' used in this section to (i) *explanation of meaning* (or specification of meaning), e.g., by means of examples or of rules of designation, and (ii) *explication* (or elucidation), in the sense of concept refinement. Recall Sec. 3.2. *Alternate Problem.* Compare explanation with definition.

9.1.3. The sentences 'I am hungry because I am hungry and thirsty', and 'He owns one book because he owns ten books' are perfectly correct as far as logic goes. Would they count as explanations? *Alter-*

nate Problem: Is it correct to say of an explanation that it is true (or false)?

9.1.4. In the text the rule "All wrong drafts must be burnt" was employed as an explanans premise. Should we regard it as an ultimate explanans or is it possible to suggest an explanation why all wrong drafts ought to be burnt?

9.1.5. If scientists were to limit themselves to making, recording and describing observations and abstained from asking for explanations, would they introduce hypotheses? And if they did not, would they go far in gathering scientifically valuable data?

9.1.6. Elucidate the concept of *partial explanation* such as it occurs in the following example. Explanandum: A certain intelligent child misses many classes. Explanans premises: "Unmotivated children often miss classes" (generalization) and "The given child is not interested in school" (circumstance). We may regard this as part of the explanation (or as a partial explanation) of the given fact. What would complete the explanation? And is it possible to ever get a complete explanation of an actual fact? Be careful to distinguish the actual fact from the statement that expresses it.

9.1.7. Examine the thesis "There are no final explanations in science". What kind of proposition is this: is it an empirical generalization or a hypothesis proper? In case it is a hypothesis: of what kind? What can its ground be, if any? And is it refutable?

9.1.8. A number of philosophers have taught that the schema of scientific explanations is "*C*⊢*E*", where '*C*' stands for the causes producing the effects denoted by '*E*'. Examine this view.

9.1.9. During centuries, until the early 1920's, certain fossil molars were regarded in China as dragon bones with definite medical virtues and so they were sold in the apothecaries, until anthropologists interpreted them as remains of specimens of *Sinanthropus*. This illustrates the thesis that fact interpretation depends on background knowledge (in science, on systematic background knowledge). Expand on this thesis. *Alternate Problem:* Must explanation kill beauty?

9.1.10. Examine the opinion that statistical explanations (those involving statistical laws and/or data) are, by contrast with deductive explanations, cases of "probabilistic reasoning" in the sense that the explanandum is in them a probable rather than a certain consequence of the explanans. *Alternate Problem:* Discuss the opinion that the generalization-circumstance pattern fails to fit explanation by motives and goals, which is prominent in the sciences of man, and accordingly the latter calls for a logic of explanation essentially different from the one adopted in the text.

9.2. Nonscientific Explanation

Every rational explanation of a fact is a hypothetico-deductive sequence involving generalization(s) and information (see Sec. 9.1). If the generalization(s) and the information are scientific and the argument is correct (logically valid) we speak of *scientific* explanation—or perhaps of *nomological* explanation to emphasize the role of scientific law in it, or even of *theoretical* explanation to underline the occurrence of the argument in a body of theory. If the generalizations and the data are technological, we call the explanation *technological* and occasionally *nomopragmatic* (rather than nomological) to underline the function of nomopragmatic statements, or grounded rules, play in it. Every other explanation, even being rational by our definition in Sec. 9.1, will be called *nonscientific.* Needless to say, a given nonscientific explanation can be more correct than thousand scientific explanations: but this is beside the point, since science has not the monopoly of truth but only the monopoly of the means for checking truth and enhancing it. Nonscientific explanation will be of interest to us both because its study will give us additional information on what science is not and because it occurs in the practice of science.

But before dealing with nonscientific explanation we should dispose of a kind of argument that often passes for explanation although it fails to meet the requirements laid down in our definition (see Sec. 9.1)—namely, *labeling.* Labeling, or name-calling, is a pseudoexplanation and it should be dealt with in an account of explanation much for the same reason that unauthentic ("false") pearls must be mentioned in a treatise on pearls. Labeling is conspicuous in ideological debate, in pseudoscience, and even in the early stages of science (protoscience). Here go some outstanding cases of pseudoexplanation.

(i) "Why are living beings different from nonliving bodies? Because they have a vital force" (vitalism and, in particular, H. Bergson). (ii) "Why is the mesquite [a Mexican plant] so widespread? Because of its capacity to occupy new terrain" (read in an ecological study). (iii) "Why does opium produce sleep? Because it has the *virtus dormitiva*" (Molière's physician). (iv) "Why do people often fight each other? Because they have an aggression instinct" (psychoanalysis). (v) "Why can higher animals solve certain problems rather quickly? Because they have got a special insight" (Gestalt psychology). (vi) "Why can people recall some experiences? Because of their memory" (psychology of the faculties, now rechristened 'abilities'). (vii) "Why do we seem to repress certain feelings and thoughts? Because we harbor a censor" (Freud's super-ego). (viii) "Why did the ancient Greeks excel in practically everything? Because they were a talented people" (read in a history book). (ix) "Why do some people behave intelligently? Because they have a high IQ". (x) "Why do some people make above-chance guesses? Because they have paranormal abilities" (parapsychology).

The logical property of labeling is that it is circular, i.e. of the form "*q* because *q*". The epistemological correlate of this circularity is that it consists in what has long since been called *explicatio ignoti per ignotium*—of the unknown by the unknown. Its psychological partner is its consisting in an *explicatio obscurum per obscurius*—of the obscure by the more obscure. In fact, to say that a vase broke because it was fragile or had the virtue of fragility is to repeat the explanandum, only in a slightly different and possibly more imposing way. Now, repetitions are harmless as long as they are not made to pass for explanations and as long as they do not attempt to smuggle in false or untestable hypotheses. Thus, e.g., to explain actual aggression as the outcome of an aggressive instinct is not only circular but it enthrones the hypothesis that such an instinct in fact exists—a conjecture refuted by ethologists. An example of a pseudoexplanation that smuggles in an untestable hypothesis is the Aristotelian explanation of change as the actualization or unfolding of a capacity or potentiality for change—which capacity can only be tested by change itself. With this we close the discussion of labeling as pseudoexplanation.

We shall now take up two varieties of rational explanation which are not scientific: (i) *ordinary* explanation, the argument that employs common knowledge rather than scientific knowledge, and (ii) *unscien-*

tific explanation, which includes blatantly false premises. Both are rational explanations in the sense that they fit the logical pattern "Explanans⊢Explanandum".

The ordinary explanation of a fact fits the correct pattern "$\{G, C\} \vdash E$"; only, in it G will not be a set of scientific laws and C will not be a set of scientific data. In ordinary explanation, G will be a common sense generalization—a nonlaw yet lawlike statement—such as "All ravens are black", and both C and E will describe facts known or conjectured by common sense. The data C and E involved in ordinary explanation may differ in the following respects from those occurring in scientific explanation. Firstly, C and E may express *whole facts* (e.g., events) in ordinary explanation, whereas in scientific contexts they will usually refer to *selected traits* (properties) of facts such as, e.g., the initial and final concentrations of a set of chemical reagents. In particular, in common sense contexts 'C' will often designate a cause and 'E' its effect. Secondly, in ordinary explanation C and E will designate facts, whether real or imaginary, that are supposed to be accessible with ordinary means, whereas in science no such restriction is placed: 'C' and 'E' may, in fact, designate values of inferred properties.

Ordinary explanation may, of course, be correct: i.e. the premises G and C may be true and the derivation of E from them may be valid. Thus, it will be true to say that the car stopped because it ran out of gas, if it is the case that the gas provision was exhausted—which may be tested by merely refilling the tank; the generalization involved, regarding the constant relation between the fuel and the motor output is also true. But, as often as not, ordinary explanation is incorrect either because it involves false information or, most frequently, because it uses false or even untestable generalizations. But the main shortcoming of ordinary explanation is, of course, that in the more important cases it cannot become more accurate unless it ceases to be ordinary.

Consider, for example, the question 'Why do we feel cold after a swim when we dry up in the sun?' A nonscientist might reply something like this: 'In order to evaporate, the water on our skin takes heat from the body'. This teleological explanation will be rejected by the physicist, who may offer in its stead any of the following explanations: (i) "Upon evaporation, water takes some thermal energy from the body; this heat of vaporization can be estimated". (ii) "When in

Table 9.1. Rational explanation

Ordinary	Scientific
No restriction on why-questions.	Restricted to well-formulated questions.
No restriction on reliability of either explanandum or circumstances.	Data concerning explanandum and circumstances are checkable.
The explanadum refers to a whole fact.	The explanandum refers to selected aspects of a fact.
The explanans-premises are imprecise.	The explanans-premises are precise.
The generalizations involved are extra-systematic: empirical generalizations, *ad hoc* hypotheses, or just myths.	The generalizations involved are systematic: they are law statements.
Hardly improvable accuracy and depth.	Improvable accuracy and depth.

contact with the body, the water molecules collide with the skin molecules and get an extra kinetic energy; one consequence is that the skin becomes cooler (decrease of kinetic energy), another that the water molecules overcome the surface tension (they evaporate)". The first of these scientific explanations is couched in macroscopic terms, the second in both macroscopic and microscopic terms. The two involve unfamiliar terms but both are testable and, moreover, systemic: they are applications of scientific theories. The difference between them is one of depth: the second explanation is deeper than the first in the sense that, by relating a macrofact (drop of temperature) to a set of microfacts (molecule collisions and escape), it provides a fuller account of the explanandum. In short, whereas ordinary explanation is extrasystematic, scientific explanation is *systematic,* i.e. it takes place within some conceptual system; moreover, it is testable and perfectible, which ordinary explanation may not be.

One further feature of ordinary explanation which distinguishes it from scientific explanation is that it may proceed without first critically examining what is being requested to explain. Scientific explanation, on the other hand, is an answer to a *well-formulated* problem, i.e. one which is both well-conceived and well-stated (see Sec. 4.2). There is no such severe restriction to questioning in the context of nonscience, where questions like 'Why are phantoms so noisy?', or 'Why do some people have mediumnic abilities?', or 'Why did Zeus punish Prometheus?' are not dismissed but answered with zeal. Scientific explanation, on the other hand, often consists in laughing the explanandum away by showing that is not true or that its truth value cannot be found out, so that it is not something to be explained. This is a legitimate way of *explaining away*. [Illegitimate kinds of explaining away are: (i) to dismiss a piece of evidence unfavorable to a pet hypothesis by introducing an *ad hoc* hypothesis which covers just the exception, and (ii) to refuse to acknowledge superphysical levels of reality by reducing them forcefully to the physical level. For the latter, see Sec. 9.5.] In short, whereas ordinary explanation tends to naively accept every why-problem within its reach, scientific explanation will require that the explanandum be at least verisimilar. Table 9.1 summarizes the above discussion.

A conspicuous kind of rational explanation which at first sight seems to defy the schema "$\{G, C\} \vdash E$" is the inference pattern usually called *diagnosis* or *interpretation* of facts (as opposed to the interpretation of artificial signs). Diagnosis is important both in daily life and in applied science, e.g. in medicine, where it consists in the hunting of disease on the basis of the observation of symptoms. Diagnosis is frequently ambiguous, because the relations between causes and symptoms are often many-to-one. The ambiguity can be reduced by further observation and deeper knowledge, as when the cause of recurrent headaches is traced to one of a large set of possible causes. But it can seldom be removed altogether. A second source of ambiguity is the incomplete statement of the relevant generalizations. Let us see how far we can go in reducing ambiguities of this kind.

Consider the fact expressed by the following sentence: 'President *X* has appointed Mr. *Y* as U.S. representative to the U.N.'. Any of the following six "interpretations" (diagnoses) of this fact might be given:

1. Appointment shows great (little) importance President *X* attaches Mr. *Y*.

2. Appointment shows great (little) importance President *X* attaches the U.N.

3. Appointment shows great (little) importance President *X* attaches both Mr. *Y* and the U.N.

Hypothesis 1 *presupposes* that President *X* assigns the U.N. a great (little) importance. Conjecture 2 takes for granted that President *X* assigns Mr. *Y* a great (little) value. And hunch 3 coincides with its own presupposition. We shall show that they all fit the schema of rational explanation.

1. Generalization: "Great (small) tasks, great (small) men", or "If a task is great (small) then whoever fulfils (or ought to fulfil) it is great (small)". Circumstance: "The task of representing the U.S. to the U.N.

is great (small)". Explanandum (conclusion): "The representative of the U.S. to the U.N. is a great (small) man". That this coincides with the explanandum is clear, since 'The representative of the U.S. to the U.N.' is just a definite description of Mr. *Y*.

2. Generalization: "Great (small) men, great (small) tasks", or "If a man is great (small) then whatever task he is assigned is (or must be) great (small)". Circumstance: "Mr. *Y* is a great (small) man". Explanandum or conclusion: "The task assigned Mr. *Y* is great (small)".

3. Generalization: "Great (small) men, great (small) tasks and conversely", or "A man is great (small) if and only if he is assigned a great (small) task". Circumstance: "Both the U.N. and Mr. *Y* are great (small)". Explanandum or conclusion: "Both the task of representing the U.S. to the U.N. and Mr. *Y* are great (small)".

We can see that the diagnosis-statement is just the explanandum of a rational explanation. The fact or symptom to be explained does not occur in the diagnostic save as a starter: what has been regarded as calling for an explanation is a nonobservable fact, namely, the referent of the explanandum or diagnosis. The author of any of the inferences 1–3 above jumped from symptom to diagnosis without perhaps thinking of the premises of the explanation: these were introduced or made explicit only a posteriori, in an attempt to validate the leap. As a consequence of not rendering the premises explicit, the "interpretation" of natural signs is more ambiguous than it need be. This is usually the case with historical "interpretation" (explanation), which involves commonsense generalizations and occasionally (as yet) law statements borrowed from psychology and sociology. The ambiguities of history could be decreased upon turning such presuppositions into explicit premises. But then it would attract no writers.

Having dealt with ordinary explanation and its pitfalls let us now approach *unscientific* explanation, i.e. explanation involving fallacy. This occurs not only in the context of pseudoscientific and antiscientific doctrines but even in scientific texts. An illustrious instance is a rather common answer to the question 'Why are light rays bent when they pass grazing a star?' In a number of books the following explanation is found. First, the special theory of relativity contains the law "Energy=Mass×Square of the velocity of light in vacuum" (a half-truth, because this theorem belongs to a theory of things endowed with mass and is consequently inapplicable to light). Second, the former equation means that mass and energy are the same up to a constant factor (false) or at least equivalent (another half-truth) and, particularly, that anything having energy has also a mass (false). Third, since light has energy (true), it has also a mass (false). Fourth, since light has a mass (false) and since anything that has a mass is attracted and consequently deviated from its path by a massive body (true), a body will attract light (false). Fifth, since whatever attracts deviates, a body will deviate light; in particular, a celestial body will deviate a light ray (true). This explanation is perfectly rational because it subsumes the explanandum (a generalization) under more comprehensive generalizations; but it is a wrong explanation. Moreover, it is unscientific because it hinges on an unwarranted generalization of a mechanical theorem—"$E=mc^2$"—to optics. This generalization is fallacious because, from the proposition "If the mass of a system is m, then the total energy of the system is mc^2", the converse "If the total energy of a system is E then the mass of the system is $m=E/c^2$" does not follow. The derivation has been "formal" in the sense that an arithmetical transformation (of "$E=mc^2$" into "$m=E/c^2$") has been performed without paving attention to the physical meanings of the symbols—a meaning that can be disclosed only by bringing to light the object variable of both E and m—a variable which denotes an arbitrary mass point but not a light quantum. In this way the condition of semantic closure (see Sec. 7.2) has been violated, because the concept of mass of a light ray has been smuggled into a theory that does not contain it to begin with.

The case of unscientific explanation we have just discussed is merely a case of misapplying a considerably true scientific theory, owing to a misunderstanding of one of its formulas—a misunderstanding, incidentally, made possible by a neglect of logic. Such mistakes are common and, in the long run, harmless; moreover, a misapplication of a scientific theory may eventually suggest its extension to a new field. A different kind of unscientific explanation is the one involving outright false, definitely untestable or even meaningless premises belonging to some pseudoscience. Such doctrines mix scrutable predicates with inscrutable ones in such a way that, by sheer juggling, they can occasionally derive an explanandum. This is an easy game indeed. Take as premises "All stars are pooh" and "Everything pooh is hot", where 'pooh' designates whatever inscrutable property you wish. From the two premises it correctly follows that "All stars are hot", which is probably true. But we would not regard this explanation of the hotness of stars as scientific: the initial assumptions of the pooh theory are both untestable and groundless even if they are meaningful in the context of poohism.

In order to have a scientific explanation of a fact it is necessary, though insufficient, to be able to deduce the propositions expressing the given fact from data and generalizations belonging to a scientific system. This condition will be qualified in the next section. But before going into it let it be noted that the mere existence of unscientific games like poohism makes a strong case against the doctrine that theories are nothing but conventional devices or games permitting the systematization of observational propositions.

Problems

9.2.1. Pose a why-question and find both an ordinary and a scientific answer to it. *Alternate Problem:* A physician notes that the white of a man's eyes are yellow. What is his diagnostic? And is it unambiguous? If not, how does he remove the ambiguity?

9.2.2. Analyze the following case; pay particular attention to the testability of the hypotheses involved. Explanandum: "The pain produced by a given stimulus diminishes by the interposition of a further stimulus". (Fact, summary of observations, intermediate level hypothesis?) Explanans-generalization 1: "The subject's attention is distracted upon the application of a second stimulus, so that he feels the pain less intensely (but the pain is still there)". Explanans-generalization 2: "A positive stimulus applied to a nervous system may attenuate or even altogether inhibit the activity of other nervous centers of the system" (Pavlov's law of negative induction).

9.2.3. Some historians, sociologists and psychologists have often offered explanations in terms of human nature. For example, "it is in human nature to try to improve (whence progress)"; or "It is in human nature to become easily conformed (whence stagnation and decline)". Study the logic of this kind of explanation. *Alternate Problem:* Since so many historical accounts are beyond logic, some philosophers of history have concluded either that there is no historical explanation or that, if it exists, it must be radically different from what we have been calling rational explanation. For the "deductive model" of explanation as applied to history, see K. R. Popper, *The Open Society and its*

Enemies, 4th ed. (1945; London: Routledge & Kegan Paul, 1962), Ch. 25 and *The Poverty of Historicism,* 2nd ed. (1944; London: Routledge & Kegan Paul, 1960), Sec. 30, and C. G. Hempel, "The Function of General Laws in History", *Journal of Philosophy,* **39**, 35 (1942). For the cleavage between how-historians and why-historians see V. Mehta, *Fly and the Fly Bottle* (London: Pelican, 1965), Ch. 4.

9.2.4. A famous physicist once said: "Right after a nonelastic collision with another particle, the atomic nucleus tries to get rid of its excitation energy in the simplest possible way". Is this an explanation of nuclear decay? If so, what kind of an explanation? And could you mention other instances of explanation in the same vein found in scientific publications?

9.2.5. Discuss the thesis—maintained by ordinary language philosophers—that the right way to approach the problem of scientific explanation is to survey and analyze the uses of the verb 'to explain' in ordinary parlance.

9.2.6. The apparently spontaneous noises heard in certain old buildings in bad condition have often been explained in terms of ghosts; there have even been international conferences on "spontaneous phenomena" attended by some scientists. Are such explanations scientific? In case they are not, is it reasonable to accept them or to conclude that they concern phenomena impervious to scientific explanation?

9.2.7. The explanation of actual facts in terms of predispositions, abilities or potentialities has been branded as circular, hence as pseudoexplanation} in the text. Does it follow that things cannot be assigned predispositions, abilities, or potentialities?

9.2.8. The hermeneutic school claims that everything social "is a text or like a text", whence it must be "interpreted", and more precisely in terms of intentions. Check whether such interpretation is anything other than hypothesis, and whether macrosocial facts, such as economic recessions and civil wars, can be accounted exclusively in terms of the (guessed) intentions of individuals

9.2.9. Teleological explanations are mostly wrong in reference to non-mental happenings and are therefore to be avoided if possible. (See Sec. 5.9: principles of level parsimony, level contiguity, and source level.) Does the same hold for teleological questions such as 'What is the function (or the use) of x?' If so, how is it possible for a teleological question to elicit fruitful research and for the teleological answers to the same question to be unenlightening and even obstrusive?

9.2.10. Should teleological explanation be banned from every science or may it be accepted *in extremis* and as long as it concerns facts on the highest levels and on condition that it be not regarded as ultimate but as in need of explanation in non-teleological terms?

9.3. Scientific Subsumption

Since fully stated explanations, whether scientific or not, fit all the same logical pattern (see Secs. 9.1 and 9.2), logic will be unable to tell the differences between a scientific and a nonscientific explanation. In order to pin down this difference it will suffice to recall that every deduction is made in some context and that the peculiarity of scientific deduction is that it takes place in the bosom of scientific theories, whether adult or not (see Sec. 7.3). Ordinary explanation, by contrast, is made in the body of ordinary knowledge, which is a rather poor and incoherent mosaic. This suggests adopting the following *Definition 2:* A *scientific explanation* of a formula q is an answer to a well-stated scientific problem of the why-kind, consisting in an argument showing that q follows logically from a scientific theory (or a fragment of scientific theory, or a set of fragments of scientific theories), auxiliary scientific hypotheses, and scientific data, not containing q.

The data taking part in a scientific explanation will be bits of information about certain properties characterizing the fact that is being explained. Not *all* the properties, which would be impossible and useless even if possible, but only those which are relevant to the generalizations employed. For example, in the explanation of E by the argument "$\{G, C\} \vdash E$", 'C' will not refer to a whole fact but to a set of properties of a fact. Very often, C will be a set of particular values of some of the variables occurring in G. Thus, e.g., in the falling body problem C will be the couple: <initial position, initial velocity>. These are called the *initial conditions* of the problem. And in the more complex problem of an ideal vibrating string, C will be the set of the

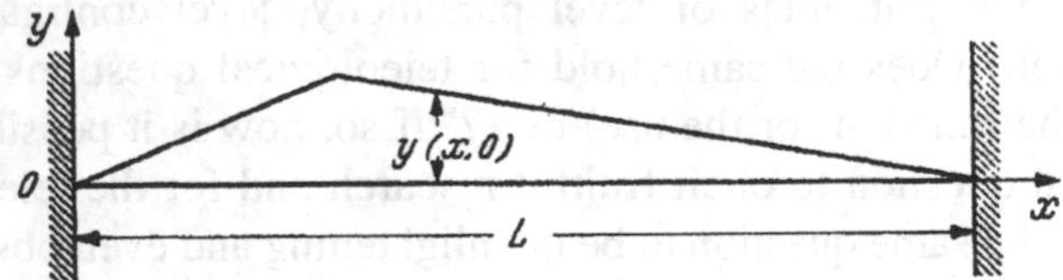

Fig. 9.1. Initial and boundary conditions in the vibrating string problem. Initial conditions: $y(x, 0)=f(x)$ and $\frac{\partial y}{\partial t}(x, 0)=g(x)$. Boundary conditions: $y(0, t)=0$ and $y(L, t)=0$.

following items: the form of the string and its velocity when released (the initial conditions) and the string's amplitude at the fixed ends (the *boundary conditions*). (See Fig. 9.1.) They are conditions in the sense that they must be satisfied by the general solution; otherwise they are data (actual or assumed).

In general, in problems amenable to quantitative treatment, *C* will be made up of information regarding both initial conditions and boundary conditions—very often purely conjectural rather than read on instruments. At any rate '*C*' does not denote a full event but a set of properties sufficient to characterize a certain state of affairs (e.g., the initial state of an elastic band). *C* could not be the complete description of a whole event, first because no such complete description is possible; second because, if it were possible, it would be a nuisance: we need those and only those data that are relevant to the generalization employed, and scientific laws are not about whole facts but about selected properties of them (see Sec. 6.1). In nonscientific contexts, on the other hand, '*C*' may refer to a whole fact—for example, to the cause of *E*.

In order to better understand the peculiarities of scientific explanation let us analyze a case within the reach of common sense, so that it can be approached either with or without scientific theory. Suppose someone asks 'Why did my nose bleed when I climbed a 5,000 metres mountain?' A number of ordinary knowledge answers may be conceived, as many as kinds of cultural background. One of them might be: 'Because most people's noses bleed when they climb 5,000 metres or higher'. This enables us to say that it is very likely that anyone's nose will bleed when on the top of a 5,000 metres mountain. It will be an explanation of sorts but not a scientific one because it is not inserted in a scientific theory. A scientific answer to the original question would begin by checking the explanandum and, upon finding it

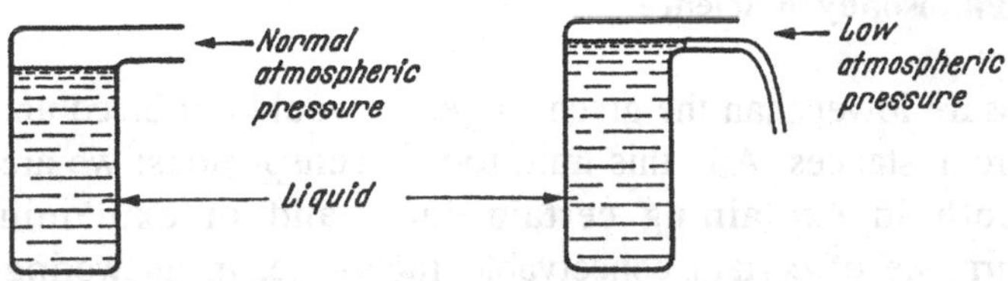

Fig. 9.2. Physical model of nose: (i) normal; (ii) at high altitude.

true or verisimilar, would proceed to build a rough *ideal model* of the relevant parts of the body, i.e. the nose and the circulatory system. It would, for example, idealize the system as a vessel with an opening at the top and nearly filled with a liquid (see Fig. 9.2). This model would then be treated by ordinary physical theory (hydrostatics), in compliance with the metaphysical principle of level parsimony. The overflow of liquid is easily explained by the decrease in atmospheric pressure, which checks the liquid level. Once the explanation of the physical fact has been achieved we apply it to nose bleeding on the assumption that this is nothing but the overflow of a liquid as a consequence of the liquid expansion upon pressure decrease. In this way a biological fact has been explained in physical terms.

Notice the following traits of the above scientific explanation of an ordinary life fact. In the first place, far from being *ad hoc,* the explanation applies to an unlimited number of similar instances, such as the dripping of pens when flying in non-pressurized planes; this is, of course, a consequence of the embedding of the given problem in a universal theory. In the second place, the explanation has involved the building of an ideal model of a real system, and precisely a model that could fit in an existing theory—to which end most accompanying circumstances had to be eliminated from the picture. In the third place, the explanation is perfectible. For example, with a few additional premises, among them the altitude-atmospheric pressure law and the subject's arterial pressure, we might have explained why his nose began to bleed exactly when he reached a given altitude: that is, we might have supplied a quantitative explanation (and even a prediction) of the given fact instead of a qualitative one if we had used quantitative generalizations and data. These three characteristics—*universality, use of a model,* and *perfectibility*—are peculiar to scientific and technological explanation.

Moreover, with the addition of the above-mentioned quantitative premises we might have explained why all those people with an arterial pressure lower than the given subject's would *not* bleed under the same circumstances. And this trait, too, is conspicuous: we are interested both in explaining certain facts and in explaining the nonoccurrence of certain conceivable facts—i.e. in answering questions of the type 'Why not-*p*?'. A particularly interesting case in which the explanation of a *non-fact* is called for happens when, according to our theories, *p* should be the case, yet experience stubbornly clings to not-*p*. Thus, e.g., a physicist may wish to explain why a certain transition, which according to his calculations should be possible, is not observed to occur; the military historian will ask why the Mongols stopped their successful invasion of Europe, why the Nazis failed to carry out their plans for the invasion of a relatively unprepared Great Britain, and so on. The explanations of non-facts may be as important as the investigation of what is the case, either because it points to a lacking necessary condition for the occurrence of the corresponding fact, or because it points to a deficiency in our knowledge; such a deficiency may consist in the neglect of a factor which is sufficient to prevent the predicted fact from occurring. In the latter case the explanation of the non-fact will call for a new set of premises, i.e. it will stimulate the correction of the relevant theory and/or of the relevant information.

Consider now, instead of an isolated fact, a time sequence of singular facts: for example, the successive events forming the history of a biological species or of a human community. When the sequence exhibits some pattern or order rather than a random character, we speak of a historical *trend*—and often mistake trend for law. (For the difference between trend and law, see Sec. 6.6.) The follow-up study of a system, whether elementary or complex, yields its *history* rather than its laws. That a history—e.g., the spacetime diagram of an electron in a field—can be analyzed, hence explained in terms of the system's laws, is true—but another story. Anyhow certain histories, far from being random time series, exhibit definite trends. What does the explanation of trends consist in? It is essentially the same as the explanation of single facts. For example, evolutionary trends are partly explained by the theory of evolution: this theory explains them with the help of certain laws of genetics and ecology. [It is a controversial point whether the theory of evolution contains *laws of evolution* of its own. Anyhow, it is an unfulfilled desideratum of theoretical biology (i) to state exactly what the initial assumptions of the theory of evolution are and

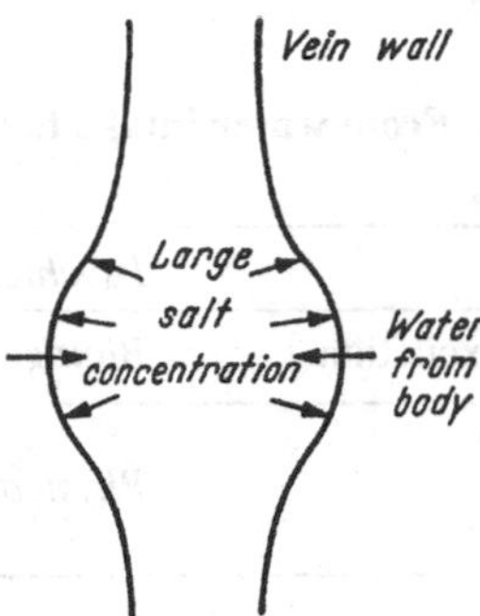

Fig. 9.3. A model explaining dehydration of bodily tissues. The excess salt molecules are not diffused through the vein walls and exert a pressure against it; this extra pressure is balanced by the pressure of water diffusing in the opposite direction.

(ii) to deduce those axioms from the more fundamental laws of genetics and ecology.]

Let us now go from facts and sequences of facts to generalizations concerning either. As a matter of fact most of the explanations that pass for explanations of facts are explanations of generalizations, and this simply for the linguistic infraction consisting in calling *facts* all those generalizations we would like to be sure about. When we asked a while ago why do noses bleed at great altitude we should not have looked for the explanation of the *fact* that noses bleed in such circumstances, as there *is* no such fact: what we can be after is an account of the empirical *generalization* that noses bleed at high altitudes. The same is true of many of the so-called *scientific facts:* they are nothing but reasonably well-established (or else lazily untested) low level generalizations.

A scientific explanation of such low-level generalizations, however familiar they may be, can be complex or even unavailable. Take, for instance, the familiar truth that thirst increases upon drinking sea water: the current explanation of this generalization requires a physiological model. When sea water is drunk part of it goes into the blood; the large concentration of salt produces, as a reaction, a diffusion of water from the nearby tissues to the blood (osmosis), as shown in Fig. 9.3. Consequently the blood becomes more watery and the body's protoplasm dries up. This event excites the hypothalamus; the psychological partner of this neural condition is the thirst sensation. The

Table 9.2. From water intake to thirst

Explanation ↑		
	Thirst	*Psychical level*
	Hypothalamus excitation	*Biological level*
	Cell dehydration Water diffusion Osmotic pressure	*Physical level*

explanation of a familiar "fact" (lowlevel generalization) has required the construction of a chain of determiners starting at the physical level and ending at the psychical level—in accordance with the metaphysical principle of level contiguity or non-skipping (see Sec. 5.9). Table 9.2 summarizes the model involved in the explanation of the empirical generalization. It will be realized that the chaining of events occurring on the various levels goes up from the lower levels but does not amount to a *reduction* of thirst to osmosis or even to its consequence, protoplasm dessication: far from being a reductive explanation proper, the foregoing appeals to certain peculiarities of various levels: it is a multilevel explanation. (More on this in Sec. 9.5.) Our example exhibits another interesting characteristic: it explains the familiar by the unfamiliar, the known by the conjectured; but this, which is peculiar to scientific explanation, will concern us in Sec. 9.4.

Science, then, is in principle in a position to explain empirical generalizations, whether those of ordinary knowledge or those of an early stage of scientific research. But it can do better than this: it can explain *laws* proper, not only of the qualitative kind but also quantitative ones. The pattern of the explanation of laws is slightly different from the logical form of the explanation of singular facts but it fits into the Definition 2 put forward at the beginning of this section. In fact, a scientific law is explained by being *deduced from other law(s),* eventually with the help of *subsidiary hypotheses.* Consequently the process of scientific explanation has, within each theory, a limit: it cannot rise beyond the axioms of the theory. These axioms could be explained only by subsuming them under an even more comprehensive theory; but at each stage in the history of science explanation is limited by the richest theory.

An historically important case of explanation of law was the derivation of Galilei's law of falling bodies from Newton's mechanics, a

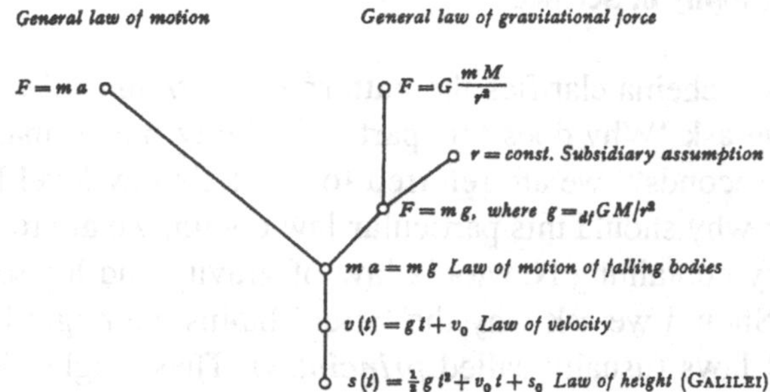

Fig. 9.4. Deduction of low-level laws of falling bodies from high-level laws plus subsidiary assumptions.

derivation which provided one of the latter's chief tests. The derivation is summarized in the tree above. It will be noticed that the two highest-level laws involved are not sufficient for the deduction of Galilei's law. Two subsidiary assumptions have been added: the tacit assumption that the sole acting force is gravity and the explicit assumption that the body travels a distance which is negligible compared to the earth's radius, so that the distance *r* between the body and the center of the earth may be regarded as constant. Both are pretences: the first, because in real situations there are additional forces; the second is obviously false. The first is not an explicit premise but rather a presupposition: it fails to occur in the derivation but it does appear in the choice of premises. The two are pretences true to a first approximation; they are *subsidiary assumptions* (not laws) that regulate the application of the basic laws. Generalizing and summarizing, the patterns of scientific explanation we have met so far are:

Fact explanation {Theory, Data}⊢Singular factual statement, [9.2]

Law explanation {Theory, Subsidiary assumption(s), Data}⊢Law. [9.3]

The tree illustrating the derivation of Galilei's law can be continued upwards a few steps along the left branch: Newton's general law of motion can be deduced from Lagrange's equations, which can in turn be derived from Hamilton's principle of stationary action, which can in turn be derived from d'Alembert's principle of virtual displacements. (The derivation of Galilei's law from general relativity follows a different pattern: a single equation, a sort of synthesis of the law of motion and the law of force, is the first element of the tree in this case. But here too subsidiary assumptions—minor branches—must be added.)

The above schema clarifies the pattern of *upstream scientific questioning* If we ask 'Why does this particular body fall so many metres in so many seconds?' we are referred to Galilei's low-level law. If, in turn, we ask why should this particular law obtain, we are referred to a whole theory containing Newton's law of gravity and his second law of motion. Should we ask why the latter obtains we might be offered higher-level laws (usually called *principles).* These higher-level laws are justified because they yield a rich set of lower level laws and these in turn are justified by their agreement with empirical data and their compatibility with the remaining formulas of the given theory. Upstream scientific questioning stops at the highest-level assumptions attained at the moment: it is a finite regress with a receding starting point.

What characterizes every step in an explanatory chain, whether of the type [9.2] or of the type [9.3], is the *subsumption* of formulas under stronger formulas. Whether this is all there is to scientific explanation, will be discussed in the forthcoming section.

Problems

9.3.1. Give a scientific explanation of an everyday life fact.

9.3.2. A geologist noticing horizontal striae in the walls of a canyon may put forward the following hypotheses. (i) "In the past there was a period during which all the water bodies in this region were frozen; in particular the water in the canyon was frozen (it was a glaciar)." (ii) "Above a certain bulk the pressure exerted by the glacier on the bottom of the canyon was such that the layers in contact with the rock began to melt and the glacier began to slide down the canyon." (iii) "During this sliding the glacier left the scars still visible on the rocks." Show the logical pattern of this explanation and examine the nature of the hypotheses involved in it.

9.3.3. Examine the scientific explanation of the rainbow (or any other well-studied natural phenomenon). Point out the basic laws and the levels involved in it. *Alternate Problem:* Compare the contemporary account of explanation with Aristotle's explanation in terms of the four causes as expounded in his *Metaphysics,* Book A (various editions).

9.3.4. How is it possible for the jellyfish to have a definite shape notwithstanding that it is made up of 99 per cent of water? Recall that the protoplasm is, physically speaking, a gel (like jello). Show what the model in the explanation is.

9.3.5. The elementary laws of the ideal pendulum, such as "The oscillation period is independent of the oscillation amplitude", follow from the single law: "$T=2\pi\sqrt{L/g}$". In turn, this law follows from Newton's law of motion on the simplifying assumption that the amplitudes are small (up to, say, 20°). Show what the pattern of the derivation is.

9.3.6. Perform a logical and epistemological analysis of the derivation of some scientific law. *Alternate Problem:* Exhibit one specimen of *reductive* explanation (like that of nose bleeding) and another of *nonreductive* explanation (like that of thirst).

9.3.7. Examine the following explanations of the historical generalization that the cradles of all civilizations had arid or semiarid environments. (i) "An achievement is possible if the material means are available. The cradles of civilization were sorrounded by soft ground which was easy to till and irrigate. Geographical factors were easier to control there than in tropical forests or even in plains." (ii) "Every achievement is an answer to some challenge: there is no advance without difficulties. The arid or semiarid environments of the cradles of civilization was a challenge that stimulated work and ingenuity" (A. Toynbee). Are these explanations mutually exclusive? Do they state the conditions both necessary and sufficient for the emergence of civilization? What do they take for granted?

9.3.8. Scientific explanation is very often called *causal.* This name would be justified if either the generalizations or the informations involved referred to causes. Is this the case with every scientific explanation or even with most scientific explanations? See M. Bunge, *Causality,* 2nd ed. (Cleveland and New York: Meridian Books, 1963), Ch. 11.

9.3.9. In scientific questioning a point is reached where a set of postulates and a set of data must be accepted at least for the time being. Does this prove that scientific postulates and data are no better than religious dogmas?

9.3.10. The philosophy of a given scientific theory can be regarded, in a sense, as an explanation of that theory. In precisely what sense? In the sense that the given theory is logically deduced from philosophical principles, in the manner of Descartes? *Alternate Problem:* Questioning children and beginning students are frequently offered metaphors instead of genuine explanations which they would fail to understand. For better or for worse, every successful textbook teems with analogies the sole function of which is supposed to be pedagogic. Examine the claim that science proceeds in the same way, whence the so-called deductive model of scientific explanation (i.e., explanation as subsumption) is inadequate. See M. Black, *Models and Metaphors* (Ithaca, N. Y.: Cornell University Press, 1962) and M. Hesse, *Models and Analogies in Science* (London: Sheed and Ward, 1963).

9.4. Mechanismic Explanation

Suppose someone asks 'Why did *c* lie to me?' and gets the reply: 'Because *c* lies to everybody'. No doubt, this is a cogent explanation since it consists in the subsumption of the explanandum "*c* lied to *d*" under the generalization "For every *x*, if *x* is a person different from *c* then *c* lies to *x*". The subsumption is made via the information that *d* is a person (circumstance-premise). But the answer, namely '*c* lied to you because *c* lies to everybody', is identical with '*c* lied to you because *c* is a liar'—by definition of 'liar'. And this is an instance of labeling or namecalling, i.e. of pseudo-explanation (see Sec. 9.2).

Scientific explanation can be subjected to a similar criticism, namely that, although it does *account for* the explanandum it hardly *explains* it—except in the Pickwickian sense of subsumption of particulars under generals. In fact if we ask, e.g., 'Why did the resistance opposed by the bicycle pump increase upon pushing the piston down?', we might be informed that the air pump approximately "obeys" Boyle's law. We might as well be told that the pump behaved the way it did because it is a Boyler or nearly so. Here, too, we are offered a logical subsumption of the explanandum under a generalization but we remain in the dark as to the mode of action or mechanism of the piston's interaction with the air. Subsumptive explanation, i.e. the subsumption of a singular under a generalization, smacks of *explicatio obscurum per obscurius:* it does not assuage our thirst for intellection. This does not entail that science is incapable of providing understanding and that, consequently, we must either remain in the darkness or rely on something else—perhaps irrational intuition—in order to get an insight into the "inner workings of things". Far from this, science is the only mode of knowledge capable of supplying the deeper explanation we miss—only, this is a kind of explanation different from subsumptive explanation.

Every scientific theory provides a subsumptive explanation of a set of data but not every scientific theory can provide what in science is often called an *interpretation* of such data. A phenomenological theory, one not representing any "mechanism", will provide subsumptive explanations; but only representational theories, i.e. theories purporting to represent the *modus operandi* of their referents, can give deeper explanations. These shall be called *mechanismic explanations.*

What scientists usually mean by 'explanation' is not just subsumptive explanation but mechanismic explanation—like the explanation of the constant acceleration of a falling body in terms of a constant gravitational field, or the explanation of the odors of things in terms of the shape of their molecules and the way these molecules fit the olfactory receptors. In a nutshell, just as

Representational theory =Deductive systematization +Mechanism,

so

Mechanismic explanation = Subsumption +Modus operandi.

Take again the explanation of the doubling of gas pressure upon halving its volume. Saying that this fact fits Boyle's phenomenological law, correct as it is, will not be regarded as a satisfactory explanation in science but just as an account and systematization of the fact. An explanation proper, or mechanismic explanation, will involve the molecular or kinetic theory of gases, which includes the following basic hypotheses (axioms) that go far beyond Boyle's law: "A gas consists of a collection of particles (molecules) of a size negligible in comparison with their mutual distances"; and "The gas molecules are ceaselessly moving at random". From these assumptions the following consequences (theorems) can be derived. First, if an obstacle is interposed in the path of the molecules, collisions will result; in particular, the molecules locked in a container will collide with its walls exerting a force on the latter; the resulting specific force (force/area) is the gas pressure. Second, if the available space decreases, the number of collisions per unit area and unit time will increase; in particular, if the

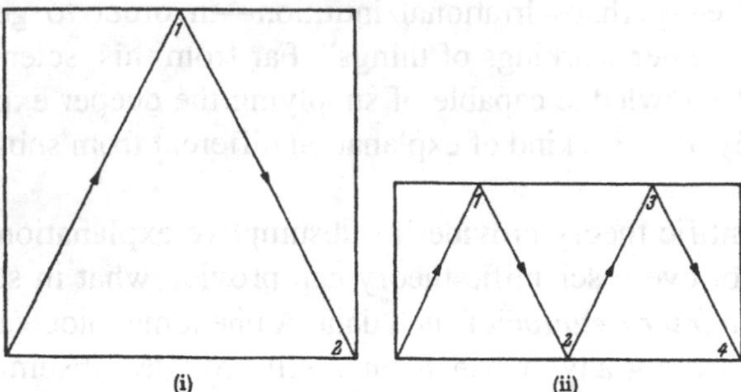

Fig. 9.5. The pressure of a gas doubles as its volume is halved because, when the space available for the molecules' flight is halved, the number of collisions of the molecules with the walls doubles. (i) Volume V: 2 collisions per unit time; (ii) Volume V/2: 4 collisions per unit time.

volume is halved the number of collisions will be doubled and, accordingly, the pressure will double (see Fig. 9.5).

The difference between this mechanismic explanation and the corresponding subsumptive explanation is due to the *interpretation* of one of the relevant variables, viz. the pressure, in microscopic terms instead of keeping it as an unanalyzed observable trait. This interpretation or hypothetical analysis of an external (observable) variable in terms of a hypothetical construct ("molecular collision") enables us to understand why a gas should exert a pressure at all on the walls of its container—a fact that must be accepted as such in the phenomenological account. This psychological bonus of mechanismic explanation does not follow from some peculiar logical trait of this kind of explanation: the logical pattern of mechanismic explanation is identical with that of subsumptive explanation. What distinguishes mechanismic explanation from subsumptive account is that the former is contrived with the help of some theory about the mechanism or *modus operandi* of facts. In short, the pattern of interpretive explanation is

$$\{\textit{Representational theory, Information}\} \vdash \textit{Explanandum}. \qquad [9.4]$$

That representational (nonphenomenological) theories are preferable to phenomenological theories because they are logically stronger, semantically richer (they have a greater content) and methodologically more exacting, was argued in Sec. 8.5. We shall now argue that representational or mechanismic theories supply the *deeper explanations.* The reason for this will emerge in the examination of a couple of cases.

The slowing down of a solid body moving in a liquid can be explained either in terms of the liquid's viscosity (or, equivalently, in terms of a coefficient of resistance to motion), or in terms of the momentum transfer of the molecules of the liquid to those of the solid. Both explanations are correct but they are not equivalent: the latter, a mechanismic explanation, is the deeper because it reaches a *deeper level* of reality. The explanation in terms of a resistance coefficient is little more than an exact description, whereas the molecular explanation supplies the *modus operandi;* moreover, in principle at least, the resistance coefficient can be deduced from molecular theory.

Consider poisoning as another example. "That man died because he took an overdose of strychnine" is a subsumptive explanation calling for a generalization like "Strychnine is lethal in high doses". (Caution: Do not define "high dose" as lethal dose.) Moreover, we might invoke a precise dose-effect relationship, i.e. a quantitative phenomenological (input-output) law, thereby producing an *exact* explanation. But as long as we ignore the mechanism of the drug action our knowledge will not advance much. The demand for a mechanismic explanation will lead, in a first stage, to finding out that strychnine inhibits the respiratory centers in the brain; the second task will be to find out the biochemical processes responsible for this paralysis; and this second task is presumably performed stepwise but never entirely finished. Incidentally, if we remained satisfied with subsumptive explanations we would not go beyond establishing the dose-effect relationship. Not only pure science but also technology (in this case pharmacology and medicine) is interested in discovering the mechanism of the drug action—if only to manufacture antidotes.

Our third and last example will be taken from economic history, in order to forestall the impression that mechanismic explanation requires a physical mechanism. Let the explanandum be the average upwards trend of prices in Europe during the second half of the 16th century—a rise that broke an age-long price stability. The rise of the bread price during that period could subsumptively be explained by reference to the general trend, but this would hardly be regarded as a satisfactory explanation. The mechanism responsible for the trend itself should be taken apart. At least two hypotheses, which are not mutually exclusive, have been proposed to explain the whole inflationary trend at that time: one is the introduction of huge quantities of gold and silver from Spanish America, the other is the reconstruction of the countries devastated by the Reformation and the peasant wars. Each of these hypotheses, in conjunction with economic theory, affords an *interpretation* of the price rise and each of them points to something which lies deeper than the referent of the explanandum.

The reason that mechanismic explanation is deeper than subsumptive explanation is that the former involves a *deeper analysis,* one often involving a reference to levels of reality other (usually deeper) than the level to which the explanandum belongs. Subsumptive explanation, by contrast, involves relations among items standing on the same level. Thus, Newton's direct interparticle theory of gravitation will subsumptively explain the gravitational interaction among any two bodies without referring to anything but the bodies' masses and their mutual distances; this is why Newton did not profess to know "the cause of gravity". On the other hand, Descartes would explain gravitational interaction in terms of a subtle fluid aether, Le Sage in terms of invisible particles colliding with the macroscopic bodies, and the classical field theory of gravitation invoked the gravitational field (a thin substance after all) as the medium through which bodies act: all these were representational theories, each providing a mechanismic explanation of gravitational interaction—and imageable explanations at that. (The modern theory of gravitation employs, in addition to the field concept, the further invisible inner stresses within the interacting bodies, particularly to explain gravitational repulsion, incorrectly called antigravity.) In mechanismic explanation the gross properties of real systems are explained in terms of the hypothesized nature, composition, structure and behavior of its low level constituents. Such an *atomistic analysis* underlying mechanismic explanation is the fulfilment of Democritus' program, i.e. the strategy and desideratum of explaining the observable in terms of the unobservable, the datum in terms of the transcendent hypothesis.

The deeper a theory the deeper the explanations it will supply. Just as there are degrees of theory depth (see Sec. 8.5), so there are degrees of *explanation depth.* The explanation depth attained with a semiphenomenological theory will be intermediate between the minimum reached with a phenomenological theory and the (temporary) maximum reached with a representational theory. This maximum depth attainable with a given theory need not be an absolute bottom: it will most likely be temporary because any given representational theory can conceivably be subsumed under a richer theory going into finer mechanism details. That no such deeper theory is in sight at a given moment does not entail that there can be no such deeper analysis disclosing, so to say, the atoms within the atoms.

At any rate, between the superficial explanation (account) supplied by the phenomenological theory, and the deep explanation (interpretation) supplied by the mechanismic theory, we often find the intermediate degree explanation afforded by a semiphenomenological theory. Take, for instance, cultural diffusionism. The central hypothesis of this anthropological and historical view is that every culture trait or function (e.g., irrigation technique, priesthood, number system) has a single origin (monogenism) and that it spreads out from its original center. This view certainly supplies a mechanism of cultural change, namely, invention-adoption; moreover, some cultural changes are indeed brought about by importation. But this, perhaps the simplest theory of cultural change, fails to explain why there is such a variety of cultural levels, why some cultures are impervious to certain cultural currents while others assimilate them eagerly, why certain traits appear simultaneously or nearly so in distant and unrelated regions, and so on. Diffusionism, though not a phenomenological theory, is not altogether representational either because it is much too simplistic and superficial: it deals mostly with secondary characters rather than with basic cultural traits and accordingly fails to provide the basic socioeconomic and psychosocial mechanisms responsible for the acceptance or rejection of any new trait coming from the outside, as well as the mechanisms inherent in independent reinvention.

We are now prepared to approach the psychological aspect of explanation, i.e. *understanding*. In ordinary life we say that something has been understood if it has been reduced to familiar terms. Thus, we say we understand the emission of a photon by an atom if we can picture it as the firing of a bullet, i.e. by analogy with a familiar experience; similarly we feel we understand the domination of one nation by another because we ourselves are relata of the domination relation. But in neither of these cases have we set up an explanation proper when just pointing to familiar similes; that is, cogent explanation is unnecessary for understanding. "Understanding" is a psychological concept rather than a metascientific one: it is essentially relative to the subject. An explanation will be satisfactory for an individual at a given moment if and only if that individual finds at that moment that the given explanation fits his scheme of things—a scheme that is largely composed of images and is inevitably laden with bias and with relics of dead theories. The explanation achieved in this way is a trivialization, a reduction of the unexperienced or unfamiliar, hence surprising, to the experienced and familiar, hence reassuring and, by the same token, limited to the subject's experience. This is the kind of explanation we seek both in everyday life and in science teaching, when we draw diagrams of abstract ideas and visual models of invisible entities.

The chief aim of scientific explanation is to extend, deepen and rationalize ordinary knowledge. In this process anything but a reduction to popular ideas takes place. Scientific explanation makes no concession to understanding in the sense of reducing the unfamiliar to the familiar, the deep to the shallow or the new to the old. Quite on the contrary, to explain familiar phenomena science will invent high-brow, transempirical constructs. Thus, chemistry will explain why carbon has such a great power to combine with other elements, but it is unable to "bring home" why this should be so. Scientific understanding, in short, does not secure understanding in the ordinary sense of this word. Only the specialist who has become acquainted with the technical ideas involved in scientific explanation, who has "internalized" them, may claim that he has achieved an intuitive understanding of some (not all) of the facts and ideas. Conversely, intuitive understanding may be obtained without genuine explanation. Thus, e.g., we may "understand" other people's behavior in the sense that, far from being surprised by it, it seems "natural" or "obvious" to us although we ignore the behavior mechanism. What we ask from science is *intelligibility* rather than intuitive understanding. And to obtain intelligibility it is enough to secure subsumptive explanation, which may involve abstract and complex ideas.

However, there are *degrees of intelligibility.* A representational theory, even if it is (epistemologically) abstract, will supply a higher degree of intelligibility than a phenomenological theory, particularly if it involves an imageable model. However, since the turn of the century it has become fashionable to abominate representational theories in physics and, particularly, theories involving pictorial models, such as Bohr's planetary model of the atom. Accordingly many scientists have been persuaded that their sole task is to execute observations or to set up phenomenological theories capable of summarizing such observations and to predict further possible empirical results without caring for an interpretive explanation of what goes on in between observations or inside what is being observed. When asked to explain a fact they may hasten to compute certain variables in terms of other variables: they may supply subsumptive explanations; if pressed for a deeper explanation in terms of structures, interactions, or motions that are out of sight, they may dismiss the question as metaphysical. Yet many a big theory has involved some mechanism or other and, as our boxes become less and less black, a deeper intellection of their inners and their inner workings is attained. Why should one reject whatever intuitive understanding science can legitimately provide? In particular, why should we refuse pictorial or semipictorial models as long as they are shown to be approximately true and not merely didactic tricks? Why should we, in short, refuse the intuitive understanding that sometimes comes with a deep and complex theory? Just in order to gratify thoughtless computers and observers as well as obscurantist philosophers who would like to conceive of the whole universe as a black box? Only untestable or patently false representational theories and pictorial models—and the explanations they lead to—should be rejected.

In sum, whenever possible we should strive for the deeper explanation, which is the mechanismic explanation, because it is supplied by the deeper theory. Such a deeper explanation will always provide a deeper intellection than the corresponding subsumptive explanation, and occasionally it may even produce an intuitive understanding. Notice, however, that this intuitive understanding, if attainable at all, will not stem from either a repeal of reason or an oversimplification, but from a more rational and complex argument going deeper into the nature of things. This complexity and depth culminates with the mechanismic explanation of laws and theories, to which we now turn.

Problems

9.4.1. C. Bernard explained the toxicity of carbon monoxide as an outcome of its combination with haemoglobin, as a result of which haemoglobin is no longer capable of transporting oxygen. Analyze this case in terms of the distinction between subsumptive and interpretive explanation.

9.4.2. According to an influential account of explanation, it would always consist in rendering familiar what is unfamiliar, in identifying the new with the old—identity being in turn the core of reason. See, e.g., E. Meyerson, *De l'explication dans les sciences* (Paris: Payot, 1921). Examine this doctrine.

9.4.3. What is the position of wholism (the metaphysics that accompanies vitalism, gestaltism, and phenomenalism) with regard to the atomistic analysis demanded by Democritus' program?

9.4.4. Associationist psychology provides no mechanism whereby two or more stimuli become associated in an animal. Is an interpretive explanation of association desirable and possible? If so, of what kind? If not, why? *Alternate Problem:* Report on J. A. Fodor, "Explanations in psychology", in M. Black, ed., *Philosophy in America* (Ithaca, N. Y.: Cornell University Press, 1965).

9.4.5. Examine the contention that the human personality can only be understood intuitively and accordingly human psychology must be a "comprehensive" (*verstehende*) discipline rather than an explanatory *(erklärende)* science.

9.4.6. What distinguishes the specialized understanding achieved through familiarity with deep (representational) theories from the intuitive understanding (*Verstehen*) demanded by certain schools in psychology and sociology?

9.4.7. It is widely held that explanation is requested just when a conflict arises between a new, surprising fact and prior belief—as when an ordinarily meek person is observed to act aggressively or gallantly. Does this account for scientific explanation?

9.4.8. Explanation is a kind of analysis. Now, every analysis reduces (logically) an analysandum to certain items (ideas, things, motions, etc., as the case may be) which are not further analyzed in the given context. Are these items (factual or conceptual "atoms" of some sort or other) absolutely atomic, i.e. unanalyzable?

9.4.9. Study the relevance—favorable, unfavorable or nil—of the various kinds of simplicity to explanation. In particular, find out whether the simpler explanation is always to be preferred. But before performing this task make clear what kind(s) of simplicity (formal, semantical, epistemological, pragmatic) you have in mind. In case formal simplicity alone is concerned, state whether it refers to the explanans premises, to the derivation, or to both. *Alternate Problem*: Is reference to another level necessary or just sufficient to produce an interpretive explanation?

9.4.10. There are essentially two accounts of the alleged phenomenon of thought transmission. One is the assumption of a direct inter-mind action without any physical intervening medium or mechanism. The other is the "emission" or "field" hypothesis, according to which a substance or wave of some unspecified kind is involved. The former is the orthodox doctrine; the latter is rejected by most parapsychologists. Why? And should we prefer the second hypothesis just because it is representational? *Alternate Problem:* There are essentially two kinds of electromagnetic theory: the Ampèrian (direct interparticle action) and the Faradayan (action by contact through a field). Examine the kinds of explanation they supply, find out whether they have the same coverage, and whether the action at a distance theories are actually independent of the field theories.

9.5. *Mechanismic Explanation and Reduction of Laws

What has so far been said about the subsumptive and the mechanismic explanation of facts can be extended, *mutatis mutandis,* to the explanation of law statements and systems of such, i.e. theories. The derivation of a law from a set of laws and subsidiary assumptions (see Sec. 9.3) may be regarded as a subsumptive explanation of the given law. If this deduction makes no reference to a lower level of reality, then it is a purely logical process that fails to explain how the given pattern came into being. In other words, the subsumptive explanation of a law formula does not account for the *emergence* of the nomic pattern which the formula is supposed to refer to. Such an account of the birth of patterns is performed by the mechanismic explanation of laws.

It might be thought that the task of explaining the emergence of nomic patterns is hopeless or even "meaningless": that laws are what they are and this is all there is. Yet a critical mind should regard ontological ultimates as suspect, and might justify in the following way the search for mechanismic explanations of scientific laws. If objective laws are patterns of facts, and if they do not tower above reality but are the canvas of reality, then it does make sense to inquire how a given objective law has come to be—e.g., how a given pattern of chemical composition has come about. This is part of the question how the world came to be what it is now. (Which is an entirely different question from the ill-conceived query 'Why is there a world?') In short, once a law has been discovered it seems legitimate to search for the *mechanism* of its emergence, if any. It would be as dogmatic to deny a priori that there might be such an emergence (at least in the case of supraphysical laws) as to maintain that all laws, including the physical laws, ought to be given a mechanismic explanation.

Now, emergence can be from a coexisting (lower or higher) level or from a prior stage in a process. Accordingly, as regards its ontological import the mechanismic explanation of laws can be of either of the following kinds: (i) in terms of different coexistent levels, or (ii) historical. An interpretation of the first kind obtains when a law is shown to emerge from patterns characterizing a *coexisting* lower or higher level, as when laws of social tension are explained in terms of conflicting economic and cultural objectives, or when patterns of the relief of social tension are explained in terms of the acceptance of new systems of habits and ideas. Historical explanation, on the other hand, consists in the tracing of the *evolution of a law,* by showing how it arose in the course of time from patterns characterizing earlier stages in an evolutionary process, as when a new pattern of social behavior is given a historical explanation. Let us examine these two kinds of explanation.

The most usual expretation of laws is in terms of lower levels; this is, in particular, the case of the explanation of macrolaws in function of microlaws. Important cases have been: (i) the explanation of the laws of light rays in terms of the laws of interfering wavelets, i.e. the explanation of geometrical optics by the wave theory of light; (ii) the explanation of some laws of thermodynamics in terms of the statistical mechanics of molecules; (iii) the explanation of the laws of chemical binding in terms of the laws of atomic physics; (iv) the explanation of the laws of heredity in terms of the laws "governing" the behavior of genes, and the latter in turn in terms of the laws "governing" the behavior of *DNA* molecules. In all these cases the lawful behavior of a whole (light ray, thermodynamical system, molecule, organism), or rather of some of its aspects, is explained as the outcome of the lawful behavior of all or some of its parts. Briefly, macropatterns are shown to emerge from micropatterns.

A particularly important and characteristically modern kind of law explanation in which two coexisting levels occur is *statistical explanation.* This is, in fact, the explanation of a regularity at a given level as the outcome of the random interplay of a large number of items on a lower level, as when the magnetization of a piece of iron is explained as a macroscopic outcome of the force exerted by an external field on the initially randomly oriented tiny magnets that constitute the large scale body. Statistical explanation requires both an *assumption of randomness* (resulting from the mutual independence or nearindependence of the individual components) and the knowledge of the *nonstatistical* laws referring to the individual components. In other words, statistical explanation consists in the derivation of collective laws from the individual laws "governing" the behavior of the members of a statistical aggregate.

The laws of the members of the statistical aggregate may in turn be nonprobabilistic ("deterministic") or probabilistic ("indeterministic" or stochastic)—but not statistical, because there is no statistics unless there is a statistical aggregate. An example of explanation of statistical laws in terms of nonprobabilistic individual laws is afforded by the derivation of some thermodynamical laws on the basis of the laws of classical mechanics applied to a large number of nearly independent molecules. (The complete and rigorous derivation of classical thermodynamics from mechanics and statistical hypotheses is so far an open problem.) An illustration of the explanation of statistical laws in terms of individual probability laws is offered by the laws of averages derived on the basis of quantum mechanics; among such laws of averages we find Newton's second law of motion. (It should be pointed out that not every law of quantum mechanics is probabilistic or stochastic; thus, e.g., the formulas for the energy levels of stable systems are not probabilistic. The probabilistic laws of quantum mechanics acquire a statistical interpretation when applied either to an assembly of similar systems or to a set of similar observations on a single system. The point here is that Probabilistic≠Statistical.)

In either case a statistical explanation consists in showing how collective regularities come about from individually lawful behavior on a different level—whether the patterns of this individual behavior are "deterministic" or stochastic. In other words, a statistical explanation consists in discovering how collective (statistical) patterns result from the interplay of numerous individuals of a certain kind (molecules, persons, etc.). Such patterns emerge when certain individual variations cancel out and build up *stable averages*, so that the former can be neglected (at the macrolevel) and attention can be focused on the latter. In short, statistical laws arise when there is randomness in some respect, e.g. as regards the individual trajectories of the molecules in a gas or the distribution of genes in the fertilization of an egg.

For the purpose of illustration let our statistical generalization (the explanandum) be "Nearly half of the new-born children are girls". A simplified statistical explanation of this law could be along the following lines:

*A*1. Every human female body-cell contains a pair of identical sexual chromosomes, *XX*. (Individual law linking three levels: organismic, cellular, and chromosomic.)

*A*2. Every human male body-cell contains two different sexual chromosomes: *XY*. (The same as before.)

*A*3. Every human sexual cell (gamete) results from the division of a body cell in two parts, with one sexual chromosome each. (Individual law linking the cellular and the chromosomic levels.)

*A*4. The union of a male and a female gamete into a zygote is independent of their chromosome content, i.e. it is genetically random. (Hypothesis of random genetic shuffling.)

*T*1 (from *A*1 and *A*3). Every female sexual cell (gamete) contains one *X*-chromosome.

*T*2 (from *A*2 and *A*3). Of the two male sexual cells (gametes) resulting from the division of a body cell, one contains an *X*-chromosome and the other a *Y*-chromosome.

*T*3 (from *A*4, *T*1, *T*2 and elementary probability theory). The probability of an *XX* egg is the same as the probability of an *XY* egg and either equals $\frac{1}{2}$. (See Fig. 9.6.)

*T*4 (from *T*3 and the Bernoulli theorem on large numbers). Nearly half of the fertilized eggs produced by a large population is of the *XX* (female) type and the other half of the *XY* (male) type.

Let us emphasize that *A*1 and *A*2 are *three-level* laws concerning individuals; *A*3 is a *two-level* law concerning individuals; and *A*4 is a two-level law concerning the building of an aggregate of microentities.

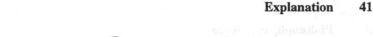

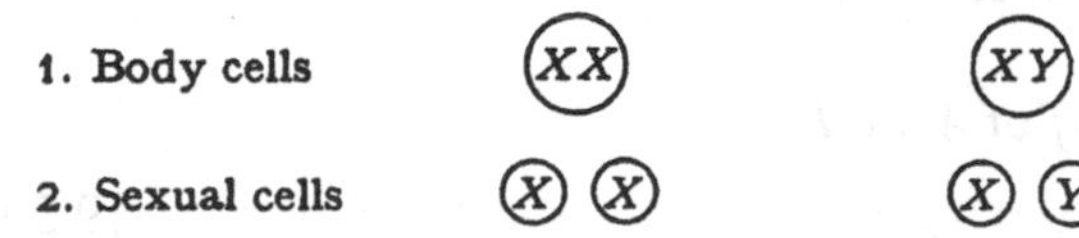

Fig. 9.6. Statistical explanation of the half-female and half-male composition of a human population.

*T*1, *T*2 and *T*3 are individual inter-level laws; whereas *T*1 and *T*2 are nonprobabilistic, *T*3 is probabilistic. Of the four axioms only one, *A*4, contains the randomness concept, and this makes it possible to apply the mathematical theory of probability. Finally, *T*4 is a statistical interlevel-law. The word 'nearly' occurring in it is typical of statistical law statements; it means that the percentage of fertilized eggs of the female type is not exactly $\frac{1}{2}$ but fluctuates around this value. Any departure from this value should be temporary; but if a systematic long-run deviation from the fifty-fifty composition were found, an additional mechanism would have to be superimposed on genetic shuffling. Anyway, *T*4 yields almost what we wanted. In order to obtain the explanandum—which refers to babies rather than to zygotes—we need only introduce the embryological law "Every human individual develops from a fertilized egg". This and *T*4, two inter-level laws, yield our one-level statistical generalization.

Statistical explanation is, then, *inter-level mechanismic explanation.* It is a kind of explanation of laws, not an explanation of individual facts with the help of statistical laws. This point has to be stressed in view that philosophers often suppose that, when scientists speak of statistical explanation, they think of the subsumption of a singular proposition under a statistical generalization according to the schema

$$\begin{array}{l} \text{Almost every } A \text{ is a } B \\ c \text{ is an } A \\ \hline c \text{ is almost certainly a } B \end{array} \qquad [9.5a]$$

or, perhaps, according to the more refined pattern

$$\frac{\begin{array}{l}\text{A fraction } f \text{ of } A \text{ are } B\\ c \text{ is an } A\end{array}}{\text{The probability that } c \text{ is a } B \text{ equals } f} \qquad [9.5b]$$

This schema, the so-called *statistical syllogism, is* analogous to the explanation schema "$\{G, C\} \vdash E$" but it is not an explanation proper because it is not a valid deductive argument. In fact, the would-be explanandum does not follow from the given premises. It cannot, if only because the explanandum is a theoretical statement containing the theoretical concept of probability, whereas the pseudoexplanans premises might, for all we know, be an empirical generalization ("A fraction *f* of *A* are *B*") and a singular empirical proposition ("*c* is an *A*"). The explanandum is a probability statement, and probability statements are entailed only by probability statements.

The schemas [9.5a] and [9.5b] are not explanations; nor are they "inductive arguments" either, because they do not consist in generalizations on the strength of observed instances. Those schemas are patterns of *statistical inference* and, more precisely, [9.5b] exemplifies the following inference pattern:

Rule 1. If an individual belongs to a certain class [the reference class *A*] which overlaps with another class [the attribute class *B*] to the extent *f* [i.e., the measure of $A \cap B$ over the measure of *A* equals *f*], assign that individual the probability *f* of being in the attribute class. In other words: if the observed relative frequency of *B*'s in a class *A* is *f*, assign every member of *A* the probability $P(B|A)=f$. This rule of inference encourages us to go from an *observed frequency* (a collective property) to a *hypothesized probability* (an individual property): it is a bridge between observation and theory. Its dual, which allows us to predict observable frequencies from calculated probabilities, is

Rule 2. If an individual of the class *A* is assigned the probability $P(B|A)=p$ of being also in the class *B*, take *p* as a fair estimate of the observable frequency *f* of its occurrence in the intersection of *A* and *B*. Rule 2 bridges the gap between interpreted stochastic hypotheses and theories, on the one hand, and their statistical test on the other. But, just as in the case of Rule 1, we have to do here with a *fallible rule of nondeductive inference*, not with an explanation.

What scientists call a *statistical explanation* of a law is the interpretive explanation of a law in terms of the behavior patterns of the members of a statistical aggregate. In other words, a statistical explanation of a law is an argument showing that, far from being fundamental the law concerned is derived and, moreover, owes its existence to the random interplay of individually determinate entities. (Incidentally whenever a macrolaw is derivable from microlaws, it is possible to compute the probability of any given deviation of the actual course of nature from the average behavior predicted by the macrolaw. Such a probability may then be regarded as a measure of the error of the macrolaw relative to the corresponding microlaws.)

Now a few words on the historical explanation of laws, i.e. on the explanation of the emergence of certain objective patterns as the outcome of a gradual transformation of patterns characterizing prior stages of a process. This kind of explanation is less frequent than inter-level explanation, yet more common than might be thought. In fact, the ultimate goal of the study of the evolution of anything is the discovery of the evolution of nomic patterns. Thus, e.g., the study of the evolution of learning is essentially an inquiry into the transformation of learning patterns, i.e. the modification of old patterns and the emergence of new learning patterns. The ontogeny of such patterns in the human being, from automatic adjustment to trial and error and back, and from these to conceptual learning, is familiar to every parent; what is far from familiar is the mechanism of such an individual evolution. Even less known, but under active research, is the phylogeny of behavior patterns. That this study must parallel the study of biological evolution seems obvious, especially since learning is a feature of animal life and a prerequisite of survival. That the evolution of learning cannot be reduced to biological evolution is less obvious yet equally likely, particularly since we know that not every major biological progress has resulted in improved learning patterns. Thus, e.g., amphibians and reptiles do not seem to have higher learning capabilities than fishes; on the other hand, although the anatomy of man differs but little from that of apes, we seem to be justified in believing ourselves equipped with learning (and unlearning) capabilities not only quantitatively higher but also qualitatively different from those of our distant cousins.

We have just suggested that the evolution of learning patterns does not seem to be *reducible* to biological evolution. We saw earlier, in Sec. 9.3, that other patterns, such as that of nose bleeding at high altitude, are reducible: quite apart from the biological consequences of nose bleeding, this is basically the same as pen dripping at high altitudes: it has nothing peculiarly biological in itself. Generally, a *reductive explanation* is an argument showing that a fact (or a generalization) that apparently belongs to a given level of reality (e.g., the level of life) is nothing but a fact (or a generalization) belonging to a different (usually lower) level.

Reduction, then, consists in explaining *away,* as purely apparent, the peculiarities assigned to facts or generalizations that, at first sight, had been regarded as exclusive of a given level but which, on closer examination, are shown to belong to a different level. Thus, e.g., when we account for animal behavior in terms of reflexes we explain animal mind away or, equivalently, we reduce it to a system of automatisms: we perform a *reduction to a lower level.* And when we argue that automatic computers look intelligent just because men have designed, built, and programmed them in an intelligent way, we explain automata intelligence away: in this case we perform a *reduction to a higher level.*

The class of physicalist ontologies may be regarded as a set of programs for achieving a reductive explanation of every fact in terms of facts belonging to the physical level via a reduction of properties and laws. Properties would be reduced by way of explicit definitions of the higher-level properties in terms of physical properties, and laws would be reduced via the deduction of nonphysical laws from physical ones. The class of spiritualistic ontologies, on the other hand, may be construed as the set of programs for the reductive explanation (or, rather, obscuring) in terms of the highest level. This reduction, however, is not made via a careful reduction of properties and their mutual relationships (laws) but in a wholesale and confused way.

A look at the history of science should leave little doubt that, of the two extreme reductionist programs, physicalism and spiritualism, the former has been fertile whereas the latter has been barren if not obstructive, and has thrived on the failures of physicalism. Fertility, however, is just *a* symptom of truth. The failures of any cognitive program may be just temporary difficulties but, in the case of physicalist reductionism, we can explain its failures as deriving from a fatal mistake. This mistake is believing that every *logical* reduction—such as a definition or a deduction—warrants an *ontological* reduction, i.e. an elimination of levels. From the numerical equality of the average kinetic energy of an assembly of molecules and the assembly's temperature

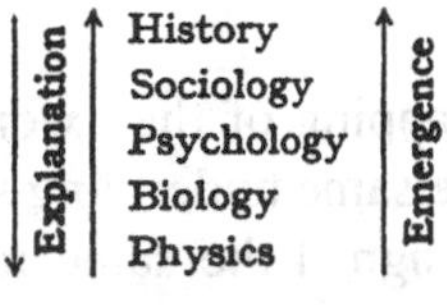

Explanation ↓ | History
Sociology
Psychology
Biology
Physics | ↑ Emergence

Fig. 9.7. Interlevel explanation, or linking of laws and theories on one level to laws and theories on one or more different levels.

we cannot conclude that "at bottom" temperature is "nothing but" a microscopic mechanical property, just as from the interchangeability of 'Plato' and 'Socrates's best disciple' we are not allowed to infer that "at bottom" Plato was already contained in Socrates. To establish relationships among concepts is not the same as to eliminate some of them in favor of the remaining properties.

For example, one might attempt to explain the reader's interest in philosophy—a mental property ψ—as a function F of both a complex of biological properties β and another complex σ of sociocultural variables. If ψ belongs to the n-th level, β belongs to the $(n-1)$-th level and σ to the (n+1)-th level. But the discovery of such a relation would not entitle us to say that we have *reduced* ψ to β and σ; we might say, on the other hand, that we had achieved a *nonreductive interlevel explanation*. And establishing such a relationship among different levels is surely more rewarding and exciting that either leaving them unrelated or crushing them down to the lowest level or up to the highest. (See Fig. 9.7.)

Genuine reduction must therefore be distinguished from interlevel relation. Properties and laws are genuinely reduced to "other" properties and laws if they are shown to be in fact *the same* despite appearances to the contrary. And a theory is genuinely reduced to another theory if the reduced theory is proved to be a *subtheory* of a richer theory (the reducing theory), so that the laws of the reduced theory turn out to be among those of the reducing theory. (For the concept of subtheory, see Sec. 7.3, Theorem 1.) In other words, full reduction obtains just in case one theory is included in another, whence the former can be deduced from the latter—as in the case of statics in relation with dynamics, and a few other cases. Every other case of intertheory relation requires additional assumptions.

Some important cases of theory interrelation, which are often mistaken for reductions, are isomorphism, partial overlapping of the predicate basis, and partial overlapping of the axiom basis. If two theories are *isomorphic,* they have the same underlying structure but their primitives are not all of them assigned the same interpretation in the two interpreted theories or models. (The converse is not true: it often happens that an abstract theory, such as group theory, has nonisomorphic models. Only the interpretations of *categorical* theories are isomorphic.) A case in point is the relation between matrix mechanics and wave mechanics: they are isomorphic, but some of their primitive symbols have not the same meaning—hence they are mathematically equivalent but physically inequivalent.

Consider next the case of two theories which do not stand in the relation of inclusion but share some basic concepts *(partial overlapping of their predicate bases*). In principle, relations between the two predicate sets can be established, with the common concepts serving as links: a whole third theory bridging the two or even embracing them can in principle be built. This is the case of electrodynamics with respect to dynamics and electromagnetic theory. Finally, two theories may have a partial overlap not only because they share some predicates but also some statements *(partial overlapping of their axiom bases).* For example, quantum mechanics retains in its basis certain classical law statements (regarding, e.g., the relation of energy eigenvalues to the momentum eigenvalues). Neither of the three cases is one of genuine reduction; in none of them can we properly say that one of the theories is reduced to the other. Of the three relations we have indicated only the first—isomorphism—has been studied in detail owing to its mathematical and logical interest. The determination of the actual formal and semantical relations among existing factual theories is an open problem and has barely been approached despite its importance.

In brief, it is often possible to frame a mechanismic explanation of laws. Such an explanation brings in the ontological hypothesis of the multilevel structure of reality (see Sec. 5.9). And this hypothesis alone seems to be able to explain (i) why we have separate sciences for dealing with physical systems, organisms, and societies respectively, and (ii) that these separate sciences are related to one another.

Problems

9.5.1. Exemplify same-level and new-level interpretive explanations.

9.5.2. Would you regard the following statistical generalizations as fundamental (irreducible or at least unreduced) laws, or would you search for the mechanism of their emergence? (i) "In every country the ratio of new-born boys to new-born girls increases during wars or immediately after them." (ii) "Children born in autumn or in summer tend to be more intelligent than any others." (iii) "Children of talented (stupid) parents tend to be less talented (stupid) than their parents."

9.5.3. Kirchhoff's first law for electric networks, "The algebraic sum of the current intensities at each knot in a mesh is zero", follows from the principle of conservation of electric charge, which holds in turn for elementary charges (electrons). Examine the derivation and comment on it. *Alternate Problem:* Do the same with any other law statement.

9.5.4. Comment on the following remarks by C. A. Coulson in "Present State of Molecular Structure Calculations", *Reviews of Modern Physics*, **32**, 170 (1960), p. 173: whereas some chemists arc primarily interested in accuracy, others are chiefly interested in explanation. For example, the latter "want to know why the *HF* [hydrogen-fluorine] bond is so strong, when the *FF* [fluorine-fluorinej bond is so weak. They are content to let spectroscopists or physical chemists make the measurements; they expect from the quantum mechanician that he will explain *why* the difference exists". The explanation must not be that the electronic computer shows that the binding energies are stronger in the case of the *HF* bond than in the case of the *FF* bond. "Any acceptable explanation must be in terms of repulsions between nonbonding electrons, dispersion forces between the atomic cores, hybridization and ionic character." Construe this text in terms of representational theory and mechanismic explanation.

9.5.5. Psychologists have often contrasted what have been called two types of scientific explanation: *reductive* explanation (by reference to lower levels) and *constructive* explanations (in terms of higher-order constructs). See, e.g., M. H. Marx, "The General Nature of Theory Construction", in M. H. Marx, Ed., *Psychological Theory* (New York: MacMillan, 1951), pp. 14–15. Are those two kinds of explanation? Is not every scientific explanation of everyday individual behavior done both by reference to two different levels (the physiological and the social) and in terms of theoretical constructs?

9.5.6. Examine the contention of holists and intuitionists that the emergence of novelty cannot be explained but must be accepted blindly. Is this thesis consistent with contemporary chemistry and biology?

9.5.7. Perform a logical and epistemological study of the chemical explanation of the basis of biological reproduction (gene replication) put forward by J. Watson and F. Crick in 1953. In particular, examine the double-helix model of chromosome structure in relation with the conventionalist commandment "Thou shalt not make thee any graven image of things unseen".

9.5.8. According to thermodynamics, isolated systems approach a state of thermal equilibrium characterized by maximum entropy. Statistical mechanics, in attempting to explain this law in terms of lower level entities and processes, reaches the conclusion that the entrops of an isolated system remains constant. Examine this unsolved conflict which stands in the way of the complete logical reduction of thermodynamics to statistical mechanics—a reduction taken for granted by the philosophers who write on reduction. See J. M. Blatt, "An Alternative Approach to the Ergodic Problem", *Progress of Theoretical Physics*, **22**, 746 (1959). *Alternate Problem:* Study the algebra of intertheory relations. Make use of filter theory, sketched in Sec. 7.3.

9.5.9. Examine the statistical explanation of the unlikelihood (next to impossibility) of the inversion of processes such as the mixing of coffee and milk or the transformation of men into hominids. *Alternate Problem:* Report on the present state of the problem of reduction. See J. H. Woodger, *Biology and Language* (Cambridge: University Press, 1952), pp. 271 ff.; J. G. Kemeny and P. Oppenheim, On Reduction", *Philosophical Studies,* VII, 6 (1956); E. Nagel, *The Structure of Science* (New York and Burlinghame: Harcourt, Brace & World, 1961), Ch. 11; Q. Gibson, *The Logic of Social Enquiry* (London: Routledge & Kegan Paul, 1960), Ch. ix; M. Bunge, *Philosophy of Physics*, Chap. 9 (Dordrecht-Boston: Reidel, 1973) and *Finding Philosophy in Social Science* (New Haven: Yale University Press, 1996); E. Agazzi, Ed., *The Problem of Reductionism in Science* (Dordrecht-Boston: Kluwer, 1991).

9.5.10. Expand on Peirce's remarks concerning the "natural history of laws of nature", i.e. the account of laws of nature as results of some evolutionary process. See C. S. Peirce, "The Architecture of Theories" (1891) and letter to Lady Welby (1911), in P. P. Wiener, Ed., *Values in a Universe of Chance* (New York: Doubleday Anchor Books, 1958). Note that Peirce may not have had in mind an autonomous evolution of laws, since he conceived of a law in a way reminiscent of Heraclitus, namely "as a reasonableness energizing in the world" (loc. cit., p. 300). In short, according to Peirce "it is the law that shapes the event" (ibid.). Speculate on the possible genesis of laws in a self-sustaining universe. *Alternate Problem:* Examine the attempts to explain (reductively or otherwise) the behavior patterns and the thought patterns as brain processes. D. O. Hebb, *Essay on Mind* (Hillsdale, N.J.: Lawrence Erlbaum, 1980); M. Bunge and R. Ardila, *Philosophy of Psychology* (New York: Springer, 1987); S. M. Kosslyn and O. Koenig, *Wet Mind: The New Cognitive Neurosciences*, 2nd ed. (New York: The Free Press, 1995).

9.6. Explanatory Power

Any given fact can, in principle, be given an unlimited number of mutually incompatible explanations: think of the ways in which a politician can excuse his defeats. The possible explanations of a generalization are less numerous yet not single either. How shall we pick out the most adequate explanation? Although there seems to be no foolproof battery of criteria, if we rest content with fallible criteria we shall find them by examining the actual practice of weighing alternative scientific explanations.

To begin with, we have a criterion for spotting suspect explanations: namely, their degree of ad hocness. An explanation is called *ad hoc* if the generalization it contains has been tailored to suit the given explanandum alone. Suppose someone comes out with the alleged generalization: "Women are no good at the medical profession." A number of *ad hoc* explanations of this "fact" can immediately be invented: women are absorbed by family life and consequently they neglect their other duties, are less intelligent, tenacious and ambitious than men, etc. Should someone point out that some of these alleged defects would disqualify nurses—who are known to do well—and that in certain countries a large percentage of doctors are women, and even proficient, another set of explanatory hypotheses can readily be invented: women are more sensitive to human suffering, they are more patient and comprehending than men, they have an acute intuition suitable for clinical guesswork, etc. As can be seen, *ad hoc* explanations are very much like excuses.

The chief traits of *ad hoc* explanation are these. First, it is *post factum,* in the sense that the generalizations it contains fit the data but could not have predicted them: in other words, *ad hoc* explanation is wisdom after the event. Second, it is rigid or *incorrigible:* having been contrived to fit a given explanandum, it must be given up—or else shielded by further *ad hoc* conjectures—as soon as the explanandum itself is found defective. Third, it is *not generalizable* except by accident: there would be no ground whatsoever to extend it to other cases. These features of *ad hoc* explanation stem from its original sin: it is not built with the help of genuinely general and systemic generalizations—i.e. with law statements proper. *Ad hoc* explanation, in other words, is not theoretical explanation, i.e. derivation performed in the bosom of a theory.

Isolated generalizations will provide opportunistic "explanations" in any number. Only generalizations included in theories will supply principled explanations and indeed in a limited number—in fact, exactly one per theory. There still will be more than one possible explanation of either a singular or a general factual statement, but the set of possibilities is now as numerous as the set of relevant rival theories—and this is always a manageable set. The reason for the drastic limitation on the number of possible explanations when the requirement of systemicity or theorification is enforced, is simply this: in a factual theory we are not allowed to introduce extra general premises that will yield the wanted explanandum but have to count on the theory's initial assumptions alone; in particular, we are prevented from introducing premises containing predicates foreign to the theory. (See Sec. 7.2 on semantical consistency.) In other words, the application of the addition rule,which countenances opportunistic deduction, is severely limited in factual theories by the requirement of semantical consistency. Consequently, only a scientific theory can have an appreciable explanatory coverage: one that is not accidental but derives from the universality of the laws included in the theory.

Within bounds, the best theory is the one with the greatest coverage. (We shall see in Sec. 10.4 what such bounds are.) The historical sequence of theories in a given field should then exhibit, roughly, an increasing coverage. A good illustration of this trend is afforded by

Table 9.3. Stages of optics

Typical facts and laws	Ray optics (Hero)	Hydrodynamic aether theory (Descartes)	Corpuscular theory (Newton)	Longitudinal wave theory (Huygens)	Transversal wave theory (Fresnel, Cauchy, Green)	Electromagnetic theory (Maxwell)	Electromagnetic theory without aether	Electromagnetic theory with curved space	Electromagnetic theory and electron theory	Quantum electrodynamics and nonrelativistic Q.M.	Quantum electrodynamics and relativistic Q.M.
1. Rectilinear propagation	×	×	×	×	×	×	×	×	×	×	×
2. Reflection	×	×	×	×	×	×	×	×	×	×	×
3. Refraction	×	×	×	×	×	×	×	×	×	×	×
4. Extremal time travel	×	×	×	×	×	×	×	×	×	×	×
5. Dispersion	×	×	×	×	×	×	×	×	×	×	×
6. Superposition				×	×	×	×	×	×	×	×
7. Double refraction				×	×	×	×	×	×	×	×
8. Decrease of speed in transparent media				×	×	×	×	×	×	×	×
9. Diffraction				×	×	×	×	×	×	×	×
10. Interference				×	×	×	×	×	×	×	×
11. Doppler effect				×	×	×	×	×	×	×	×
12. Polarization					×	×	×	×	×	×	×
13. Radiation pressure			×			×	×	×	×	×	×
14. Anomalous dispersion									×	×	×
15. Invariant speed light							×	×			×
16. Change of frequency in gravitational field								×			
17. Light scattering									×	×	×
18. Backbody spectrum										×	×
19. Photoelectric effect										×	×
20. Compton effect										×	×

the sequence of theories of light—a sequence that is far from finished. Look at Table 9.3. From this table we see that the most advanced theories of light account for nearly all the typical facts and low-level laws, but no theory explains them all. In particular, the increase (decrease) in the frequency of a light wave when going down (up) a gravitational field is explained only by the intersection of the electromagnetic theory and the relativistic theory of gravitation, which on the other hand fails to explain the quantum effects 17 to 20. This hole suggests that one of the next tasks should be the building of a theory meeting at the same time the fundamental requirements of quantum mechanics (abbreviated to 'Q.M.' in the table) and general relativity. Although a number of attempts have been made, and partial successes have been achieved, no satisfactory synthesis has so far been performed. Light is still somewhat obscure.

The above example shows, first, that the older (and usually simpler) theory *can* very often be used to account for a number of facts, namely the more familiar or known of older. It would be foolish to employ a quantum theory of light to compute the angle of refraction from the angle of incidence and the refractive index—just as foolish as to ignore the quantum theory in the derivation of the phenomenological refractive index from lower-level properties and laws. (Quantum theories of light are needed only: (i) to explain the typical quantum effects that can be expected when a few photons at a time are at stake or when they act on electrons, and (ii) when the emergence of macroscopic light patterns out of quantum patterns is to be explained.) A second lesson derived from Table 9.3 is that, although in most cases a new theory will absorb the older theory, in some cases the overlapping of the validity domain of successive theories is not complete and, as a consequence, some of the older theories *must* still be used to account for its peculiar explananda.

In view of the number of theories that often fight over a single domain of facts, it would be convenient to have a measure of the actual explanatory efficiency of a theory. A first rough measure is suggested by the maxim "Explain the most with the least", which is one of the formulations of the rule of thought economy. In fact, let a theory T start with n mutually independent initial assumptions and suppose T absorbs (explains) N otherwise mutually independent generalizations—the founder hypotheses. Then the following function ρ may be regarded as a measure of the *explanatory efficiency or yield* of T:

$$\rho(\mathrm{T})=1-\tfrac{n}{N}. \qquad [9.6]$$

If T happens to be just an assemblage of the N founder generalizations, i.e. if T is an *ad hoc* systematization, then since $n=N$ its yield will be nil. But if T is a theory proper, it will subsume a very large (possibly infinite) number of generalizations, whence its explanatory yield will approach unity. Clearly the preceding elucidation of the concept of explanatory efficiency, or performance, is too crude, for it takes for granted that all the statements concerned are completely, rather than partially, true. In other words, the above formula accounts for the *range* not for the *accuracy* of scientific explanations.

Indeed, one thing is the *range* or extension of the set of facts and low-level laws covered by a theory and quite another thing is the *accuracy* or adequacy of such a coverage. Different theories may account for a given set of explananda with different degrees of accuracy. If their range is roughly the same the more accurate among them will be preferred, whereas if their accuracy is roughly the same the one with the wider range will be preferred. The range of a theory can be extended almost without limit if the accuracy standards are relaxed. Conversely, the accuracy of a theory can be enhanced by narrowing down its range: a theory tailored to suit a single fact can be as accurate as desired. But accuracy without range is worthless because it can be attained in a number of ways with *ad hoc* constructs. And sheer range is worthless too: all-purpose constructions, such as those revolving on vague and inscrutable concepts ("repression", "gestalt", "cultural diffusion" and the like), are much too accomodating. Accordingly, (i) the coverage of a theory must be analyzed into its range and its accuracy, and (ii) range and accuracy must be requested jointly rather than separately: the requisite of range should be checked by the requisite of accuracy and conversely. We will not get both universal range and perfect accuracy, but we may attain maximum range compatible with reasonable accuracy.

The range of a hypothesis or theory may in turn be regarded as a set made up of the following subsets: (i) the set of known explananda, i.e. of available data and low-level constructs; (ii) the (unknown) set of unknown explananda of the same kind as those of the previous subset; (iii) the set of known explananda of a different kind, not taken into account when the construct was originally proposed; (iv) the (unknown) set of as yet unknown explananda of a different kind, not originally envisaged (e.g., the radiation pressure, explainable on Maxwell's electromagnetic theory but not known at the time of the latter's inception); and (v) the set of logically possible explananda, conceivable on alternative constructs but impossible according to the given construct: i.e. the set of non-facts accounted for by the construct (see Sec. 9.3). Clearly, only the subsets (i) and (iii) are limited and might accordingly be assigned a definite measure. The remaining sets are open: they may or may not expand from the time the estimate of the theory's coverage is made. Consequently, although the range of a theory may in a way be regarded as given along with the theory itself, the determination of the range's measure will change in time. Of course, we do not "discover" the elements that make up the various subsets but stepwise build these sets ourselves. At any rate, the range of a theory is not determined once and for all. What can be determined at the time of its inception is the *minimum* range, consisting in the subset (i) of facts and generalizations known at that time to be covered by the theory; and, later on, the *actual* range of the theory, equal to the logical sum of the subsets (i), (iii) and eventually (v), can be determined. Similarly, the accuracy assigned to a theory will vary with time since it depends on the available techniques of empirical test and on the other theories taking part in the test of the given theory. (For the last point see Sec. 8.4.) Accordingly, the coverage of a theory will be judged differently at different times, and at any given time we can estimate the *actual* coverage of the theory, never its *potential* coverage—which need not overlap with the actual coverage. The evaluation of the potential coverage of a theory—of all that the theory could yield—would require having access to all the testable consequences of the theory and to every possible fact relevant to the theory.

The concept of *actual coverage* of a theory can be quantified in terms of the concept of truth. Let T be a qualitative factual theory and $\{t_i\}$ the set of its testable consequences. The truth value of the t_i will be judged on the antecedent knowledge A (not including T) and whatever empirical evidence E may be relevant to the t_i. This evaluation will rest on the test of a sample of the total set $\{t_i\}$. Suppose a number C of t_i are confirmed and a number R of them are refuted. If we just add the individual truth values of the theory's tested consequences, we get $C.(+1)+R.(-1)=C-R$ on assigning the number +1 to truth and −1 to falsity. We may take this as a measure of the theory's *empirical* truth value—the truth value assigned to the theory on the strength of empirical tests. We average the empirical truth value with the sum of the theoretical truth values of the t_i, which in this case are all 1 since the t_i are assumed to be validly deduced in T. We obtain thus the total truth value or *coverage* of T:

$$C(\mathrm{T})=(\tfrac{1}{2})(C-R+N)=C, \text{ with } N=C+R \qquad [9.7]$$

The coverage thus defined coincides with the actual (not the potential) *range* of the theory. When the number N of tests is large it makes sense and it is convenient to refer to the fraction of hits or *relative coverage* $C_r(\mathrm{T})$:

$$N>>1,\ C_r(\mathrm{T})=\frac{C}{N}, \qquad [9.8]$$

a number between 0 and 1. We may also regard $C_r(\mathrm{T})$ as the *degree of confirmation* of T—as long as we do not feel tempted to equate it with a probability just because it is a fraction between 0 and 1. Here go some typical cases:

$$\begin{array}{lll} C=N,\ R=O & (\textit{all-round confirmation}), & C_r(\mathrm{T})=1 \\ C=O,\ R=N & (\textit{all-round refutation}), & C_r(\mathrm{T})=0 \\ C=R=N/2 & (\textit{fifty-fifty}) & C_r(\mathrm{T})=\frac{1}{2} \end{array} \qquad [9.9]$$

*The foregoing formulas can easily be generalized to a quantitative theory tested with quantitative accuracy. Calling $V(t_i \mid \mathrm{AE})$ the degree of truth of t_i as judged on the antecedent knowledge A and the empirical evidence E, and $V(t_i \mid \mathrm{T})$ the theoretical truth value of t_i, we may put

$$C(\mathrm{T})=(\tfrac{1}{2})\Sigma[V(t_i \mid \mathrm{AE})+V(t_i \mid \mathrm{T})], \qquad [9.10]$$

where the capital sigma designates summation over i between 1 and N. Clearly, [9.7] is a particular case of [9.10], namely when every t_i is either completely true or completely false judged on the basis of A, E and T. If a theory accounts fairly well for all N empirical situations with roughly the same error $\varepsilon(0\leqq\varepsilon<1)$ in every case, we have $V(t_i \mid \mathrm{AE})=1-2\varepsilon$ and consequently

$$C(\mathrm{T})=(\tfrac{1}{2})[N(1-2\varepsilon)+N]=N(1-\varepsilon) \qquad [9.11]$$

In this particular case, then, the coverage equals the range N times the accuracy $1-\varepsilon$. If, on the other hand, the theory is all wrong, i.e. $V(t_i \mid \mathrm{AE}) =-1+2\varepsilon$, then

$$C(\mathrm{T})=N\,\varepsilon\geqq 0 \qquad [9.12]$$

which tends to zero with ε. As long as the inaccuracy of the theory—which depends both on the tested statements t_i and on the testing techniques—does not vanish, the theory has a nonvanishing coverage. That is, inaccuracy shields poor or even nil range. (In the last case the theory's range was $C=0$.)

*Just as in the case of qualitative theories (cf. [9.8]), it is convenient to introduce the concept of *relative coverage* $C_r(\mathrm{T})$:

$$N \gg 1, \qquad C_r(T) = (1/2N)\,\Sigma\,[V(t_i \mid AE) + V(t_i \mid T)] \qquad [9.13]$$

Typical cases:

$C=N, R=0$ *(all-round confirmation within error* ε*)*, $C_r(T)=1-\varepsilon$

$C=0, R=N$ *(all-round refutation within error* ε*)*, $C_r(T)=\varepsilon$ [9.14]

$C=R=N/2$ *(fifty-fifty)*, $C_r(T)=\frac{1}{2}$

*If the theory is moderately complicated, simplifying assumptions may have to be made in the derivation of the t_i. These simplifications will introduce additional (theoretical) errors $\delta_i \geqq 0$, so that $V(t_i \mid T)=1-2\delta_i \leqq 1$. Supposing (again an ideal case) that all these theoretical errors are the same, i.e. $\delta_i=\delta$, instead of [9.14] we get in the same cases

$C=N, R=0$ *(all-round confirmation within errors* δ *and* ε*)*,
$C_r(T)=1-(\delta+\varepsilon)$

$C=0, R=N$ *(all-round refutation within errors* δ *and* ε*)*, [9.15]
$C_r(T)=\varepsilon-\delta$.

$C=R=N/2$ *(fifty-fifty)*, $C_r(T)=\frac{1}{2}-\delta$.

The second formula of the triplet [9.15] covers the case, not uncommon in scientific practice, in which experimental errors mask or compensate for theoretical approximations. Further formulas, linking coverage with inaccuracy and predictive performance, will be evolved in Sec. 10.4.*

A high coverage is an important symptom of the truth value of a theory's degree of truth. Yet maximal coverage is not an absolute desideratum: maximal explanatory *depth* is an additional desideratum (see Sec. 9.4), and one that can conflict with coverage. Thus, a preliminary representational theory with an as yet restricted coverage may be worth while being worked out although the corresponding phenomenological theory may have a still larger coverage: we are interested both in comparatively inexpensive truths concerning a given level and in more adventurous conjectures concerning the relations of the given level to other levels. A theory capable of furnishing many-level explanations will clearly have a greater explanatory level than one restricted to single-level explanations: the former will supply interpretive explanations of both facts and generalizations (see Secs. 9.4 and 9.5), whereas the latter will yield subsumptive explanations

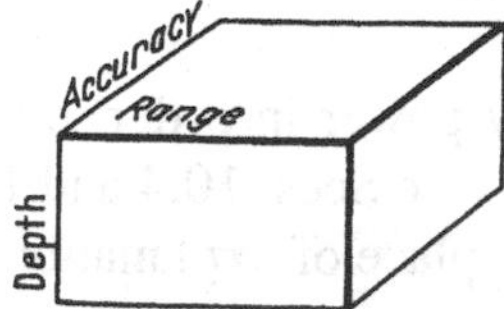

Fig. 9.8. The volume of a theory: E(T)=Range (T) × **Accuracy** (T) × **Depth** (T).

only. This suggests introducing the concept of *explanatory power* E(T) of a theory as the product of the coverage and the depth of it:

$$E(\mathrm{T})=C(\mathrm{T}) \cdot D(\mathrm{T}) \qquad [9.16]$$

The dot in this formula cannot be interpreted as an arithmetical product unless we succeed in setting up a measure of depth. This may eventually be done in the context of a theory of levels. For the time being we may adopt the naive procedure of assigning numbers (on an ordinal scale) to the various levels crossed by the theory concerned. Thus, e.g., a purely behavioral theory would be assigned depth 1, a theory containing both behavioral and psychological variables depth 2, and one containing behavioral, psychological and physiological concepts depth 3.

*For example, the explanatory power of an n-level theory accounting for all N relevant data with the accuracy 1–ε is, according to [9.11] and [9.16],

$$E(\mathrm{T})=N \cdot (1-\varepsilon) \cdot n \qquad [9.17]$$

i.e., the product of the range, the accuracy and the depth of the theory. We may accordingly picture the explanatory power of a theory as the theory's volume. (See Fig. 9.8.) The minimum value, determined by the minimum range, is all we can estimate at a theory's birth. From then onwards the volume may expand or it may shrink, but either development is hardly predictable.*

In sum, the concept of explanatory power may be analyzed into coverage and depth. In turn, coverage is determined by range and accuracy, which are determined by the degree of truth of the theory's tested consequences. Since the latter are just a sample, and not even a random sample of the total set of the theory's testable consequences, a high explanatory power is not a guarantee of the theory's truth but just one symptom of it. Further symptoms of truth must be taken in conjunction with explanatory power in order to formulate an adequate—yet corrigible—diagnosis (see Secs. 10.4 and 15.7).

Let us finally study the place of explanation in science.

Problems

9.6.1. During and after wars more boys than girls are born: this is an empirical generalization of which the following explanations have been offered. (i) "Nature, which is a self-regulating whole, compensates for the loss of men in the battlefield". (ii) "War accelerates the engagement process and accordingly favors early motherhood. Now, young couples tend to conceive more boys than girls". Examine the two explanations as to ad hocness.

9.6.2. Two explanations were given of the Boyle and Mariotte law for ideal gases. The central assumption in one of these explanations is that a gas is a collection of molecules moving rapidly at random. The other explanation assumes that a gas is a continuous fluid in rotational motion. The law can be derived accurately from either the molecular or the continuous medium theory and accordingly the two explanations were regarded as equivalent during a long period. What were the initial range) accuracy, and depth of the two theories? And what happened subsequently, i.e. how was a decision finally reached?

9.6.3. Is range desirable in itself?

9.6.4. Examine the widespread belief that the coverage of every new theory is larger than that of its predecessor's or otherwise it would not be taken into account. *Alternate Problem:* Is it possible to identify the explanatory power of a theory with its entailment (or deduction) power?

9.6.5. Construct a table similar to Table 9.4 for some field other than optics.

9.6.6. Compare the explanatory power of a pseudoscientific theory with that of a corresponding scientific theory.

9.6.7. Compare any two rival scientific systems as to explanatory power, both at inception and after some crucial tests have been performed.

9.6.8. A desideratum for any new theory is to account for both the successes and the failures of the theory it attempts to supersede. Expand this theme and illustrate it.

9.6.9. The perihelion of Mercury has a small secular motion unaccounted for by Newton's theory. A number of explanations of this discrepancy between theory and observation were put forward in the 19th century before Einstein explained it with the help of his field theory of gravitation. Those explanations fell into two sets: those which assumed and those which did not assume a law of gravitation different from Newton's. Among the former the following may be cited: (i) the exponent in Newton's law is slightly different from 2, namely n=2.00000016; in this case Mercury's anomaly is explained but then the Moon's motion becomes anomalous; (ii) a corrective term is added to Newton's law, namely $k \cdot r^{-n}$, with n=3, 4, or 5; but no value of k is found to fit every planet; (iii) Newton's law is multiplied by an exponentially decreasing function, e^{-kr}; but then, again different k-values must be assumed for the various planets. These proposals, then, were inconsistent with the total evidence; moreover, they were inconsistent with the (classical) field theory of gravitation based on the Poisson equation Yet they were seriously proposed by competent investigators (the exponential correction by Laplace). Moreover, in our days antirelativists have proposed similar patchings up of Newton's law without caring whether the proposed new formulas were independently testable or whether they matched at least the classical theory of the gravitational field. Explain this behavior. For details on those various ad hoc explanations, see M.-A. Tonnelat, *Les principes de la théorie electromagnétique et de la relativité* (Paris: Masson, 1959), pp. 253ff.

9.6.10. Introduce the concept of growth rate of a theory—a rate that could be negative. Hint: start with a measure of theory coverage. *Alternate Problem:* Propose a measure of theory *ad hocness*.

9.7. Functions and Reach

Why should scientists ask for explanations? Should descriptions and predictions not be enough? According to several influential philosophical schools science should not try to answer why-questions but only how-questions, i.e. it should limit itself to producing as complete and economical descriptions of actual and possible phenomena as possible. Explanation, in this view, would be dispensable or even misleading. We shall show that this opinion, which may be called *descriptivism*, is inadequate. But before this a more detailed characterization of descriptivism will be needed.

There are important differences among descriptivists. We may distinguish three species of descriptivism: (i) radical descriptivism, which accepts data alone and, at most, data-packaging hypotheses; (ii) black-boxism, which tolerates transempirical hypotheses and theories as long as they are not assigned any meaning beyond what can be observed, and (iii) antipictorialism, which objects to explanations in terms of imageable models and regards them in all cases as metaphorical or didactic tricks rather than as pictures of reality

Radical descriptivism means what it says about description and explanation in science, namely, that scientific research consists or should consist of observations and inventories of observation reports. *Black-boxism* adds that such inventories should be economical rather than true, to which end they must consists of compressed empirical information or of data-churning devices; in other words, black-boxism allows for laws and theories as long as they are phenomenological (non-representational). Or, as we may also say, whereas radical descriptivism bans explanation, black-boxism allows subsumptive explanation while calling it description and rejecting interpretive explanation. Finally, *antipictorialism* does not object to the introduction of constructs, however far from sensible experience, as long as they are not employed in building visualizable models. In other words, the explanations acceptable to antipictorialism involve theoretical models that are not iconic but rather symbolic.

Radical descriptivism and black-boxism may be traced to the quest for certainty or infallibility, which feeds the prejudice against hypotheses. Blackboxism is, in addition, linked to conventionalism, fictionalism, and instrumentalism, according to which science cannot represent reality—a metaphysical misfit—but may systematize experience and foresee it. Antipictorialism, too, has often these same philosophical roots but it has independently been invigorated by the failures of the mechanical models of the aether and the atom. (See Alternate Problem 8.5.1.) In the case of many scientists, antipictorialism is an iconoclastic reaction against naive mechanism and naive realism rather than a systematic philosophy: it is a sound but exaggerated rebellion against the building of naive models the main function of which is to bring home what essentially lies beyond the reach of the senses.

There can be little doubt that mechanism, the world outlook of the mechanical engineer, has been superseded by a richer outlook involving nonmechanical entities and nonmechanical changes on a number of levels. But it is also true that mechanism can often provide a first approximation, if only because so many of the real objects do have mechanical properties. Kelvin's aether machinery is untenable, but Kekulé's molecule models have only been refined by quantum mechanics, and Bohr's atomic orbits remain as average distributions. The interests of science are best served by steering a middle course between iconolatry and iconoclasm. There is no reason to reject visualizable models as such: after all, the referents of factual theories exist in spacetime, i.e. they are somehow localized, and whatever is localized—even with no sharp boundaries—is up to a point imageable. Accordingly rather than smashing icons we should object to untestable and to patently false models, whether pictorial or not (see Sec. 9.5). A nonvisualzable model of a material system, such as Aristotle's—in terms of form, potency, and final cause—is not better than a mechanical oversimplification; the attempt to force reality into untestable or just outdated mental schemes is not necessarily tied to a pictorial language. In short, then, antipictorialism may be regarded as an overzealous warning against the temptation to trivialize ideas, and in this capacity it will be useful. On the other hand the other varieties of descriptivism have not been a sound root and can be dangerous. This we shall try to show next.

In the first place, it is impossible to give a complete and simple description of every fact. The description of all the facts, both macro and microscopic, happening in our classroom during one second would—even if it were technically possible—take centuries. Not even all the phenomena proper (experienceable facts) occurring in the same lapse could be described accurately, first because of their huge number and complexity, second because any intelligent and accurate description requires theories, and theorizing is never completed. Second, and more important, even if a complete description of any given set of phenomena were possible, it would be useless for science: in good science only facts relevant to some interesting idea are usually recorded and described; that is, the facts deserving to be described are selected among a host of facts and this selection, no less than the description itself, is made with the help of hypotheses and theories. Moreover every scientific description, far from being complete, is deliberately partial in the sense that it involves a limited number of variables and purposefully neglects several others: it concerns those and only those variables that could presumably be relevant to some view and some aim. The addition of irrelevant information, however accurate and interesting for another aim, would be a nuisance blocking the invention of adequate models (see Sec. 8.1). In other words, in science we are interested just in relevant information and its theoretical processing. But relevant information is information relevant to a set of views and aims; all else is, with regard to that particular set, sheer noise to be discarded. Third, detailed descriptions of that which happens are insufficient to attain scientific knowledge proper: firstly because what might happen and what cannot happen are important, too—and both possibles and impossibles can only be surmised on some theory or other; secondly, because knowledge is not called scientific unless it can bind together the items that are separately given in experience. In conclusion, radical descriptivism is preposterous.

Radical descriptivism does not seem to have been advocated by scientists but rather by philosophers; what several scientists have advocated is the liberal version of descriptivism we have called *black-boxism* (see Sec. 8.5). It would be instructive to draw a list of the scientific advances frustrated by the strictures of black-boxism. Here we shall only recall the discovery of the thermoelectric effect, a fact that remained practically buried during decades just because its discoverer refused to make a leap beyond what he observed. In 1821 T. J. Seebeck found out that, on heating a metallic circuit consisting of two different conductors, the circuit became magnetized as tested by the deviation of a nearby magnetic needle. He thereupon summarized his findings in the following generalization: "The heating of a circuit containing two different metals is accompanied by the magnetization of the circuit". This was a lowlevel hypothesis compressing and generalizing the observed conjunction of two fairly directly observable properties: heat and magnetism. The hypothesis was strictly phenomenological: it supplied no mechanismic explanation of the fact. This generalization remained isolated, hence sterile during many years, even worse, the refusal to go beyond it misled Seebeck into thinking that the terrestrial magnetic field was due to the temperature difference between the poles and the equator. Eventually a deeper hypothesis was formulated, namely: "The temperature difference in a circuit containing two different metallic conductors generates an electric current [unobservable] which in turn produces a magnetic field [unobservable] that acts upon the magnetized needle". The mechanism of the building up of the current as a result of heating a welding joint was not thereby entirely cleared up: the new hypothesis was semiphenomenological. But the addition of the invisible link—the electric current and the accompanying magnetic field—was sufficient to rescue thermoelectricity from its isolation. Had Seebeck been bold enough to formulate that semirepresentational conjecture instead of sticking to the observable facts, he might have opened an entire new field of physics. As it was he just added one more stamp to the stamp collector's world view—the view of the world as a heap of unrelated facts. Other researchers, eager to explain the effect—i.e. to explain the initial empirical generalization—advanced the subject.

Another example of the limitationist policy of descriptivism is the injunction to abstain from inquiring into the origin, development and mechanism of social institutions, and the associated recommendation that anthropologists and sociologists should limit themselves to summarizing the social events occurring in the communities at hand, and to describing the functions of habits and institutions. It is argued that this can be done on the basis of observation, whereas questions of origin, evolution and mechanism require hypotheses. But not even this modest program (functionalism) is exactly fulfillable. First, because institutions as such are unobservable; second, because observation does not supply the ideas suggesting us to pick up certain traits of society rather than others. Thus, e.g., the gathering of data regarding economic and social history did not begin until it was explicitly hypothesized that production and commerce generate certain basic social relations and that these constitute the tissue of every society. No social fact, let alone, social function, is ever observed by a theoretically unprejudiced scientist: all we do observe is a few active individuals and crowds; the rest is conjectured or inferred. An adequate description of a social institution requires a handful of ideas just as an adequate description of a cloud requires a handful of theories; and the more representational the theories used for the description the more profound the latter will be.

In conclusion, descriptivism has two harmful effects: (i) it blinds people to any new fact beyond the surface of things, thereby blocking the gathering of information suggested by the deeper theories, and (ii) it makes people miss the opportunity and the joy of new thought that every new fact affords. By virtue of (i) descriptivism is self-defeating; by virtue of (ii) it encourages superficiality and prevents the transformation of protoscience into science.

The barrenness of descriptivism does not entail that science should dispense with description: it just forces us to revaluate description. Two kinds of description are needed in science: nontheoretical and theoretical description. A *nontheoretical description is* an account not involving theoretical concepts. A *theoretical description,* by contrast, presupposes a temporary commitment to a set of theories. In the quantitative branches of science theoretical description is made with the help of low-level law statements. Thus, e.g., the narrow beam of light which the layman may describe in nontechnical terms, can be described by the physicist with the help of a solution to an equation of propagation of light. (Only an explanation of the same thing will require the high-level equation itself.) Whereas nontheoretical description bears always on some singular fact which is supposedly real, theoretical description bears on an admittedly idealized model of a class of facts: theoretical description is hypothetical rather than actual, because it uses law statements and simplifying assumptions. Moreover, there is no reason to sharply distinguish theoretical description from subsumptive explanation (see Sec. 9.3). In this sense, accurate theoretical description is a kind of explanation.

The above does not erase the difference between description and explanation. Firstly, nontheoretical description (of singular actual facts) is sharply distinguished from theoretical description (of models of facts of a kind). Secondly, theoretical description or, if preferred, subsumptive explanation, is equally distinguished from interpretive explanation (see Sec. 9.4). What is often called a *descriptive theory* is either no theory at all (e.g., the radical behavioristic programme) or a black-box theory supplying subsumptive explanations (e.g., thermodynamics). Theories proper enable us to perform, at the very least, subsumptive explanations, i.e. theoretical descriptions; representational theories can yield, in addition, interpretive explanations.

Are there limits to scientific explanation? The traditional view is that facts can be classed into explainable and unexplainable: the former are to be explained by causes (Aristotelian approach) whereas the latter are to be dreaded or revered (Goethe's *das Unerforschliche ruhig verehren*). The dominant contemporary view is rather the following. Everything except the existence of the universe—which is presupposed by any question—is explainable to some extent even if it has not yet been explained. A few facts can be explained rather detailedly, most only in outline, and whatever can be explained by science at any moment is explained with the help of the laws and theories available at the moment. Not every explanation is a causal explanation: there are as many kinds of explanation as kinds of theories, and causation is just one of the determiners involved in scientific theories. Defects in explanation are due either to an insufficient (but improvable) knowledge of laws and circumstances, or to practical difficulties in the application of basically correct theories to highly complex systems—such as, e.g., the polyester fibre. In no case is explanation limited by an alleged opacity of reality to human reason and human experience: after all, being a part of reality we have some inside information about it. The assumption that nature and, to a lesser extent, culture are irrational is refuted by the advancement of science. Only some individuals are irrational—and a few to the point of looking for reasons to validate irrationalism. What is true is that no *perfect*—completely exhaustive and accurate—explanation is ever achieved; consequently there are no *final* explanations. If we remember how theories are built (see Sec. 8.1) we must grant that this imperfection is essential rather than circumstantial. At the same time we surmise that every explanation is improvable and, in general, that the explanatory power of scientific theories can be increased without a priori limits precisely because it is possible to eventually pin down the defects of any theory. The possibility of progress is suggested by the history of science and by the way theories are built and critically examined, but there is no law of scientific progress: there is just a progressive trend, and nothing warrants its continuation.

*It has repeatedly been suggested, though, that in physics no further basic progress is possible: the present quantum theory would be final and the limitations of it would not be contingent but insurmountable. For instance, the present stochastic laws would remain fundamental, i.e. underivable from more basic laws referring to lower-level things and events. Those laws, being stochastic, are said not to explain individual events and processes but only the collective outcome of a number of such facts—e.g., the distribution of spots on a photographic plate subject to bombardment by subatomic particles diffracted by a crystal. (This presupposes the adoption of the frequency interpretation of probability to the exclusion of its propensity interpretation—which is just as legitimate.) Such laws being final, we shall never be able to explain the individual fact. Consequently we should not ask why such and such individual fact happens as it does: this is a meaningless question. Thus far the claim of the man who, being unable to predict what will happen to an individual electron when shot through a slit system, is wise enough to forecast that no men in the future shall ever come to know what he does not. We may not know most things and we cannot even state certain problems within our present theories: true. *Ignorabimus*—we shall never know—what we now ignore: a dangerous ungrounded prophecy. We do not know whether the individual atomic and subatomic events will eventually be explained in more detail by future theories, but we should encourage rather than dogmatically decry any serious attempt to enrich the problem system that can be attacked within the present theoretical framework, especially since the latter is ridden with internal difficulties (such as the famous infinite quantities) and it does not cope with much of the experimental material regarding the so-called elementary particles. If we do not hope that such difficulties will eventually be overcome by a more nearly true yet still imperfect theory, we had better give up fundamental research altogether. Fortunately few believe the dogmas they proclaim: they keep trying to explain the unexplained.*

The choice before us is not a choice between exhaustive and exact explanation (naive rationalism) and complete lack of explanation (mysticism, systematic skepticism, and traditional positivism), but rather among theories with different explanatory powers (see Sec. 9.6). In particular we must choose between remaining content with phenomenological theories and going beyond them. If we care for possible new and deep truth rather than for safety we shall prefer the bold testable hypothesizing of mechanisms underlying the observable. But this preference for representational theories is far from universal: it must be recommended by the metascientist not because it is fashionable—which it is not at the time of writing—but because it is the most fertile (see Sec. 8.5).

Those who hold that scientific explanation is inherently limited argue that this could not be otherwise because science seeks only relations and does not inquire into the nature of its objects. Transcendentalists add that every object has an immutable essence beneath its properties and relations and that essences, which escape reason as well as experience, are apprehensible with nonscientific means: with intuition of some sort or other. Phenomenalists refuse to countenance essences just because they are nonphenomenal but they agree with transcendentalists that science is limited to establishing relations, whether among phenomena or among fairly directly observable traits of facts, but that it never gets to the things themselves, which are mere metaphysical constructs. These two relationist views are mistaken and they rest on a superficial analysis of scientific law and scientific theory, one that misses no less than the object variable and the set of its possible values (i.e., the universe of discourse).

Science does not investigate properties and relations in themselves, apart from things and events: what we are given is just an outward glimpse of entire things and events, and it is only by analysis and invention that we separate their various aspects (variables) and detach the invariant relationships among them (laws). Whatever relations are established among properties (variables) in factual science, they always refer to some concrete system or other. (See Secs. 7. 1 and 8.4.) This is taken for granted in scientific contexts, where one writes, say, '$R=kS^p$' (the basic psychophysical law) rather than 'For every x, if x is an organism subject to the stimulus S, then x reacts with the strength $R(x)=kS^p(x)$'. The absence of the object variable x may mislead the layman into believing that a ghostly relation among variables is meant; but it behoves the philosopher to spell out what the scientist hints at. Secondly, science certainly ignores all about ideal essences or even about real but inscrutable, immutable and ultimate essences—but why should this be a shortcoming? In every set of properties we can eventually recognize those which are source properties from those which derive from them; and similarly for laws. In science, essences are not conceived as the souls of things but as certain properties and laws: namely those which, on a given theory, are fundamental at a certain level. This is why the qualifier *essential*, applied to both variables and laws, can be found in the scientific literature whereas the noun *essence*, in its metaphysical sense, is nearly absent from it. Science has, accordingly, no use for metaphysical explanations in terms of essences; but on occasion scientific research does hit on properties and laws that may be regarded as essential on a given level (rather than ultimately essential)—until new notice. Thus, e.g., the atomic number is an essential property at the chemical level, but it is a task of nuclear theory to account for it in terms of still profounder properties, in which the chemist is not interested because they do not directly show up at the chemical level.

To conclude. Despite the prohibitions of philosophers of the descriptivist persuasion, scientists have not ceased to ask whys and to answer them, i.e. to offer explanations both subsumptive and interpretive. Only, in contrast to traditional philosophers, scientists rarely claim that their explanations—or, at any rate, those of their colleagues—are final: they usually recognize that every explanation *must* be defective because it is built with imperfect theories, simplifying subsidiary assumptions, and more or less inaccurate information. At the same time, they are usually confident in the unlimited improvability of the range, the accuracy, and sometimes also the depth of scientific explanations.

The chief motive of this unceasing quest for improved explanation—for theories with increasing explanatory powers—is that every new explanation widens the horizon of the known and the reach of reason. We might go as far as saying that explanation consummates the marriage of reason and experience, since it is a logical argument requiring empirical data and tested (or at least testable) generalizations. Furthermore, every new explanation increases the systemicity of our knowledge since it transforms the isolated datum, or the stray generalization, into a piece of a conceptual whole. Consequently, to demand the suppression of scientific explanation, or even its restriction to subsumptive explanation, is very much like killing the golden egg hen of science.

Problems

9.7.1. Discuss the theses of descriptivism as expounded in some of the following writings. (i) G. R. Kirchhoff, *Vorlesungen über mathematische Physik: Mechanik* (Leipzig: Teubner, 1883), pp. iii and 1. (ii) E. Mach, "Beschreibung und Erklärung", in *Populär-Wissenschaftliche Vorlesungen,* 4th ed. (Leipzig: Barth, 1910). (iii) P. Duhem, *The Aim and Structure of Physical Theory* (1914; New York: Atheneum, 1962), Part I. (iv) K. Pearson, *The Grammar of Science*, 3rd ed. (London: Adam and Charles Black, 1911), p. v. (v) L. Wittgenstein, *Philosophical Investigations* (New York: Macmillan, 1953), I, p. 47. (vi) P. Frank, *Philosophy of Science* (Englewood Cliffs,

N. J.: Prentice-Hall, 1957), p. 23. Kirchhoff's programme for mechanics, extended by himself to electrodynamics, was "to describe natural motions in a *complete* and in the *simplest way*" (op. cit., p. 1).

9.7.2. Point out the differences between the following propositions. (i) "If two bits of aluminium paper are touched with a lucite rod that has previously been rubbed with a piece of fur, they separate". (ii) "Two bodies charged with electricity coming from a third charged body repel each other". *Alternate Problem:* The behavior of a black box can be explained by many different hidden mechanisms. Can this be held against explanation in depth? See W. Ross Ashby, *An Introduction to Cybernetics* (London: Chapman & Hall, 1956), Ch. 6.

9.7.3. Would Darwin have proposed his theory of evolution through natural selection had he not attempted to explain likenesses—which he explained as kinships? *Alternate Problem:* Examine the limits of scientific explanation according to E. Mach, *History and Root of the Principle of Conservation of Energy* (1872; Chicago: Open Court, 1911), p. 55. (Thesis: explanation can be pursued no further than the simples into which it resolves the explanandum)

9.7.4. Examine the defense of metaphysical or "concrete" (as opposed to the scientific or "abstract") explanation as provided by empathy. Evaluate its validity claims and establish its relation with the universal sympathy of the Hermetic and Stoic doctrines, as well as with the sympathetic understanding (empathy, *Verstehen)* of W. Dilthey, R. G. Collingwood and other intuitionists. Find out whether 'understanding' is used by those philosophers in a logical or in a psychological sense. See H. Bergson, *L'évolution créatrice* (1907; Paris: Presses Universitaires, 1948), p. 177, and M. Bunge, *Intuition and Science* (Englewood Cliffs, N. J.: Prentice-Hall, 1962), pp. 10ff, and *Finding Philosophy in Social Science* (New Haven: Yale University Press, 1996), pp. 150–157.

9.7.5. Among naturalists and historians it is a widespread conviction that description or narration is more important or at least methodologically prior to explanation: that no *why* can be answered until all the *hows* have been secured. Discuss this view in relation to the selection of evidence and its interpretation, that even the most candid fact-gatherer is forced to perform.

9.7.6. Compare the following descriptions. (i) "The substances *A* and *B* react and give rise to the substances *C* and *D*, i.e. $A+B\dot{\rightarrow}C+D$." (ii) "Let *A* be a monatomic gas which reacts with a diatomic gas every one of the molecules of which has the composition *B–C*, where '*B*' and '*C*' symbolize the component atoms and '–' the bond among them. When *A* comes close to *B–C*, the two interact and constitute a short-lived intermediate complex *A . . . B . . . C*. The bond *A–B* eventually predominates and the new molecule *A–B* emerges, whereas *C* remains free. The net reaction is then: $A+(B-C)\dot{\rightarrow}(A-B)+C$."

9.7.7. A run of observations conducted to test a parapsychological thesis, if done with a minimum of care, will rely on objective facts concerning the subject's behavior, such as his writing down a series of conventional marks. Both a parapsychologist and a psychologist may agree over the description of the subject's overt behavior. Will they necessarily agree over the interpretation (explanation) of it? And would it matter if they did not? *Alternate Problem:* In 1766 Priestley made by chance the surprising discovery that charcoal was an excellent conductor of electricity. This empirical generalization and others that were discovered thereafter—e.g., that the conductivity of carbon increases with temperature—remained unexplained until the 1950's, when the quantum theory was applied to semiconductors. Priestley himself tried to explain his discovery but to no avail. We read in Joseph Priestley *Selections from his Writings* (University Park, Pa.: Pennsylvania State University Press, 1962), p. 229: "But notwithstanding my want of success, I make no doubt but that any person of tolerable sagacity, who has an opportunity of making experiments in a laboratory, where he could reduce to a coal all kinds of substances, in every variety of method, might very soon ascertain what it is that makes charcoal a conductor of electricity". Did he have the right strategy, i.e. was he right in hoping that purely empirical operations would supply the wanted explanation?

9.7.8. Does science reject explanations in terms of occult entities and properties? If not, what is the difference between modern science and scholasticism or even the occult "sciences" such as astrology and parapsychology? Begin by specifying what shall be meant by 'occult': not directly observable, or inscrutable (belonging to untestable hypotheses)? *Alternate Problem:* Skeptics claim that explanation has unsurpassable limits. Very often their rationale, far from being obscurantist, is illuministic: in fact, they usually oppose the claims of nonscientific speculation, such as vitalism, which professes to know what the essence of life is. Make a study of agnosticism and skepticism as a critique of unscientific speculation.

9.7.9. Leibniz' *principle of sufficient reason* can be construed as a programmatic hypothesis stating that everything is explainable (in principle). Formulate, illustrate and analyze the principle and ascertain whether it is logical, epistemological, or ontological—or whether it can be stated in either of these forms. See M. Bunge, *Causality,* 2nd ed. (Cleveland and New York: Meridian Books, 1963), pp. 229–239 and the bibliography cited therein.

9.7.10. Outline an account of philosophical explanation. Take into account that the generalizations involved in philosophical explanation seem to fall into the following categories: (i) arbitrary generalizations of a high level, not connectible with testable formulas; (ii) inductive generalizations taken from either ordinary or scientific knowledge and promoted to the rank of principles, and (iii) regulative principles, proposals, and norms. *Alternate Problem:* Analyze the concept of ultimate explanation. Remember that the Aristotelians believed they had the ultimate explanation of every fact in terms of the four causes, the Newtonians in terms of mechanical laws and gravitation, and in our own days optimists hold that almost every physical fact has its ultimate explanation in the laws of quantum electrodynamics.

Bibliography

Beckner, M.: The biological way of thought. New York: Columbia University Press 1959.

Braithwaite, R. B.: Scientific explanation. Cambridge: University Press 1953.

Bunge, M.: Causality, 2nd ed, Ch. 11. New York: Dover 1963.

——— Finding Philosophy in social science. New Haven: Yale University Press, 1996.

——— Mechanism and explanation. Philosophy of the Social Sciences 27. 1997.

Campbell, A.: What is science? (1921). New York: Dover 1952.

Craik, K. J. W.: The nature of explanation. London: Cambridge University Press 1952.
Feigl, H., and G. Maxwell (Eds.): Minnesota studies in the philosophy of science III. Minneapolis: University of Minnesota Press 1962. Particularly: Explanation, reduction, and empiricism, by P K. Feyerabend; Deductive-nomological vs. statistical explanation, by C. G. Hempel; Explanations, predictions, and laws, by M. Scriven; and Explanation, prediction, and 'imperfect' knowledge, by M. Brodbeck.
Gardiner, P.: The nature of historical explanation. Oxford: University Press 1952.
——— Theories of history. Glencoe, Ill.: Free Press 1959. Particularly the 8 papers on historical explanation.
Harvey, D.: Explanation in geography. London: Edward Arnold, 1969.
Hempel, C. G.: Aspects of scientific explanation. New York: Free Press 1965.
———, and P. Oppenheim: Studies in the logic of explanation, Philosophy of Science, **15**, 135 (1948).
Mill, J. S.: A system of logic, 8th ed. (1875), Book III, Chs. XI to XIV. London: Longmans, Green 1952.
Nagel, E.: The structure of science, Ch. 2. New York and Burlinghame: Harcourt, Brace & World 1961.
Pap, A.: An introduction to the philosophy of science, Chs. 18 and 19. New York: Free Press 1962.
Popper, K. R.: The logic of scientific discovery (1935), Sec. 12. London: Hutchinson 1959).
——— The aim of science. Ratio **1**, 24 (1957).
Symposia of the Society for Experimental Biology, No. XIV: Models and analogues in biology. Cambridge: University Press 1960.

10

Prediction

Forecasts are answers to questions of the form 'What will happen to x if p is the case?', 'When will x happen if p obtains?', and the like. In science such answers are called *predictions* and are contrived with the help of theories and data: scientific prediction is, in effect, an application of scientific theory. Prediction enters our picture of science on three counts: (i) it *anticipates* fresh knowledge and therefore (ii) it is a *test* of theory and (iii) a guide to action. In this chapter we shall be concerned with the purely cognitive function of prediction, i.e. with foresight. The methodological aspect of prediction (its test function) will concern us in Ch. 15, and the practical side (planning) in the next chapter.

10.1. Projection

The concepts of prediction and retrodiction belong to the family of ideas of anticipation and retrospection. The following members of this family may be distinguished: (i) *expectation*: an automatic attitude of anticipation found in all higher animals; (ii) *guessing*: a conscious but nonrational attempt to figure out what is, was or will be the case without any ground whatever; (iii) *prophecy*: large scale guessing on the alleged ground of revelation or other esoteric source (occult "science", untested "laws" of history, special powers of the leader, etc.); (iv) *prognosis* or informed guess or common-sense prediction: forecast with the help of more or lest tacit empirical generalizations, and (v) *scientific prediction* (or *retrodiction*): forecast (or hindcast) with the help of scientific (or technological) theories and data.

Expectation is the biological basis of prediction. Animals act with certain goals, the highest of which is self-preservation at the species level. They adjust their behavior to both a fixed set of goals and varying circumstances; and the higher animals preadjust themselves to some extent on the basis of prior experience: they have expectations or anticipations of sorts. A dog shown his promenade leash can become frantically excited at the prospect of a walk. It has been conditioned to that: he has learned to automatically associate the leash to the walk and it rejoices beforehand; the association and the ensuing expectation will fade away if the animal is consistently cheated. In prehuman animals the goals are built into their organisms by natural selection and, so far as we know, the adjustment of their behavior to goal achievement is entirely automatic. Man alone can become fully aware of some such goals and can accordingly alter them deliberately to some extent and design the most adequate means for attaining them— and this involves prediction. Biologically speaking technological prediction is, then, the highest type of adaptation: through it man preadapts himself to new conditions he shapes himself. But scientific prediction fulfils no biological function, at least not directly: it satisfies our curiosity and it validates and invalidates knowledge claims.

Guessing is superior to expectation only because it is a conscious operation; but it is groundless or nearly so, whereas expectation is rooted to reflexes and is therefore more likely to be successful. Guessing is, at best, a nonrational trial justifiable in the absence of knowledge— after all it can be successful, particularly when the possibilities are scarce and not equally likely. Guessing can be an amusing game, the lesser of evils, or a dangerous superstition turning people off from honest and hard scientific work. What foresight can be obtained is attained on the insight into the present and the hindsight of the past; without some research we cannot even know what the present circumstances are. And even founded foresight is at best plausible. The belief in the possibility of precognition without prior knowledge is just a relic from magic. There are, to be sure, procedures that yield information of certain kinds and are called guessing, such as the successive partitionings of a set into mutually complementary subsets all the way down to the desired set or individual (see tree questioning in Sec. 1.3). But these are all questioning techniques, none of them yields knowledge out of ignorance, and none supplies knowledge about the future.

Prophecies, or large scale guessworks, such as those of the *Apoca-*

lypse, Nostradamus, and of certain politicians, are every bit as unfounded as guesses. Yet they, too, can often be right—surprisingly so for the layman. To be effective a prophecy need just be either obvious or vague and ambiguous: in short, it must not be exact, hence risky. Gipsies are, of course, experts at uttering obvious prophecies ("You will make a trip", "You are waiting for news") as well as vague ones (" Something marvelous will happen to you some day", "Great worries are in store for you"). As to ambiguity, oracles have always been known to excel in it. When Croesus asked the Delphi oracle what would happen if he were to attack the Persians the answer was: 'A great kingdom will be destroyed'. Croesus did not realize the ambiguity and attacked. As a result a kingdom—his own—was destroyed as prophesied.

Prognoses or common sense forecasts are of an altogether different sort: they rely on generalizations and trends. The weather forecast of the experienced peasant, the horse-racing wager of the bettor and the prognostic of the medical practitioner are all based on definite empirical generalizations even though these are not always made explicit. Sometimes the anticipation consists of a mixture of prophecy and prognosis. This is the case of science fiction forecasts made by well informed writers such as Verne and Wells. Guessing and prophesying yield *unconditional* sentences of the form '*P* will happen'. On the other hand prognoses are *conditionals* or hypotheticals of the forms 'Since *C* is the case *P* will (or may) happen', or 'If *C* happens then *P* will (or may) occur'. Hence, whereas the prophet and the gipsy, the politician and the prescientific psychologist state *that P* will happen, the conscientious practitioner of a craft will cautiously say that, *if C* happens (or since *C* has happened) *P* will or may occur. His forecast is iffy and yet has some ground, however feeble. This ground is usually an empirical generalization concerning the regular conjunction or else the succession of facts, such as "If faced with *x*, *y* will do *z*".

Finally, scientific prediction is as iffy as common sense prognosis, only more refined than it. In scientific prediction the generalizations are explicit rather than tacit and they consist in laws or theories; as to the data, they can be checked and eventually improved with the help of scientific techniques. The generalizations themselves may have been guessed at but the data have been found and possibly checked, and the derivation of the predictive statement has been a piece of rational work. For example, in order to forecast (or to retrodict) what will (or

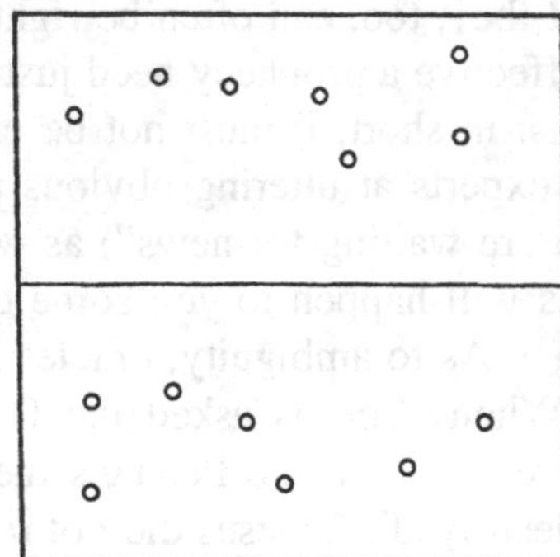

Fig. 10.1. The Big Dipper constellation now (above) and 100 centuries from now (below).

did) happen to the pressure of a gas upon halving its volume, we might invoke the law "pressure × volume=constant" to infer that the pressure will (or did) double. The prediction (or the retrodiction) will in this case be the following argument:

$$\{p_1V_1=p_2V_2,\ V_2=V_1/2\}\vdash(p_2=2p_1)$$

The premises of the argument are a law statement, namely "$p_1V_1=p_2V_2$", and a bit of information regarding specific circumstances, namely, that the volume of the gas in state 2 (future or past) is half its volume in state 1 (present). In this particular case state 2 can be later or earlier than state 1, hence the inference to the future is the same as the inference to the past. This is because an exactly reversible process has been assumed to occur; in other words the ideal gas law is time-independent, hence invariant under the exchange of future and past, whence retrodiction could not differ from prediction. In general, however, the two projections, the one to the future and the one to the past, will not be the same. (See Sec. 10.2)

The previous example exhibits the essential characteristics of scientific projection (prediction or retrodiction): (i) it is *conditional*, in the sense that it says what will (or may) happen or when something will (or may) happen *if* such and such laws are "obeyed" and such and such circumstances take place; (ii) it is *theoretical* or nomological, i.e. it employs law statements— which, incidentally, need not contain the time variable; (iii) it accordingly refers to sets of *properties* or things and events rather than to whole things or whole events; and (iv) it can be as *accurate* as our knowledge about laws and circumstances, whereas prophecies are inherently indefinite and prognoses can be accurate but there is no means of improving their accuracy—except to replace them by scientific predictions made with corrigible law statements and data. Improvable accuracy, rather than accuracy, is a hallmark of scientific prediction: it can be accurate as in astronomy (see Fig. 10.1) or it can consist of rough estimates like those of the behavioral sciences. The point is not that scientific prediction is accurate in an absolute sense but that it is founded and, for this reason, it can be improved.

The logical structure of scientific projection is then the same as that of explanation, namely, a *deductive inference from laws and data:*

$$\{Law(s), Circumstance(s)\} \vdash Projectandum$$

The premises of the argument may be called the *projectans premises* and the conclusion the *projectandum* (what is predicted or retrodicted). Sometimes the argument or inference, at other times its conclusion, is called the prediction (or retrodiction). The name is immaterial as long as one keeps distinct the concepts of predictive inference (or argument) and predictive proposition, the argument's conclusion. Although prediction does not differ logically from explanation, it does differ from it in other respects. To begin with, the projectandum is not any proposition but a singular proposition with a factual reference—never a law statement. Consequently the law statements occurring in the projectans premises are *low-level* laws. For example, in order to predict the time of flight of a bullet we shall not use its equations of motion but a solution to the latter. We shall return to the extralogical differences between projection and explanation in Sec. 10.5.

To become better acquainted with scientific projection let us examine some instances; the first of them will employ some geological knowledge. In 1820 three Alpinists got lost in a Mount Blanc glacier and it was predicted that their bodies would be found 40 years later at the glacier's foot— and so it happened. The generalization involved was that the glacier speed was constant and about 75 meters per year; the circumstance datum was the approximate location of the bodies. The prediction was, of course, of the same kind as the arrival time forecasts we make in motoring. The projection would have been falsified if the information had been wrong or if, owing to some important climatic change, the sliding speed had changed. This latter risk, of the possible change in the pace of some characteristic of processes, is peculiar to prediction in contrast to explanation.

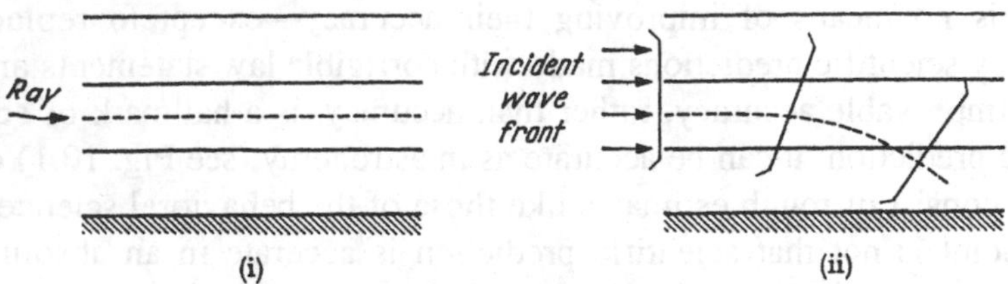

Fig. 10.2. The path of light in a three-layer flat atmosphere, (i) according to ray optics and (ii) on wave optics.

Our second example will be taken from optics. Suppose we wish to predict the path of a light ray entering the atmosphere at dusk, i.e. horizontally. To this end we proceed in a way typical of science: we begin by building an *ideal model* of our object (the earth atmosphere) and pick up a theory to deal with the model. To a first approximation a set of three horizontal air layers (see Fig. 10.2) should do as a model of the atmosphere. We further suppose that the medium layer has a refractive index intermediate between the maximum of the bottom layer and the minimum of the top layer. These assumptions are, of course, simplifying hypotheses—almost pretences—making a simple theory application possible. This model should suffice for a *qualitative prediction*; a quantitative prediction will call for a more realistic model, but still an ideal model involving, e.g., the perfect sphericity of the earth. We next try the simplest optical theory, namely ray optics; on this theory a light ray traveling horizontally will remain horizontal because in this direction the refractive index is constant and the light simply does not exist but along its ray. But, as we know, this prediction is not borne out by observation: in fact our light ray is bent downwards—something which, by the way, is counterintuitive if we are used to thinking of light as made up of rays or of corpuscles. On the other hand any of the wave theories of light will predict the observed downwards bending, just because the refractive index changes vertically. The upper portions of the incident wave front will travel faster than the lower portions and, since the trajectory (ray) is perpendicular to the wave front, the light ray will bend downwards, i.e. towards the region of higher refractive index. Once the qualitative projection has been performed, a *quantitative* projection will be tried and, to this end, the preliminary model of the atmosphere will be complicated step by step.

Speaking of the atmosphere, we should say that the rapid progress of meteorological theory is rendering obsolete our pet jokes about the weather man. The transition of meteorology from the empirical stage to the theoretical stage is accompanied and in turn fostered by the shift from empirical prognosis to theoretical prediction. Empirical weather forecasts are made on the strength of empirical generalizations regarding chiefly the pressure distribution and the cloud type. This is the least expensive method and, for short-term forecasts, quite safe. Meteorology is abandoning this empirical and phenomenological approach and is beginning to regard pressure and cloud type as symptoms to be explained in terms of internal variables. It is taking apart the weather mechanisms (of, e.g., snow and hail formation) and is analyzing the direction of travel of weather and the changes along its travel, as large scale outcomes of smaller scale events. In short, whereas classical meteorology was separated from physics, contemporary meteorology is being approached as the physics of the atmosphere in interaction with the oceans. By abandoning the phenomenological approach it has gained in external consistency and in depth and is producing grounded and reliable rules for weather forecasts.

On the next level, that of chemistry, we find prediction problems like this: Given the kinds and quantities of reactants as well as the physical conditions—such as temperature and pressure— predict the reactions that can take place and, if possible, the times they will take to be completed. The inverse problem, of retrodiction, would be to find out what the initial reactants and physical conditions were, given the final reaction products—a harder problem indeed. Another typical problem of chemistry is this: Predict whether a given formula will correspond to a stable compound obtainable in the laboratory. Such problems are seldom solved with accuracy at the present time, owing to the adolescent state of chemical theory. The chemist, like the biologist, can still make important discoveries with much flair and little theory—not because intuition is a better guide than calculation but because the available theory is still insufficient to give intuitive insights and exact projections. A dramatic case was the discovery (1962)—excluded by the available theory—that "noble" gases can form stable compounds such as xenon tetrafluoride. The theory of chemical binding available at the time could predict that such compounds would not exist and gave good reasons for it. On the other hand, the choice of fluorine in searching for compounds of noble gases was not accidental: it had been known that fluorine is the most reactive of elements, so that, if the "noble" gases could react at all, they would combine with fluorine. The discovery was then preceded by a half-scientific prognosis stimulated, in part, by a heterodox doubt concerning the ultimate perfection of the present theory of chemical binding.

In the sciences dealing with the higher levels of reality prediction is making progress but in many areas it still is very inaccurate, hence unreliable, or just impossible. Unscientific philosophers have claimed that such difficulties are inherent in the subject matter: that, e.g., biological prediction is impossible because life processes are spontaneous and indeterminate. These gloomy prophecies have not been fulfilled: accurate predictions are possible in physiology and genetics, less accurate ones in ecology, much less so in learning theory, and even less in social psychology. Quantitative predictions are being made in fields where qualitative forecasts only were possible a few years ago, and qualitative predictions are now succeeding where none at all were in sight in subjects that were devoid of theory. The more mature a science—the richer and better tested theories it includes—the more accurate projections it can make and the more it can learn from both successes and failures in projecting. Whereas pseudoscience and protoscience either make no projections at all or state irrefutable prophecies on no clear grounds, mature science exhibits the grounds of its predictions—namely, laws and data—and consequently can correct or even altogether abandon them if they lead to false projections. Consequently there is no ground for pessimism regarding the future of scientific projection in the presently underdeveloped branches of science—particularly if we realize that not every scientific prediction is of the type of astronomical prediction, as we shall see in the next section.

Problems

10.1.1. Examine Aristotle's treatment of prophesying by dreams and compare it with Freud's. See "De divinatione per somnium", e.g. in R. McKeon, Ed., The Basic Works of Aristotle (New York: Random House, 1951), pp. 626ff. Note, in particular, that Aristotle is distrustful of oniromancy because he can find no explanation for it, and he fails to explain it because no mechanism for it seems conceivable: "the fact that our seeing no probable cause to account for such divination tends to inspire us with distrust". *Alternate Problem:* Analyze the question "What does the future hold in store for us?" Does it presuppose any metaphysical doctrine?

10.1.2. Analyze F. Bacon's essay "On Prophecies", e.g. in *The Essays of Francis Bacon* (London: Thomas Nelson, s.d.). Bacon maintains here that most prophecies "have been impostures, and by idle and crafty brains, merely contrived and feigned, after the event past", and that the credit which some prophecies enjoy is partly due to the fact "that men mark when they hit, and never mark when they miss" (pp. 179–180). *Alternate Problem:* Make a list and an analysis of prophecies such as "You will benefit from a recent experience" (found in a Chinese cookie).

10.1.3. If we know that someone is a good calculator we can predict that he will be able to solve a given arithmetical problem. From this we might be tempted to conclude that predictions do not all require generalizations. Is this true? *Alternate Problem:* Examine some of the forecasts for the next millenium currently issued by scientists, technologists, and historians. What is their ground?

10.1.4. Propose examples of scientific prediction, both qualitative and quantitative, and point out what could falsify them and how they could be improved. *Alternate Problem:* Examine how unsuccessful predictions are sometimes reinterpreted as confirmations of a theory. See, e.g., C. E. Osgood, *Method and Theory in Experimental Psychology* (New York: Oxford U.P., 1953), pp. 370ff

10.1.5. Most analyses of prediction assume that it requires laws containing the time variable (e.g., differential equations with time as the independent variable). Look for counterexamples and explain the origin of the fallacy. See M. Bunge, *Causality*, 2nd ed. (Cleveland and New York: Meridian Books, 1963), pp. 312ff.

10.1.6. Study the pragmatic side of prediction, i.e. regard it as an activity performed before the predicted event occurs—an activity that takes time and is done with a definite aim. Take into account the case in which the predictive operation takes more time than the predicted event.

10.1.7. Make an analysis of market forecasts. In particular analyze the grounds on which investment councellors give advice: establish whether they act on predictions and, if so, of what kind and on what basis.

10.1.8. Assess the claim that certain privileged subjects have the faculty of precognition either because they infer future events from looking at certain present facts (augurs, oracles) or because they see the future directly (the visionaries reported on by parapsychologists). *Alternate Problem:* We forecast events that will eventually take place in time, or within a given time interval. Would it be possible to forecast the time instants or intervals themselves?

10.1.9. Can prediction say anything about the structure of a theory, as held by G. Ryle in "Predicting and Inferring", in S. Körner, Ed., *Observation and Interpretation* (London: Butterworth; New York: Academic Press, 1957)?

10.1.10. Predictions are sometimes made with the help of theoretical systematics rather than with explicit theories. How is this possible? (For theoretical systematics see Sec. 2.5.) *Alternate Problem:* Analyze the following statement by cosmologist W. H. McCrea, "Information and Prediction in Cosmology", *Monist*, **47**, 94 (1962): "We can predict that the universe will not change very rapidly. For the cosmical background forms part of the boundary conditions of any experiment, and when we assume the boundary conditions to be reproducible we include the assumption that the cosmical background is effectively the same throughout a large number of repetitions of the experiment" (p. 95).

10.2. Stochastic Projection

The least precise of projections is the bare *enumeration of possibilities* on condition that they are not just logical possibilities, as in the case of the overcautious prophecy "Either there will be a war or there will not be a war". Suppose a system (molecule, person, social group) is presently in state 1 and all we can foresee within a given unit of time is that, during that lapse, it will either remain in state 1 or go over into either of the states 11 or 12 (see Fig. 10.3). Suppose we can also assume that, if the system does go over to state 11, then after another unit of time it will either remain in that state or make a jump to state 111 or to state 112—and similarly with regard to state 12. The description of such a tree of (physical) possibilities is all we can achieve in many daily life situations and in the social sciences. (Prima facie,

Fig. 10.3. Branching possibilities.

projections of this kind are untestable, but this is a wrong impression: the system might jump from any of the specified states to a new, unforeseen state, and this would ruin the forecast concerning the tree of possibilities.) Though qualitative, such a prediction will be scientific if the possible transitions are figured out on the basis of law statements.

Next comes the projection in which every possibility is quantified, i.e. assigned a definite probability. This assignment of numbers can be made on the basis of past experience and/or with the help of theory. In any case the various probabilities will be related to one another. In the foregoing example we would have, among others, this relation: $P(1 \rightarrow 1)+P(1 \rightarrow 11)+P(1 \rightarrow 12)=1$. If every branch in the tree of possibilities is assigned a probability and if the various probabilities satisfy the conditions imposed by the probability calculus (such as, e.g., the formula we have just written), a *stochastic projection* will have been made, and one that can be tested with quantitative accuracy (with the limitation inherent in probabilistic hypotheses: see Sec. 5.6). Most of the predictions about transitions made with the help of quantum theories are stochastic because they employ stochastic laws and, in particular, laws concerning the transition probabilities of a given system.

In many cases of interest to the social scientist and the historian the various estimated probabilities are markedly different from one another. Hence to a first approximation we shall be justified in neglecting the least probable transitions (which, if very unlikely, are called *prohibited*). Thus, in our example we might have $P(1 \rightarrow 11) \gg P(1 \rightarrow 12) \gg P(1 \rightarrow 1)$ and would accordingly conclude that *the most probable course of events* within the next time unit would be a jump of the system from state 1 to state 11. In this case we would predict the change trend, or the most likely direction of change, rather than the precise future states. Social science would benefit from a more intensive theoretical and empirical study of transition probabilities: it would allow it to formulate more precise—hence more accurately testable—projections and, in some cases, to forecast what looks as by far the most likely course of events. After all, this is what is done both in atomic physics and in everyday life—only in the latter case our stochastic projections are usually qualitative rather than quantitative, and ungrounded or feebly grounded rather than based on scientific theories. The point is that stochastic projection is scientifically respectable and that the mere awareness of this could accelerate the pace of the nonphysical sciences.

A projection will be called *individual* if it concerns an individual (physical or cultural), and *collective* if it is about an aggregate. [The distinction between an individual taken as a block and an aggregate is largely dependent on the level of analysis we care to adopt but not entirely so: in fact, an aggregate may constitute a system, hence an individual on a higher level, even if its parts are not tightly interrelated. Whereas some statistical aggregates are systemic (e.g., biological populations), others (e.g., a sequence of marriages) are not.] Whether individual or collective, a projection can be made either on nonstochastic or on stochastic laws, depending on the available knowledge and on the aim. Thus, e.g., the behavior of a single light quantum is projectible, in probabilistic terms, on the basis of the probabilistic laws of quantum optics, whereas the behavior of a large aggregate of light quanta, such as those in a light beam, will be projectible on the basis of the classical macrolaws emerging from the micropatterns of individual photons (see Sec. 9.5). A statistical aggregate need not be *actual*: it may also be *potential*, i.e. it may consist either of loosely interrelated coexistents or it may be a random sequence of successives, such as a series of measurements of a given magnitude. In either case a regular behavior may result from the nearly independent behavior of the components, and in both cases the patterns will be collective, i.e. every individual component, taken by itself, may look indeterminate but it will fit in an overall pattern.

Every statistical regularity (trend or law) refers to an aggregate, i.e. depicts a collective pattern, but not every regularity regarding an aggregate is statistical or even probabilistic. We shall call a projection *statistical* if it relies on statistical regularities (trends or laws). A statistical projection is collective in the sense that it is about a statistical aggregate, i.e. a collection of items of some kind with some dose of built-in randomness in some respect—however nonrandom they may be in other respects. The explanation of the statistical regularities themselves will call for an analysis into component parts on a different level (see Sec. 9.5), but statistical prediction will disregard such detail and will concern the whole on its own level.

The simplest statistical prediction is, perhaps, that of political behavior in a stable society. It is the simplest and also the safest if based on accurate data that can be gathered by random sampling. The chief assumption playing the role of generalization in voting prediction is so obvious that it is hardly ever mentioned: namely, the conservation "law" that the political composition of any population will remain nearly constant from the moment the poll is taken to election day, as long as this lapse is a few weeks. The circumstantial information is, of course, the set of data about the political composition, or attitude, of a representative sample of the whole population. If the prediction is not borne out by the election, the blame may be put on the sampling or on a last moment switch of loyalties, i.e. on a breakdown of the "law" used in the inference. Predictions of performance in personnel psychology follow a similar pattern: they, too, assume that the scores on the abilities on which the subject's success depends will not change appreciably in the course of time. And some of the retrodictions made in palaeontology have the same characteristic: for example, we may measure the scatter or dispersion of a certain character, such as the skull breadth/length ratio on a sample of a living animal species and may extrapolate this back to an immediate extinct antecessor population. This may prevent us from jumping to the conclusion that a given fossil skull belongs to a new species just because it has an unusual skull breadth/length ratio: if the discrepancy lies within the dispersion of the related living species we may as well count it as an ordinary specimen. The generalization assumed in this case is, of course, the approximate constancy of the breadth/length ratio of the skulls of any given species in the course of a given lapse, say one million years—a corrigible hypothesis indeed.

Prediction on the basis of definite trend lines comes next. In this case the aim is to forecast the most probable value of a variable y from the corresponding value of a correlated variable x and certain statistical parameters such as averages and standard deviations. In this case a quantitative estimate can be made with accuracy, particularly if x and y are strongly linearly correlated, i.e. if the points (data) on the coordi-

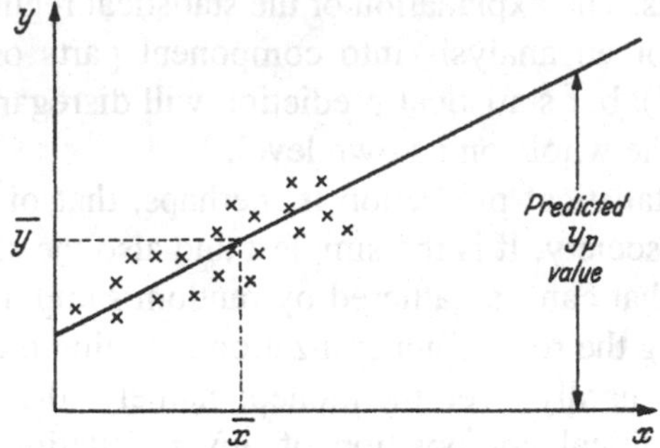

Fig. 10.4. Extrapolation with the help of the regression line equation

$$y_p(x) = \bar{y} + \frac{\sigma_y}{\sigma_x} \cdot r(x,y) \cdot (x - \bar{x})$$

where $r(x, y)$ is the correlation between x and y, and σ_x, σ_y the standard deviations.

nate plane tend to bunch up along a straight line (see Fig. 10.4). But an accurate computation is made of a probable value, i.e. of the value around which the actual value is likely to be found; the actual value will most probably differ from the computed value.

The trend line need not have a straight axis: any trend line with a regular core will serve as well. A typical case is that of human population growth. The trend of population growth is the outcome of the interplay of a number of little-known laws. Every demographic projection, whether to the future or to the past, is an extrapolation of the trend "observed" in the last century or so; in other words, the projection is made on the assumption that the present relative rate of growth is constant. But this is a pretence, as shown by the inability of demography to predict more than one decade or so in advance. Which in turn strongly suggests that there is no *single law* of human population growth, but just a trend line probably resulting from the interference of a number of laws, and which can be projected between a low and a high estimate. Thus, the world population for the year 2000 has been estimated between 5,000 and 7,000 million (see Fig. 10.5).

Trend lines are unreliable projection tools while they are not backed by some theory—for instance, a population growth theory—accounting for the mechanism responsible for the central trend. In this case the central line may be close to a line proper and the deviations from it may be assigned to random perturbations such as historical accidents. In information-theoretical terms, the central line will be the (orderly)

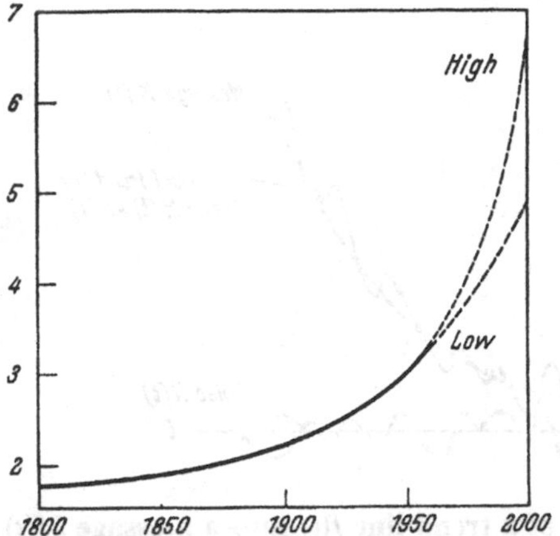

Fig. 10.5. Projection of world population. Solid line: estimated values for the recent past. Dotted lines: optimistic and pessimistic extrapolations of trend line.

message on which a (random) noise is superimposed (see Fig. 10.6). A deeper theory may in turn analyze the noise into a second message and a higher-order (smaller amplitude) disturbance, i.e. the noise that cannot be further analyzed with the knowledge at hand. In conclusion, we may trust *theoretically analyzed* trend lines because, knowing the mechanism responsible for their major traits, we have some *grounds* for assuming that "the future will resemble the past". And the reliability will be the greater, the deeper our resolution of noise into message plus a fainter noise. If no message is discerned at all, we are not in the presence of a *trend* line proper but of pure randomness, in which case only very vague statistical prediction is possible, one concerning the sequence as a whole rather than any of its components. This is the case of random time series (see Sec. 6.6).

Random time series, in contrast with stochastic laws and trend lines, have no projective power. Not every mapping or history of a course of events, however accurate, has projective power: only law statements have, and they do not describe whole events but rather selected aspects of them. The moral for history is clear: chronicling will give us no foresight, and this because it gives no insight. Projectibility (predictability or retrodictability) depends on our knowledge of laws, and such knowledge may be scanty or even nil in a given area. Consequently, failure to project events of a given kind will not prove that

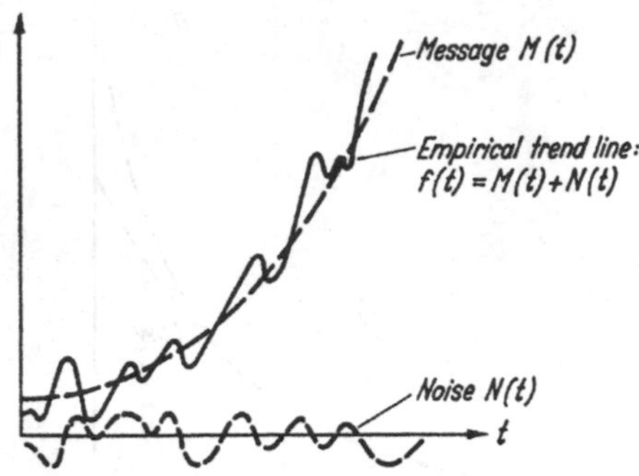

Fig. 10.6. Analysis of a trend line *f*(*t*) into a message *M*(*t*) and a random noise *N*(*t*). The noise is the unanalyzed residuum.

they are inherently unprojectible events or even chance events: the absence of a theory proper should be suspected.

Chance is much too often blamed for our incapacity to project, but unjustly so. As we have seen, there is no chance unless certain statistical regularities are satisfied (Sec. 6.6). For instance, we are not entitled to infer that a given coin is fair and has been flipped in a random way unless a rather stable heads/tails ratio is obtained in a long sequence of throws. Individual chance events are unpredictable but unpredictability does not point to randomness; furthermore, individual randomness results in collective patternedness. Moreover, what looks like a chance event in a given cognitive context may cease to look accidental in a different context, in which it will accordingly become projectible. For instance, until recently all mutations were regarded as random just because no constant relation between them and controllable agents had been discovered; to be sure, by means of certain radiations it was possible to experimentally induce mutations but never in a predictable way, because radiation acts very much like bombing. In 1944 precisely predictable mutations were induced in certain pneumococci by means of deoxyribonucleic acid (*DNA*); from then on spontaneous, hence hitherto unpredictable mutations (interpreted as "mistakes" in the duplication of nucleic acid molecules) had to be distinguished from directed, hence predictable mutations. This achievement was a first step in mastering the evolution of species, which up to now has been a random process, though one with definite trend lines. In fact, evolution is partly random because mutations, which provide the raw material for natural selection, are independent of their adaptive value, instead of being cleverly produced with the aim of enhancing the individual's fitness to its environment. The latter, by selecting the mutants, induces trend lines on a higher level. If we succeed in directing certain mutations by manipulating nucleic acids, we shall be able to steer evolution tailoring genotypes to order and perhaps decreasing the natural waste. At the time of writing genetic engineering is both a promise and a menace.

The inference is clear: there are no inherently random events, and consequently there are no inherently unprojectible events. Both randomness and projectibility are *relative*: chance is relative to the total physical context, projectibility to the state of knowledge. The expression 'x is unprojectible' should therefore be regarded as incomplete and, if taken literally, as misleading. We should instead say 'x is unprojectible on the body of knowledge y'. Change the theory or the information belonging to the particular body of knowledge used in trying to project x, and x may become projectible. The following cases should reinforce our conclusion that projectibility is essentially relative to the projection instrument, i.e. to the available knowledge.

Consider first a large collection of individually random moving molecules and such that the system as a whole is in thermal equilibrium, both internally and with its surroundings. We do not know what the velocities of the individual molecules are but we have Maxwell-Boltzmann's law for the distribution of velocities: this law is statistical in the sense that it refers to a statistical aggregate but it is far from an empirical generalization: in fact, it is a theorem in statistical mechanics, and therefore it has a larger projective power than any unanalyzed statistical regularity, such as a trend line. It predicts the distribution of velocities that will actually be measured if a small hole is made in the container; on top of this it enables one to predict the average energy, temperature, and density of the system. These are all collective predictions, i.e. they concern the system as a whole. The same theory yields also individual projections but of the probabilistic type: thus, it will not tell us what the actual velocity of a molecule taken at random is, but what the probability is that its velocity lies between such and such bounds. Retrodiction will yield the same value as prediction because time is not involved in equilibrium laws such as Maxwell-Boltzmann's. Things are quite different both in nonequilibrium situations and on the lower level of the constituents particles: here the symmetry between

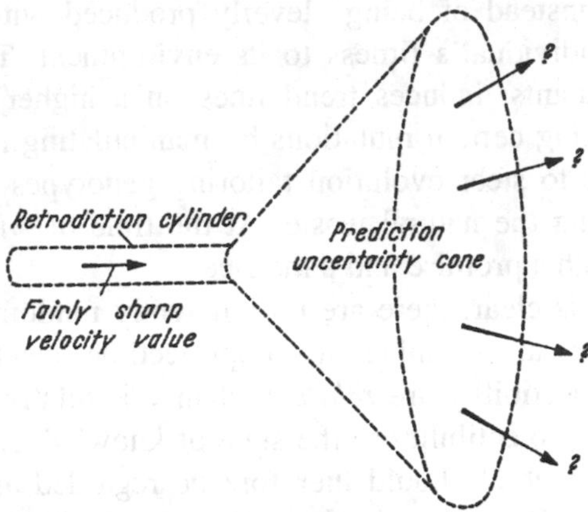

Fig. 10.7. Uncertain prediction but fairly certain retrodiction in atomic physics. From the moment a position measurement is made the electron trajectory can be retrodicted within a cylinder but can be predicted only within a cone.

prediction and retrodiction can be lost because of the irreversibility of the processes themselves.

Consider, for example, an electron fired with a given known velocity by an electron gun or by some natural source. If the electron position is measured, its trajectory can be retrodicted rather accurately; on the other hand the prediction of its future path will be rather hazy because of the blurring of its position (see Fig. 10.7). In this case, then, a fairly accurate retrodiction is possible although the prediction is next to impossible—within the present quantum theory. There is, of course, no reason to think that every future theory will retain this feature. Our point is just that there are cases in which, in a given theory, retrodiction is possible while prediction is not, and conversely.

It is only when the line between past and future is unique, that retrodiction is just the mirror image of prediction: this is the case of bullets (not of electrons) and of light rays (not of light quanta), but by no means the case of a statistical aggregate of atoms, such as a glass of water. In fact, any number of past histories of the individual components of a statistical aggregate may end up in one and the same final state; in such a case, from a knowledge of this final state and the law of the process as a whole we cannot reconstruct the past history of the system. The reason for this is that, in the case of statistical aggregates, the precise past histories of the individual components are immaterial.

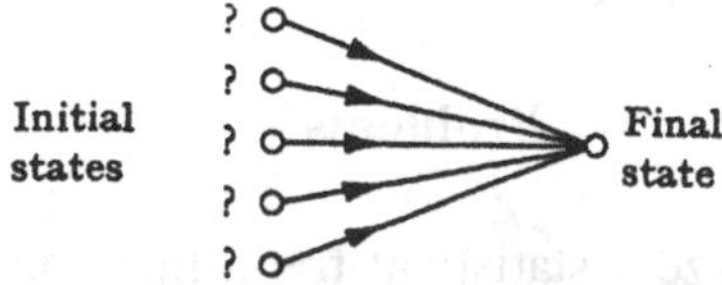

Fig. 10.8. The many-to-one relation between initial states and a final equilibrium state precludes retrodiction.

Just think of the number of trajectories of a coin all of which end up in the final state we label 'head'. A similar situation obtains in processes that are not ostensibly random although they do emerge from random shuffling on a lower level, such as the mixing of a set of bodies initially at different temperatures: a given value of the final equilibrium temperature is attainable in infinitely many ways, i.e. by starting with any of infinitely many combinations of individual temperatures (see Fig. 10.8). In all such cases, the more randomness the less memory and vice versa.

Generally speaking, given the laws of an irreversible process and the initial state of a system undergoing such a process, its final state can in principle be predicted, either fairly certainly or in probability. But, with the aid of stochastic laws alone, it will in principle be impossible to trace back or retrodict the evolution of the system from a final state to its initial state. Only a knowledge of the final state and the individual laws of the components of the aggregate would render the retrodiction possible, at least in principle, though perhaps not technically possible for large lapses owing to the uncertainties in the very knowledge of the final state and of the influences operating on the system along its history.

To sum up, (i) stochastic projection, either collective (statistical) or individual (probabilistic) is no less important and interesting than nonstochastic ("deterministic") projection, particularly if made with the help of theories rather than on the basis of empirical statistical generalizations; (ii) predictability does not involve retrodictability or conversely, and (iii) projectibility is relative to the knowledge at hand rather than inherent in facts. We are now ready to approach some of the complexities of historical projection.

Problems

10.2.1. State and analyze a statistical trend line and a stochastic law, pointing out the collective and the individual aspects of the projections that can be made on the basis of the latter. *Alternate Problem:* In daily life we make a number of indefinite forecasts and act thereupon. For example, we "save for a rainy day", assuming that it will come some day although we ignore what kind of "rain" it will be and when it will fall. Similarly, in technology provisions for events unforeseeable in detail are made (e.g., for the failure or the wearing out of machine parts). In either case we forecast sets of possible facts rather than singular facts. Elucidate this concept of indefinite forecast. Hint: use the concept of branching possibilities.

10.2.2. What is known and what is ignored when stochastic laws alone are available? And can individual behavior be predicted with the help of such laws? If so, in what way? If not: must we rely on intuition? *Alternate Problem:* Analyze the claim that, in psychology, individual (clinical) forecast, based neither on statistics nor on law statements, can be as accurate as actuarial (statistical) prediction or even more so. How are such individual prognoses made and validated? See P. E. Meehl, *Psychodiagnosis* (Minneapolis: University of Minnesota Press, 1973).

10.2.3. T. R. Malthus predicted in 1798 that mankind would end up hungry because population would eventually outgrow food supply. He assumed that population growth is accelerated (power law) whereas food production increases at a constant rate (linear law). Was this a scientific prediction? And why was it falsified? Finally: can we now do better than parson Malthus?

10.2.4. Does the statistical account of a fact, or collection of facts, prove that its referent is indeterminate? And does the existence and increasing spread of the stochastic approach establish indeterminism? *Alternate Problem:* Study how randomness in the small (e.g., genic mutation) can be smoothed out in the large (either by selection or by averaging). In other words, study the problem of how randomness on one level may lead to nonrandomness on another.

10.2.5. A car collision at a street corner might be regarded as accidental (i.e. as a chance event) by the drivers involved: otherwise they would have prevented it. But an onlooker who has been watching the collision from a terrace might have been able to forecast the accident: for him the same event was perfectly determinate. Dissolve this paradox. *Alternate Problem:* Work out the idea that, although chance may be objective (i.e., though there are random distributions and processes in reality), our judgments about the degree of randomness is relative to a set of law statements. See M. Bunge, *The Myth of Simplicity* (Englewood Cliffs; N.J.: Prentice-Hall, 1963), Ch. 12.

10.2.6. Perform a detailed study of forecasts on the strength of hypotheses of the type: $y=E(y|x)+\varepsilon$, where x and y represent key properties of a system, $E(y|x)$ is the expectation value of y given x, and ε a random variable lumping the disturbances.

10.2.7. Make a detailed study of retrodiction. In particular, examine (i) law statements that make prediction but not retrodiction possible; (ii) the converse case; and (iii) retrodiction on stochastic laws. For an example of (i), see M. Bunge, "A General Black Box Theory", *Philosophy of Science*, **30**, 346 (1963); for an example of (ii), see M. Bunge, *Causality*, 2nd ed. (New York: Dover, 1979), p. 314, fn. 5; for (iii), see S. Watanabe, "Symmetry of Physical Laws. Part III: Prediction and Retrodiction", *Reviews of Modern Physics*, **27**, 179 (1955).

10.2.8. Examine the thesis that unpredictability in a given domain warrants the inference of the objective indeterminateness of events in the same domain. Hint: distinguish determinacy (an ontological predicate) from predictability (an epistemological predicate). See M. Bunge, *Casuality*, Ch. 12.

10.2.9. Assuming that whatever is predictable by law is determined in some way, study the claim that this assumption leads to either of the following theses. (i) "Everything determined is predictable in principle". (ii) "Anything determined is predictable in practice". (iii) "Anything is determined if and only if it is predictable either in principle or in practice".

10.2.10. It is sometimes claimed that a scientific theory with explanatory power may have no predictive power whatsoever, and the theory of evolution is mentioned as an outstanding example of that. Examine this claim. Hint: Begin by ascertaining whether, as a matter of fact, no predictions are made in microbiology and no retrodictions in vertebrate palaeontology with the help of evolution theory. For a discussion of the projective performance of this theory, see Sec. 10.4.

10.3. Hindcast

We make backward projections, i.e. hindcasts, all the time. The sight of a beautiful grandmother suggests a beautiful bride—on the generalization that beauty, like the purchasing power of money, can deteriorate but not increase; an instructive book suggests the severe training of its author—on the assumption that learning and creative work are not obtained free; and so on. We wish to reconstruct the past as much as we wish to figure out where we are bound to; not only out of intellectual curiosity but also because, in order to understand the present, we must understand how it originated and, in order to forecast, we need information, and all information proper is about facts already past.

The hypothetical retrodiction of the human past is, of course, the specific task of history. But history is not the sole historical science if 'historical' is taken to meant "retrodictive". Cosmology—including the theories of stellar evolution—attempts to reconstruct the history of the universe and its components. Geology tries to retrodict the earth's past. Chemistry is concerned with hypothesizing the history of elements and compounds. And palaeontology endeavors to trace back the origin and evolution of living beings. All these sciences afford examples of scientific retrodiction, sometimes in a strict, at other times in a lax sense. We shall call *strict scientific projection* the one made with scientific theories and scientific data alone. On the other hand *lax scientific projection* will mean for us the projection made with the help of scientific theories and data plus assumptions, concerning trend lines and patterns, that are not scientific laws proper. Strict projection is typical, though not exclusive, of physical science, whereas lax projection is typical of the sciences of man in their present state.

An elementary example of strict theoretical retrodiction is the dating of geological strata by the study of sedimentary rocks. Knowing the sedimentation rate and the rock's thickness, the time taken by the

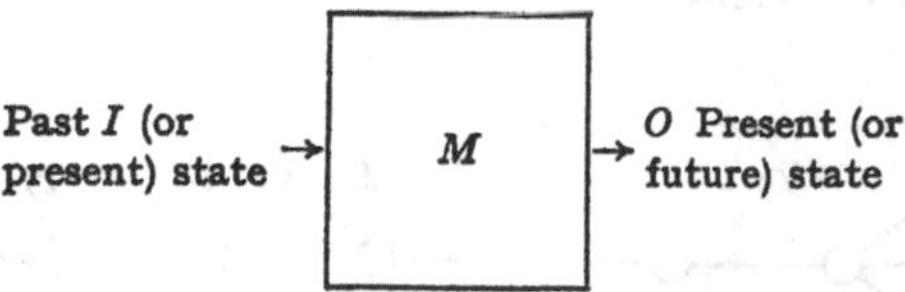

Fig. 10.9. The projecting box that relates the past inputs *I* to the present outputs *O*. The conversion operator *M* specifies the kind of box without, however, hinting at its entrails.

formation of the rock can easily be estimated (i.e. reckoned approximately) The law is in this case "v=constant", where the velocity v of sedimentation is measured in the laboratory; the circumstance datum is the rock thickness h; the inference is simply this: from $v=h/t$ we compute the time t, i.e. the age of the rock. For example, the maximum thickness of the quaternary period in Europe is about 1,000 m, and the estimated sedimentation velocity v=.03 cm/yr. Hence the maximum duration of that era was t=1,000 m/.0003 m/yr≅3,000,000 years. The dating of radioactive rocks and meteorites by weighing their lead and helium content follows a similar pattern and so does radiocarbon dating: a well known (yet in principle corrigible) law of radioactive decay is employed in these cases.

Nomological or strict prediction is possible whenever (i) there is a one-to-one correspondence between past and future states, i.e. when the system has a strong "memory", so that its history can be traced back by examining, as it were, the system's scars, and (ii) the law of evolution is known. (In other words, when the functions that map the past onto the present can be inverted.) In this case *alone* retrodiction is just the inverse of prediction. Clearly, a limited class of systems complies with both conditions; accordingly nomological or strict retrodiction is far from being universally possible.

A general—but superficial—treatment of the problem can be given by employing a black box representation of the system, i.e. one that specifies neither the constitution nor the structure of the system. Let 'I' represent an input variable, 'O' the corresponding output variable, and 'M' something that, applied to the column of input values, yields the output column. In short, $O=MI$. (See Figs. 8.16 and 10.9). The *direct projection problem* is then: Given the past (or the present) input values I and the "mechanism" operator M, find the present (or future)

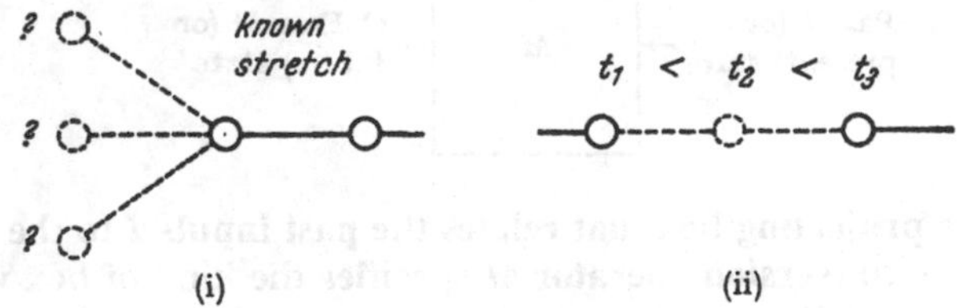

Fig. 10.10. Lax retrodiction: (i) extrapolation, (ii) interpolation.

output values O. This problem is solved by performing the operation MI, which yields O. The *inverse problem* of projection is: Given the present output values and the "memory" operator M, find the past input that brought about the present condition O. The problem, if solvable at all, is solved by finding the inverse M^{-1} of M and applying it to the data column at hand, since $I=M^{-1}O$. Whereas the direct projection problem is in principle solvable—as long as the law "$O=MI$" and the data I are available—the inverse problem may not be solvable: the inverse M^{-1} of the "mechanism" operator may not exist or it may not be unique. In the first case no retrodiction will be possible, in the second a number of mutually incompatible retrodictions will be obtained, usually without an estimate of their relative probabilities. Only under certain restrictive conditions will M^{-1} establish a 1:1 correspondence between the given output and the sought inputs (see Problem 10.2.7). In other words, strict retrodiction is possible for a limited class of black boxes alone.

If the system's history is largely immaterial to its present state, or if its evolution pattern is unknown (either because it does not exist or because we have so far been unable to unravel it), then lax retrodiction may be tried. Like prediction, lax retrodiction may consist of extrapolations or of interpolations. An *extrapolation* is a projection into an uncharted area and is therefore a risky and, if successful, very rewarding operation; an *interpolation*, on the other hand, is the more cautious inference consisting of making an estimate between two fairly reliable data (see Fig. 10.10). Speculations on the origin of social institutions are extrapolations, whereas the hypothetical reconstruction of a missing link is an interpolation.

An important case of retrodictive interpolation is the set of hypotheses on the origin of life. Evidentially considered this is an extrapolation since the projection goes beyond all fossil evidence. But if the

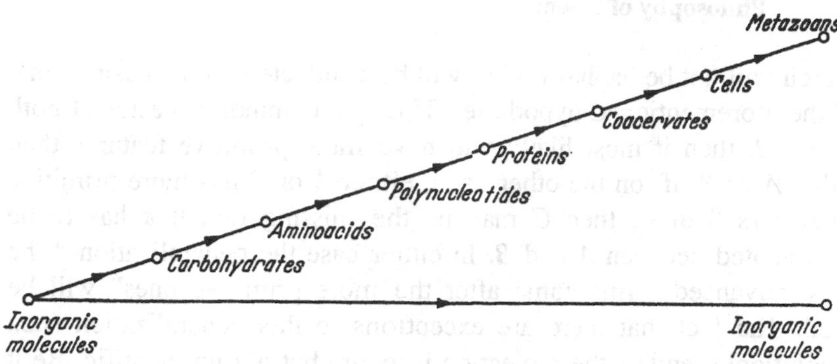

Fig. 10. 11. Biogenetic theory as a sequence of interpolations (roughly after Oparin).

hypothesis is accepted that life emerged spontaneously—though lawfully—from nonliving matter, then nonliving and living matter, as known today, become the fixed reference points between which the primaeval living things are to be interpolated. This requires, of course, the additional assumption that nonliving matter has not appreciably changed in our planet for the last couple of billion years (see Fig. 10.11).

Another typical example of backwards interpolation is offered by palaeobiology, when backed up by the theory of evolution. The doctrine of the fixity of species made retrodiction and prediction of species changes pointless; these problems became meaningful only in the bosom of evolutionary theories. Transformism does not yield accurate extrapolations, either backwards or forwards, because it does not so far contain specific *laws of evolution*, i.e. laws for the evolution of the separate genera, but rather *trend lines* to be eventually explained by the laws of genetics and ecology. Yet this is sufficient both to suggest the search for evidence of transformations and to make fairly accurate backwards interpolations on the basis of the assumption that transitions between neighboring species are almost continuous. Thus, given two neighboring species *A* and *B* of the same genus, evolution theory will suggest the following hypotheses: h_1="*A* descends from *B*", h_2="*B* descends from *A*", and h_3="*A* and *B* descend from a common ancestor". In either case transformism suggests the existence of *further* varieties *C*; and this existential hypothesis will guide the search for or at least the recognition of specimens of *C* or their fossil remains. The search will not be haphazard but will be conducted on the basis of one of the aforementioned hypotheses. If *C* is a common ancestor of both *A* and *B*, then it most likely will have more primitive features than either *A* or *B*. If, on the other hand, either *A* or *B* has more primitive characters than *C*, then *C* may be the missing link that has to be interpolated between *A* and *B*. In either case the generalization "The most advanced forms come after the more primitive ones" will be used. The fact that there are exceptions to this generalization (i.e., involutions) renders the projection insecure but not unscientific. Be it as it may, biological retrodiction is at the same time an application of the theory of evolution and its test: only, it is lax rather than strict projection.

In human history, too, backwards projection is mainly interpolative and it is possible to some extent because, unlike macroscopic mechanical systems such as the solar system, human societies are endowed with a strong "memory", not in the sense that they profit from experience but that their state at any given time strongly determines their future. In history interpolation is rarely, if ever, done with the help of laws proper but is more like puzzle solving: either a piece of evidence has to be fitted in a known or a conjectured pattern, or a gap is found in the pattern itself and the corresponding piece of evidence is sought. The prehistorian who finds a piece of fabric will set out to search for remains of looms; the archaeologist who discovers rupestrian paintings will search for human bones nearby. A set of data, properly interpreted in the light of a framework of historical knowledge, suggests a specific pattern for the human group under investigation; the pattern in turn suggests as yet unknown facts, i.e. it suggests retrodictions that the historian sets out to test in order to check the conjectured pattern and to gather fresh information. Even if he does not employ historical laws proper the historian does not proceed blindly either: he employs, besides data, (i) the common-sense knowledge—often erroneous—about "human nature", (ii) the refined knowledge supplied by biology, economics, sociology, psychology, and any other discipline he may find useful, and (iii) empirical generalizations, trend lines and other patterns of historical events and processes.

Historians, then, make lax retrodictions, mostly backwards interpolations. One occasionally indulges in forwards projections—not to speak of prophecies. Ex. 1: in the 18th century Count Aranda prognosed that the Spanish colonies in America would end up by being absorbed by the U.S.A. His was not a mere unfounded prophecy: he explicitly employed the generalization that population increases lead to territorial expansion, and the information that the young republic had a far more rapid demographic growth than the backward Spanish colonies. Ex. 2: in 1835 A. de Tocqueville prognosed that the U.S.A. and Russia would end up by taking in their hands the fate of one quarter of mankind each. Nor was this a mere prophecy: de Tocqueville extrapolated a trend line, namely the ceaseless American and Russian territorial expansion and rapid progress at the time. In both cases a general *trend* and the *mechanism* underlying it were made the projectans premises. The projections were not made on merely phenomenological generalizations of the type of "Whenever *A* happens it is followed by *B*": these are unreliable because, when no "mechanism" can be indicated, a mere coincidence can be suspected.

Many traits of the present civilization were predicted a century ago on similar grounds. Thus, the increasing division of labor, the strengthening of the state and the mounting of tradeunionism and of the political radicals were prognosed as outcomes of industrialization. On the other hand many other historical forecasts, perhaps most of them, have not been borne out by the events, and also many important events were foreseen by nobody. Thus, e. g., the increasing prosperity of the working class, World War I, the alliances during WW II, the post-war strengthening of Catholicism and militarism, the rapid collapse of colonialism and the USSR—save perhaps those who had a hand on the steer. The conclusion is obvious: so far, historical forecast has been *mostly unsuccessful.*

Two explanations of this failure offer themselves. First explanation: historical prediction is impossible, and whatever forecasts did come true were so by chance or because they were much too obvious, vague, or ambiguous. Second explanation historical prediction, in a lax sense, is possible *in outline* as long as the right and relevant generalizations, trend lines and law statements (belonging perhaps to adjacent disciplines such as the social sciences) are employed. The first explanation may be correct but does not look so at the moment, (i) because it is not followed by an explanation why historical prediction should be impossible, and (ii) because scientific prediction has not really been tried in the field of history. The second explanation looks more likely if only because, as a matter of fact, *short term* and *partial* historical predictions are successfully made all the time Only, they are not made by historians, who are more interested in the past, but by statesmen and politicians, who are in a way professional historical agents.

Making the budget of a modern nation requires forecasting the probable output in the near future on the basis of present trends and of a few laws of economics; likewise, a factory's production schedule is based on sales forecasts, for which operations research is developing more and more powerful tools. In general, the more civilized we become the more we tend to rule our present actions on forecasts: this is, indeed, the meaning of 'acting according to plan': acting in order to attain preset, hence roughly predicted goals. In the past such forecasts were made by soothsayers; nowadays they are made by administrators, economists, social scientists, and politicians. If they are often inaccurate it could be because they rely on insufficient knowledge. Wishful thinking and even untruthfulness are additional sources of failure in the prediction of social action. At any rate, failures in forecast do not prove the *impossibility* of prediction: all they do prove is the *fallibility* of prediction. It might even be argued that short-term historical forecast is in principle easier than individual performance forecast, precisely because the former is concerned with statistical aggregates of individuals whose precise behavior is not really relevant to the over-all trend. Thus, e.g., the military (economic) strategist will plan a course of action and forecast its most likely outcome without caring for the individual fate of the soldiers (workers); likewise, the tax collector will make an estimate of the total revenue of his district without caring how every one of his taxpayers earned his living. A second feature of large scale human behavior that makes prediction in outline possible is the strong "memory" of human communities. A third and last factor of predictability is the increasing intervention of the state and consequently the nearly totalitarian planning of the course of events in modern life. (Just compare Thoreau's *Walden* with Skinner's *Walden Two*, separated from the former by only a century.) For better or for worse the increasing planning of social life and the resulting rigidity of it are taking history nearer physics—which is of course just a prognosis. Historical prediction is not possible because there are *inexorable laws* of history: if there are any laws they are our own make and therefore they can be changed, i.e. they are not inexorable. Historical prediction should in principle be possible, rather, because (i) human history is the evolution of human society, and we are beginning to learn something about social mechanisms and (ii) people sometimes deliberately apply themselves to make events happen.

It has usually been taken for granted that historical forecast should either concern itself with the deeds of exceptional individuals (on the assumption that these are the chief causal agents) or with the over-all "fate" of nations (on the assumption that there are inexorable laws of history). Both extreme desiderata are next to unattainable and they are not worth while arguing about if their presuppositions are shown to be false. What the sociologically-minded historian may attempt to forecast with some success is *the most probable broad lines* of development of human groups in certain *definite respects* (e.g., economic, or educational, or religious) on the assumption that certain factors, mechanisms, and trends, will either continue or stop to operate. This kind of forecast will not start from *over-all laws*—which are anyhow not in existence—but will rather be preceded by an *analysis* of the system concerned into parts and relationships and the subsequent application of a number of hypotheses and if possible of theories concerning the parts and relationships in question.

The mistake of all historical prophets has been to make intuitive and holistic forecasts—i.e. over-all projections with no previous analysis and no underlying theory. Little wonder they should fail—unless they took the precaution to state them in an empirically irrefutable way. It is as if a mechanical engineer asked to predict the state of a car five years in advance took the machine as a whole, paying no attention to the present state of its parts and to its driver's habits, and made no definite assumptions regarding the normal wear and the probability of accidents, on the basis of known statistical generalizations. The best prognosis a competent car technologist could made on the basis of such knowledge and such assumptions would be of the form: "The probability that the car will end up in state 1 is p_1; the probability of the final state 2 is p_2, and so on, where the various probabilities add up to 1". If such forecasts of probability trees are the most that can be expected concerning cars and yet they are useful, why should so much more be required from the scientific historian of the future? (See Sec. 10.2.)

So far we have been concerned with forecasting the *most likely* course of events. What about the *least likely*, i.e. radical novelties such as advances in pure and applied science? Important scientific and technological discoveries and inventions are unpredictable *in detail*, and so are great works of art. But some such innovations can be forecast *in outline* and, moreover, they are being prognosed by every major institution where important research projects are being carried out. Everyone could now venture to make the following and even more daring prognoses concerning scientific developments in the near future. Firstly, mathematical biology, sociology and psychology will continue to make progress. This is an extrapolation of a trend initiated essentially after the second world war. Secondly, the mechanism of photosynthesis will be taken apart and subsequently mastered and applied to the large scale synthesis of food, which will in turn result in the gradual disappearance of agriculture, which in turn will complete the process of urbanization. The ground of this extrapolation is the increased insight into the process of photosynthesis and the successes of the programme of synthesizing protoplasm out of aminoacid chains. Thirdly, the chemical repair of brain dysfunctions will improve. The ground for this prognosis is that some compounds are known to perceptibly affect the brain. All three prognoses have the following characteristics: (i) they are forecasts in outline, not in detail; (ii) they are time-indefinite, hence empirically irrefutable; and (iii) they do not rely on laws proper: no laws concerning the invention and discovery of ideas and gadgets are so far known. But they are not wild prophecies either: they rest on information regarding present conditions, trends, and goals, as well as on a firm confidence in the thrust and power of scientific research. Moreover, they are not idle forecasts: they constitute the basis of certain research projects and of the hopes put in them. Without such prognoses, plans, and hopes, no funds could be raised for the corresponding research.

Historical forecast, in short, does seem possible at least in outline. But this is not to deny that there are important differences between historical prognosis and physical prediction—the two poles of scientific forecast. A few outstanding differences are these: (i) whereas the physicist is often in a position to make strict nomological predictions, the historian can so far make lax predictions only, perhaps because (ii) up to now history has disclosed only a few laws, mostly of a superficial (phenomenological) kind, such as laws of conjunction and succession, whence (iii) historical forecast is inaccurate. (iv) Historians have not cared to build scientific theories in history: they have hounded the isolated fact and the isolated chain of events and, at most, they have imagined grand but untestable or outright false philosophies of history; (v) in particular, historians have, until recently and save exceptions, not tried to disclose the mechanisms of events, which mechanisms are presumably biological, psychological, economical, social, and cultural. (vi) History has never been approached in a statistical way even though human society is in many respects more like a statistical aggregate than like an organism. (vii) Whereas the physicist need employ no laws other than physical laws, the historian needs lower level (chiefly sociological) laws—as the social and economic historians have realized long ago. (viii) Whereas the laws of the lower levels are assumed to hold for cosmic stretches of time or perhaps forever, social laws are supposed to change along with society itself; this adds complexity to historical projection but does not render it impossible: after all, perhaps the laws do not change at random and their laws of change could be disclosed. (ix) Unlike physical prediction, which by itself does not influence facts (unless it is seized upon by the engineer), prediction concerning human affairs can influence those very events it refers to, to the point of changing the course of events from the one that would have followed had the prediction not been made public—as in the cases of the wishful forecasts made by financiers and politicians, which forecasts contribute to molding the future (see Sec. 11.3).

But in this last respect historical prediction is just like the technological forecasts—about, say, machine performance and lifetime—made with the help of physics. And the causal efficacy of some historical predictions is quite a strong argument in favor of the thesis that historical prediction can be successful. Only, it has little or no test value, because we cannot experience what would have happened had the forecast not been made. In this case, the more foresight the less insight. Also, the inaccuracy of historical forecast can be contrasted to the accuracy of astronomical prediction, but it is not worse than the inaccuracy with which electron trajectories are predicted, or than the present inability to predict earthquakes and tornados. Finally, if the historian is forced to resort to lower level laws so is the geologist, who relies on laws of physics and chemistry, or the biologist, who, in order to forecast the probable evolution of a bacterial colony, uses laws of genetics and ecology. If, instead of contrasting history with astronomy, we compared it with statistical mechanics or with the study of biological evolution, many infirmities allegedly peculiar to historiography, or even to its subject matter, would seem to be shared by perfectly scientific disciplines. History is not physics but it is not literature either.

And to bar the possibility of a maturation of historiography to a science is as little conducive to progress as to feel complacent about its present protoscientific condition.

Predictions are all fallible and prognoses the more so, whether in physics or in history. Forecasts can be risky but they are not dangerous as long as they do not consist in gloomy forecasts concerning the impossibility of certain human achievements, whether social or cultural—such as, e.g., the prophecy that we shall never be able to foresee the direction of social change. The mere issuance of such a prophecy will lead the conformist to act so as to confirm the prophecy, i.e. it will inhibit his search for social theories with predictive power; and it will stimulate the nonconformist in the direction of refuting the forecast. (There are neither inherently self-fulfilling forecasts nor inherently self-defeating forecasts, but just people ready to be defeated and others unwilling to surrender.) But the causal efficacy forecasts can acquire when seized by men as guides to action will concern us in Sec. 11.3. Our next task will be to elucidate the concept of predictive power.

Problems

10.3.1. A politician, if shrewd, will be able to forecast, regarding any major international crisis, that it will eventually be settled. Why? And will he risk an opinion as to how and when will the crisis be solved? And under what circumstances, if any, could he do it? *Alternate Problem:* In his astounding science fiction *The World Set Free* (1914), H. G. Wells prophesied the discovery of artificial radioactivity for 1933 (it was found in 1919), the building of an atomic power plant for 1953 (it was made in 1951), and the blasting of the first atomic bomb for 1959 (it was exploded in 1945). On the other hand Lord Rutherford, the maker of induced radioactivity, prophesied in 1933 that nuclear energy would never be released, let alone tamed. Analyze this story in terms of prophecy, prognosis, and imagination.

10.3.2. Analyze the way in which physicist C. G. Darwin makes "average" forecasts in *The Next Million Years* (Garden City, N.Y.: Doubleday, 1953). That is, find, list and examine the basic assumptions (laws and trends) on which he bases his projections. *Alternate Problem:* Examine the advices given by S. Lilley to forecasters of scientific and technological innovation, namely: (i) Extrapolate present trends and (ii) make prognoses of the kind of "It will become possible to visit another planetary system", without specifying how and when. See "Can Prediction Become a Science?", in B. Barber and W. Hirsch, Eds., *The Sociology of Science* (New York: Free Press, 1962), p. 148.

10.3.3. Some of the particle accelerators or atom smashers were designed and built for specific purposes such as the artificial production of certain particles as yet unknown but predicted by theory. Thus, the Berkeley bevatron was designed for the production of antiprotons and the CERN protosynchrotron was designed to produce muons and pions. Discuss these cases in terms of the prediction and planning of scientific discovery.

10.3.4. Metals have some "memory", as shown by their fatigue; liquids have less "memory" but they retain for some time the vortex motions produced in them; finally, gases are almost entirely memoryless. In a number of important cases the past history of the system does not determine uniquely its future history, i.e. the system has little or no "memory"; in other cases (Markov processes) only the immediate past is relevant, i.e. the system has a short-range "memory". Examine physical "memory" as the objective basis for retrodiction. Recall Sec. 10.2.

10.3.5. What is predicted can be hoped to be observed, whereas what is retrodicted is unobservable since it has ceased to exist. How come we can test our retrodictions and what is the upshot of this difference between prediction and retrodiction for the methodology of projection? *Alternate Problem:* Study the methodology and the achievements of the social and economic historians of the new French school—such as F. Braudel and P. Vilar—grouped around the Paris *Annales*, E.S.C.

10.3.6. Biologist J. B. S. Haldane, in "On Expecting the Unexpected", *The Rationalist Annual* (1960), p. 9, has stated that "It is an essential part of the scientific method to expect the unexpected". Can this expectation be boundless or should we exclude events "prohibited" by fairly accurately known laws? And could we refine such an expectation of unpredictables into a more precise prediction? If we did, would it not entail violating the maxim that some room must always be made for the unexpected? See Alternate Problem 10.2.1.

10.3.7. Would long-term projection be possible if the objective laws kept changing? Consider the cases of random and lawful change of laws.

10.3.8. Examine H. B. Acton's argument that the prediction of individual discoveries and inventions is logically impossible because to predict them would be to make the discoveries and inventions before they had actually been made. Does the argument apply to forecasts in outline? *Alternate Problem:* Examine the view that individual discoveries and inventiones can be forecast in outline on the basis of present conditions, means and goals, and that when the means are produced "the times are ripe for the innovation", as corroborated by the fact that the innovation can be introduced simultaneously by independently working individuals. See J. R. Platt, *The Excitement of Science* (Boston: Houghton Mifflin, 1962), especially Ch. 4. Recall that Einstein told L. Infeld that, while special relativity might have been invented by someone else, general relativity would still be unknown were it not for him.

10.3.9. Examine K. R. Popper's refutation of the thesis of historical predictability as expounded in the Preface to *The Poverty of Historicism*, 2nd ed. (London: Routledge & Kegan Paul, 1960): In a nutshell, Popper's argument is this: (i) The course of human history is strongly influenced by the growth of human knowledge; (ii) the growth of human knowledge is unpredictable; (iii) hence the future course of man is unpredictable. Does the first premise hold for events such as the independence of African nations in the 1960's? And is the second premise plausible in view of the increasing planning of research and of the technological applications of scientific principles? Recall that the Manhattan project, among others, was executed roughly to schedule and that the interval between scientific innovation and application has been steadily shrinking in the last hundred years. Contrast Popper's position with that of the social scientist engaged in predicting major events and trends. See H. Hart, "Predicting Future Events", in F. R. Allen et al., *Technology and Social Change* (New York: Appleton-Century-Crofts, 1957).

10.3.10. Examine the following paradox. The statement "*x* is unpredictable" is short for "For every t, if t is an instant of time and x is an event, then it is impossible to predict x at t". For any value of t denoting a future instant, the preceding formula becomes a prediction: it predicts unpredictability. *Alternate Problem:* Collect and analyze a good number of historical forecasts currently made by social scientists and statesmen—e.g., "The agricultural population will continue to decrease all over the world", "The tertiary sector (administration and services) will continue to increase proportionally all over the world" "The educational requirements will continue to rise in all occupations" and "European science will not be able to catch up with North American science". Are they grounded? If so, on laws or on trend lines? And what could possibly refute them?

10.4. Projective Power

Predictions and retrodictions, in order to be regarded as scientific, must satisfy certain requirements distinguishing them from guesses, prophecies, and ordinary prognoses (see Sec. 10.1). And factual theories, to be regarded as scientific, should in turn yield projective formulas complying with those requisites. Ultimately, then, some of the conditions on scientific projective statements will propagate back to the premises that entail the projections. Let us study those conditions.

Quite obviously a projective proposition, if it has to have a test value or a practical value, must not be logically true: i.e. it must not be true by virtue of its form or by virtue of the meanings of its component signs. Otherwise it would be independent of fact: it would not be fact-referent. A weather forecaster would be dismissed if he issued a forecast like this: "Tomorrow it will either snow or not snow", because 'snow' occurs vacuously in this statement, i.e. it can be replaced with any other verb. And he would not be much better off if he made an announcement like this: "It may snow tomorrow", because this modal proposition is empirically undistinguishable from its denial "It may not snow tomorrow." (Only quantitative probability propositions are testable under certain circumstances.) In short, projective formulas must not only be synthetic, i.e. dependent on fact, but they must be definitely so: they must not be elusive or noncommittal. But this is insufficient: a statement may be quite definite, i.e. unambiguous and free from vagueness, yet untestable, such as "This book displeases Zeus". If a projective proposition is to be compared with observation reports it must be translatable without much distortion into an *observational proposition*, i.e. one containing directly or at most indirectly observational concepts alone; i.e. it must be, epistemologically speaking, a low level statement. In short, then, a projective proposition must have both a *high information content* and a *high degree of testability*. Or, as we may also say, a projection must be both referentially and evidentially definite if it is to count as a test of the body of hypotheses from which it derives.

Another way of saying the same thing is this: in order for a projection to count as a test of a scientific theory it must unambiguously specify the projected fact, i.e. in such a way that we may recognize it unmistakably. (This does not disqualify lax predictions or even indefinite forecasts such as those dealt with in the last two sections: it only diminishes their value as indications of the theory's degree of truth.) A practically complete specification of the projected fact will have to include the following items: (i) the *kind* of fact—e.g., that it consists in the synthesis of a certain nucleic acid molecule; (ii) the *conditions* under which the fact is produced—e.g., the reagents initially present, the pressure and temperature, etc.; (iii) the circumstances under which the fact will be or could have been *observed* by an operator handling a given experimental equipment—e.g., the quantity of nucleic acid sufficient to be detectable with the analysis techniques now available.

Both the nature of the fact and the conditions for its production and observation can be exactly specified by giving the values of its characteristic quantitative variables; but such a quantitative characterization is not always technically possible and sometimes it is not necessary. The time of occurrence of the projected fact will have to be reckoned among the values of the relevant variables if the original question that prompted the projective inference asked for it. If all these conditions are satisfied we may decide, prior to the empirical test, that whatever the truth value of the projection may be, this value will depend critically on fact alone and consequently the checking of the projection will be relevant to the theory's test. Such a dependence on fact should be ascertained prior to the test or the application: otherwise the former might be irrelevant or the latter unreliable.

*Now the requirements of information content (referential definiteness) and testability (evidential definiteness) can be satisfied to various extents. If a projective proposition excels in complying with such conditions, we may say that it has a high *projective content*. The projective content of logically true propositions is nil, that of oracular prophecies and unquantified possibility statements is nearly nil, somewhat higher that of empirical prognoses and maximal that of quantitative projections. Can we do better than just order projections with regard to their projective content: can we quantify the concept of projective content? Two proposals have been made to measure projective content, and both involve the probability concept. The first proposal is to equate projective content and probability, on the assumption that the best projections are those which are more likely to be borne out. In short, the first proposal is to set: *Projective content=Probability*. The second proposal is suggested by the remark that the gipsy's prophecies are almost as worthless as tautologies precisely because they are much too probable. Accordingly, projective content and probability would be inverse to each other, i.e. *Projective content*=1–*Probability*. The troubles with either proposal begin as soon as we attempt to measure the probability in question, i.e. as soon as we ask the following questions: (i) What is the argument of the probability function *P*, i.e. what is *P* a probability of, and (ii) how is the numerical value of the probability to be estimated on the basis of experience? Let us deal with these questions.

*Since we are speaking of the projective content of propositions it would seem that the argument of the probability function must be a statement *s*, i.e. that we should write: *P*(*s*). And, since a projective statement is about some fact *f*, we may as well write '*P*[*s*(*f*)]' for the probability of the statement *s* expressing the fact *f*. This (epistemological) probability must be kept separate from the (physical) probability *P*(*f*) of the fact *f* itself. In fact, whereas *P*(*f*) is supposed to be a dispositional property of a chunk of reality, *P*[*s*(*f*)] depends on the state of the human knowledge that suggests making the statement s. The two probabilities are then conceptually different and, in general, they also differ numerically. Moreover, the two are unrelated, as the following case will show. Consider the scattering of neutrons by an atomic nucleus. The collision of a single neutron with the nucleus is an extremely improbable (evidentially speaking, unfrequent) event owing to the small target offered by the nucleus (a target of the order of 10^{-24} cm^2). But physical theory is often in a position to make fairly accurate predictions of such improbable events; in other words, predictions with a small probable error can often be made concerning such improbable events. Suppose we equate the probability of the prediction to its accuracy. That is, calling ε the probable error, let us put: $P[s(f)]=1-\varepsilon$, where '*s*' stands for the projection that describes the fact *f*. On this assumption we have a scientific prediction, of an extremely improbable event, that has as high a probability as a gipsy's divination and, at the same time, an undeniably high predictive content. This counterexample ruins the equation of projectibility and improbability, just as the case of the gipsy ruins the equation of projectibility and probability. And it also ruins the so-called *straight rule* of inductive logic, according to which an event and the statement describing it are equiprobable. In short, both improbabilism and probabilism are unlikely.

*Moreover, we seldom determine the probability of a proposition, whether before a test or on the basis of it. First, because the phrase 'probability of a proposition' is so ambiguous that it is hardly meaningful; what can be assigned a definite meaning is the phrase 'the probability that a proposition *p be true* given (or assuming) the truth of the proposition(s) *q*'—i.e. the conditional probability $P(p|q)$ rather than the absolute probability $P(p)$. (For conditional probability see Sec. 7.5.) Second, because the determination of the numerical value of such a probability prior to test is impossible unless a definite and finite mechanism is assumed, such as the urn model (used, incidentally, in computing the probability of genetic hypotheses). What we do estimate is the *performance* of a system in doing a projective task. But, obviously, this can only be done *after* a sequence of empirical tests, because such a performance is just a measure of the extent to which the projections agree with the empirical reports relevant to them. Since we are chiefly interested in such an adequacy of our projections we shall shift the whole discussion from probability to *truth*. (That probability is a wholly inadequate measure of degree of truth was argued in Sec. 7.5.) And since scientific projections are derived in the bosom of theories we shall make a further shift, from stray projections to whole sets of projections belonging to a theory.*

We want to obtain *maximally true* projections hoping that this will be a symptom of the truth of the theories employed in computing them. After all, both for cognitive and practical purposes we seek likely (though refutable) theories. Only, for test purposes we shall not be so naive (or so dishonest, as the case may be) as to select the most likely or verisimilar among the projections afforded by our theory, because they are the more trivial and those that differ the least from the projections supplied by previously accepted theories. We are not anxious to confirm over and over the opinions countenanced by the antecedent body of belief: we want to test the new, unheard—of consequences of the theory we wish to evaluate—if this is, indeed our aim. Accordingly we must distinguish the *prior plausibility* of a projection from its *posterior plausibility*—and the same holds for the theory originating it.

In other words, for *test* purposes we shall not pick out the projections more likely to be confirmed but—in agreement with Peirce and Popper—we shall select those which, on the basis of antecedent knowledge (not embodying the theory under discussion), are the least likely to come out true. (Note that this automatically precludes the use of probability, since the set of projections we choose for test purposes is anything but a random sample of our theory: this applies to both of the proposals about projective content dealt with above.) On the other hand we shall *adopt* (until new notice) all and only such projections as are empirically confirmed, so that they have a high degree of truth on the basis of both theory and experience (posterior verisimilitude). There is no incompatibility between the former precept, which directs us to choose the boldest projections for testing, and the latter rule, which commands adopting the best confirmed projections. The first norm is applied before the empirical test, the second after it.

We shall then test those projections t_i derived in the theory T and having accordingly the maximal theoretical truth value $V(t_i|T)$ but with a small prior truth value $V(t_i|A)$ on the basis of the antecedent or background knowledge A. If t_i passes the empirical test we shall temporarily adopt it; in this case we shall assign t_i a high degree of truth $V(t_{ir}|TE)$, i.e. with regard to both theory and fresh experience E. In short, we adopt the following precepts.

Rule 1. For test purposes select those projections t_i such that the $V(t_i|T)$ are high and the $V(t_i|A)$ are low.

Rule 2. Adopt (until new notice) those t_i that, having been obtained in T, have also passed the empirical test—i.e., those with maximal $V(t_i|TE)$

These rules account for the obvious yet often neglected difference between testing a conjecture and adopting or believing it. Rule 1 should satisfy the deductivist injunction *Try the riskier hypothesis*, whereas Rule 2 should conform to the inductivist prescription *Adopt the best confirmed hypothesis*. The two rules are mutually compatible if only because they refer to different operations: test and adoption. Accordingly their joint acceptance should contribute to bridge the gap between deductivism and inductivism.

The above rules suggest several measures of projective performance, one of which shall now be presented. If many accurate projections with a low prior verisimilitude $V(t_i|A)$ are borne out by experience we can say that the mother theory T which entails them has an appreciable projective performance. (The total projective *power* of the theory is, on the other hand, impossible to compute: its determination would require the effective derivation of the infinite set $\{t_i\}$ of projections and the empirical test of every one of them.) The more numerous and the more accurate the original projections supplied by the theory, the higher its projective performance. Darwin's theory of evolution exemplifies the class of theories with a high projective performance (even though its projections are lax rather than strict, as we saw in Sec. 10.3); among others, it made the heretic projection that ape-like fossils, remains of our antecessors, would eventually be found. A theory supplying original and, at the same time, true projections, will direct the search for its own evidence. Every original projection, if confirmed, will constitute fresh evidence in favor of the theory; if refuted, against it. Moreover, in certain cases—like in Fresnel's wave theory of light, in Darwin's theory of the descent of man, and in Einstein's theory of gravitation—the projected facts constitute, at the same time, the very first evidence in favor of (or against) the theory. Clearly, if such an original projection comes out true, the degree of truth we shall assign the theory will be much greater than if the theory has been tailored after a given set of facts (ad hoc systematization).

*An adequate measure of the projective performance of a theory should then be a function of the *differences* $V(t_i|AE)-V(t_i|A)$ between the posterior and the prior truth value of the theory's projections t_i. These differences will be zero when the t_i are confirmed and the theory, far from being new in some respect, is part of the antecedent knowledge. They will be large if the t_i are foreign to A, i.e. if $V(t_i|A)=0$. And the differences will be maximal whenever the t_i are false according to A, in which case $V(t_i|A)=-1$. A simple, though not necessarily adequate formula for the projective performance $\Pi(T)$ of a theory T entailing the projections t_i that are gauged by the background knowledge A and the outcomes E of empirical procedures, is this:

$$\Pi(\mathsf{T})=\sum_{i=1}^{N}[V(t_i|\mathsf{AE})-V(t_i|\mathsf{A})] \qquad [10.1]$$

where capital sigma stands for summation over i between 1 and $N < \infty$, the total number of tests. Clearly, the preceding formula represents the sum of the differences between the posterior and the prior verisimilitudes of T's projections t_i. For large numbers of tests it will be convenient to introduce the *relative projective performance* $\Pi_r\mathsf{T}$):

$$N >> 1, \Pi_r(\mathsf{T})=(1/N)\sum_i[V(t_i|\mathsf{AE})-V(t_i|\mathsf{A})] \qquad [10.2]$$

Let us insist that, since N is the number of actually tested projections, these formulas do not measure the projective power of a theory; but they may be regarded as providing a rough *estimate* of it, much as sampling affords an approximate picture of the whole population.

We shall decide to assign as much weight to the body of our new experience E as to the antecedent body of knowledge A in the regard that is being investigated. That is, we shall rule that the truth value of a proposition t_i relative to AE be equal to the average of its truth value according to A and E separately:

$$V(t_i|\mathsf{AE})=(\tfrac{1}{2})[V(t_i|\mathsf{A})+V(t_i|\mathsf{E})] \qquad [10.3]$$

Introducing this formula into [10.2] we obtain finally

$$N>>1, \Pi_r(\mathsf{T})=(\tfrac{1}{2}N)\sum_i[V(t_i|\mathsf{E})-V(t_i|\mathsf{A})]. \qquad [10.4]$$

$\Pi_r(\mathsf{T})$ attains its minimum value, –1, when all the N projections are empirically refuted and, at the same time, they agree with the antecedent knowledge (which is then clearly superior to the new theory); in this case, in fact, every term in the sum equals $-1/N$ and the whole adds up to $-N/N=-1$. The maximum value +1 of $\Pi_r(\mathsf{T})$ corresponds to a set of projections all of which are empirically confirmed and, at the same time, contradict antecedent belief: for, indeed, in this case every term contributes $(\tfrac{1}{2}N)[1-(-1)]=+1/N$ to the total value $N/N=1$. And for an *ad hoc* theory which covers E and obeys A, $\Pi_r(\mathsf{T})=0$. In short, the projective performance defined by [10.4] is bounded between –1 (all N projections wrong) and +1 (all N projections both empirically confirmed and maximally original).

*The same result is obtained by starting from the remark that the projective performance of a theory is the greater the more original and, at the same time, the less inaccurate it is: that is, $\Pi(\mathsf{T})\propto O(\mathsf{T})-I(\mathsf{T})$. Now, the *originality* of T relative to the body of antecedent knowledge

A will be given by the distances $V(t_i)|T)-V(t_i|A)$ between the truth values assigned to the t_i on the basis of T and those assigned assuming that A holds. We may then write, again for large values of N,

$$O_r(T)=(\tfrac{1}{2}N)\Sigma[V(t_i|T)-V(t_i|A)]. \quad [10.5]$$

Since $V(t_i|T)$ is near 1 and $V(t_i|A)$ lies between −1 (heterodoxy) and +1 (orthodoxy), every addend will be nonnegative and O_r (T) will lie somewhere between 0 (or nearly 0) and 1. The minimum of O_r (T) will correspond to a theory consisting in just a rearrangement or reformulation of a system contained in A. And the maximum originality value, +1, will correspond to a theory clashing with the body of antecedent knowledge A.

*Now, boldness is valuable only insofar as it is consistent with minimum inaccuracy. And the degree of *inaccuracy* of T can quite naturally be defined (again for large values of N) as

$$I_r(T)=(\tfrac{1}{2}N)\Sigma[V(t_i|T)-V(t_i|EA)]. \quad [10.6]$$

Since $V(t_i|T)$ is near unity, $0 \leq I_r(T) \leq 1$. Subtracting this expression from [10.5] and recalling [10.2], we obtain

$$O_r(T)-I_i(T)=\tfrac{1}{2}\Pi_r(T) \quad [10.7]$$

as suspected. This intuitive result enhances the plausibility of our measure [10.2] of projective performance.

*Recalling the formula [9.13] for theory *coverage* (one of the factors of explanatory power as elucidated in Sec. 9.6), i.e.

$$C_r(T)=(\tfrac{1}{2}N)\Sigma[V(t_i|AE)+V(t_i|T)]. \quad [10.8]$$

we obtain the obvious result that the larger the coverage the smaller the inaccuracy and conversely: in fact, adding [10.6] and [10.8] we get

$$C_r(T)+I_r(T)=(1/N)\Sigma V(t_i|T). \quad [10.9]$$

When no theoretical inaccuracies are involved in the derivation of the t_i owing to the introduction of extra simplifying assumptions, i.e. when $t_i \in T$, the former formula reduces to

$$C_r(T)+I_r(T)=1. \quad [10.10]$$

which is just an elucidation of the intuitive idea that inaccuracy is the dual of coverage. And adding the expressions for coverage and originality one obtains

$$C_r(\mathsf{T})+O_r(\mathsf{T})=(1/N)\Sigma V(t_i|\mathsf{T})+\tfrac{1}{2}\Pi_r(\mathsf{T}). \qquad [10.11]$$

$$=1+\tfrac{1}{2}\Pi_r(\mathsf{T}) \text{ for } t_i\in\mathsf{T}. \qquad [10.12]$$

The interpretation of this formula is clear: projectibility is made up of coverage and originality, but in such a way that we cannot maximize both at the same time. This is one among various reasons for not confusing explanatory power with predictive power; there are further reasons (see Sec. 10.5).

*Let us illustrate the preceding formulas by applying them to a theory T, $N>>1$ projections of which are empirically tested. Suppose that, as is so often the case, approximations have been made in deriving the projections; and assume, for the sake of simplicity, that all the approximations introduce the same theoretical percentual error $\delta<<1$. Suppose further that a constant experimental error e is made in every test and that $0\leq\varepsilon<1$. A number R of these tests will be unfavorable (refutations) and a number $C=N-R$ will be favorable (confirmations). We disregard the ever-present possibility of inconclusive tests, i.e. we make the simplistic assumption that every answer of experience is either 'yes' or 'no'. As regards the relation of the projections to the antecedent knowledge we shall consider two extreme cases: (i) the t_i are all in agreement with what was known before the theory was introduced (orthodoxy); (ii) the t_i are all contradicted by the antecedent knowledge (heterodoxy). In short, our assumptions are

For every i, $V(t_i|\mathsf{T})=1-2\delta$

For every i between 1 and C, $V(t_i|\mathsf{E})=1-2\varepsilon$ (Confirmation)

For every i between C+l and N, $V(t_i|\mathsf{E})=-1+2\varepsilon$ (Refutation).

The following values are found after a short calculation.

Conformist theory: For every i, $V(t_i|\mathsf{A})=1$

Coverage: $C_r(\mathsf{T})=\tfrac{1}{2}+C/2N+(R-C)\varepsilon/2N-\delta$

Inaccuracy: $I_r(\mathsf{T})=R/2N+(C-R)\varepsilon/2N-\delta$

Originality: $O_r(\mathsf{T})=-\delta\cong 0$

Projective performance: $\Pi_r(\mathsf{T})=-R/N+(R-C)\varepsilon/N\cong -R/N$.

Nonconformist theory: For every i, $V(t_i|\mathsf{A})=-1$

Coverage: $C_r(\mathsf{T})=C/2N+(R-C)\varepsilon/2N-\delta$

Inaccuracy: $I_r(\mathsf{T})=\tfrac{1}{2}+R/2N+(C-R)\varepsilon/2N-\delta$

Originality: $O_r(\mathsf{T})=1-\delta\cong 1$

Projective performance: $\Pi_r(\mathsf{T})=C/N+(R-C)\varepsilon/N\cong C/N$.

*The best that can happen to a conformist theory is that it matches with fresh experience: in this case R=0 and $\Pi_r(\mathsf{T})$=0; the worst will

happen if the new experience is against the theory, i.e. if $R=N$, in which case $\Pi_r(\mathrm{T})=-1$. On the other hand, the worst that can happen to a nonconformist theory is that it be refuted, i.e. that $C=0$, in which case $\Pi_r(\mathrm{T})=0$; but if the theory is confirmed ($C=N$) its projective performance will be maximal: $\Pi_r(\mathrm{T})=+1$. Clearly, our preference for theories with a high projective performance is a preference for theories that point accurately to a large number of previously unknown facts. If, on the other hand, we held coverage (or inductive support) in higher esteem than projective performance, we would avoid bold new theories and seek cautious *ex post facto* systematizations.

*The desideratum of maximal projective performance, as understood in the above sense, would seem to contradict the desideratum of optimal consistency with accepted knowledge (external consistency: see Sec. 7.6). There is no contradiction if we stipulate that consistency with the *bulk* of accepted knowledge shall be required of a theory. More precisely, the new theory can clash with only a part of the body of accepted belief, namely, with that older part of established opinion that the new theory attempts to supersede. But the new theory shall not demand a simultaneous revolution in every branch of knowledge and it should not violate the usual metascientific requirements.

*Our formulas [10.2] and [10.4] for the projective performance of a scientific theory do not claim to provide a clear cut criterion for theory *acceptance* or for theory *credence* (degree of rational belief or credibility). First, because a high projective performance is not the sole valuable property a theory can have; we saw in Sec. 9.6 that explanatory depth is valuable as well. Second, because not even a high degree of truth of a theory's projections warrants the truth of the theory itself, if only because the same projections could have been derived from different assumptions. Accordingly a theory should not be definitively accepted even if its projective performance is very high. It might happen that the first N projections obtained with the help of the theory exhaust the subset of bold and true projections, all others (still not deduced and tested) being trite or outright false.

*(Mathematically speaking, our formulas for the projective performance of a theory are *not well defined*, since Π_r is "defined" on the set of projections that we happen to select for test. It may even happen that our formulas yield different values for the projective performances of two theories eventually shown to be empirically equivalent. In order for such formulas to be well defined they would have to involve

the whole set of projections; i.e. the summations over i would have to extend to $N=\infty$. In such a case we would have formulas for the projective power of theories rather than for their actual projective performance. But such formulas would be altogether useless because we cannot know all the infinitely many projections that can be derived from a set of assumptions. And even if we could deduce, test, and evaluate them, the infinite sums might not converge: for one thing, the terms in those sums obey no mathematical law, since factual truth is not an intrinsic property. In short, well-definedness, a mathematical desideratum, is a pointless condition in the case of the quantification of the projective performance. The same holds for the related formulas, in particular for the coverage.)

*All our formulas for the projective performance of a theory purport to give is a rough estimate of one of the various traits a theory should possess in order to be selected for detailed further elaboration, careful discussion, empirical test, and provisional acceptance. Moreover, an estimate of the projective performance, for being a posteriori and dependent on test techniques, can change in the course of time, as further projections are made and checked or as the previously computed projections are rechecked. Thus, e.g., the predictive value of Newton's gravitation theory, rated as maximal until about mid-19th century, has declined ever since, whereas Einstein's gravitation theory has been assigned an increasing projective power from its inception until the mid twenties. It remained constant until about 1960 and has increased ever since as new consequences of the theory have laboriously been derived and some of them empirically confirmed with astounding precision thanks to the invention of new techniques such as Mössbauer's. Yet there is reason to suspect that even this theory will eventually be found to be perfectible: one reason is the intensity with which it is again being investigated.

*In a sense, then, $\Pi(\mathrm{T})$ measures one aspect of the *growth* of theoretical knowledge; not the growth ability of a theory but rather its actual growth. (Explanatory depth is another index of the growth of knowledge.) On the other hand $\Pi(\mathrm{T})$ is not an index of *projective reliability*, as it should if theories were just tools (projection tools). Consequently, estimates of $\Pi(\mathrm{T})$ have no practical value. All the above formulas serve, in the best of cases, is the purpose of *elucidating* the concepts of projective performance, originality, and coverage. This elucidation has been made in terms of the basic concept of partial truth and has required neither the concept of probability nor the espousal of any major epistemological or ontological thesis: it is, therefore, philosophically neutral (i.e. of no interest to any philosophical *ism*).

Problems

10.4.1. Give examples of projections so imprecise that they are either sure to come out true or such that no datum could conceivably be relevant to them.

10.4.2. Mention outstanding cases of theories that have projected previously unheard-of facts. See W. S. Jevons, *The Principles of Science*, 2nd ed. (New York: Dover, 1958), Ch. xxiv, Secs. 6–11; W. B. Cannon, *The Way of an Investigator* (New York: W. W. Norton, 1945), Ch. vi; R. K. Merton, *Social Theory and Social Structure*, 2nd ed. (Glencoe; Ill.: Free Press, 1957), Ch. iii. Comment on the relevance of such facts to the popular doctrine according to which theories are just data systematizations and consequently must come always after the relevant facts have been discovered and "collected".

10.4.3. Is serendipity, i.e. the capability of predicting or retrodicting entirely new facts, a miraculous property? Hint: show that it seems miraculous only if (i) we forget that the base of a theory proper is a handful of high-level hypotheses the consequences of which are deduced only step by step, and (ii) we adopt a radically pluralistic world view, according to which no fact and no property are objectively and lawfully related to any other fact and property. *Alternate Problem:* Examine the case of the prediction of proton radioactivity, confirmed experimentally in 1963. Would it have been possible without inventing a hypothesis concerning the mechanism of decay (namely, electric repulsion)?

10.4.4. A quantitative projection should be accompanied by a statement concerning the estimate of the probable error of projection (usually prediction). Does this latter statement belong to the projective statement or is it a metastatement? If the former, how could a substantive theory produce it? If the latter, why do we say that scientific projections are derived from theories, rather than with their help?

10.4.5. Examine the proposal to measure the predictive content (of statements) in terms of probability. See (i) A. Rapoport, *Operational Philosophy* (New York: Harper & Brothers, 1954), pp. 176–77; (ii) K. R. Popper, *Conjectures and Refutations* (New York: Basic Books, 1963), Addenda 1 and 2. Even assuming that such a measure were adequate, could it be extended to theories, taking into account that the various projections derived in a theory are far from being all of them logically independent?

10.4.6. Propose a measure of the accuracy of theories. Hint: Try $A_r(\mathrm{T})=1-I_r(\mathrm{T})$. See whether the replacement of $I_r(\mathrm{T})$ by $1-A_r(\mathrm{T})$ in the formulas in the text sheds any clarity.

10.4.7. Is [10.4] applicable to the comparison of the projective performances of two rival theories?

10.4.8. Try to set up a formula for the projective content of a projection such that it increases with information content and testability.

10.4.9. Elucidate the notion of bulk of accepted knowledge and the concept of compatibility of a theory with the bulk of accepted knowledge. Hint: Decompose A into a set A′ that remains unquestioned for the time being, and a set T′ of questionable theories.

10.4.10. Quantify the concept of external consistency in such a way that it becomes the dual of originality.

10.5. Riddles

Let us now deal with some puzzling features of projection which do not characterize explanation. Let us, in the first place, ask ourselves what the truth value of a projective proposition can be. Suppose the weather bureau announces: 'Rainy tomorrow'. What is this sentence today, i.e. before it is tested by the observation of tomorrow's events: true or false? How can we know without waiting for tomorrow? But if we do not know the truth value of that sentence, can we say it expresses a proposition, i.e. a formula with a definite truth value? If we deny that the forecast has a truth value before it has been tested, we must conclude that it is not a proposition. But then we must account for the miracle that a set of projection premises (generalizations and data), which are supposed to be propositions, can generate something that is not a proposition. If, on the contrary, we assert that our predictive statement has a truth value before being tested and, at the same time, we grant that we do not know that truth value, we seem to consecrate a Platonic realm of ideas that cannot be wholly our make since we do not know whether they are true or not. In either case we seem to drown ourselves in mystery.

A second riddle is this. We grant that, when the weatherman announces 'Rainy tomorrow', he means that the forecasted event is possible, nay highly likely, yet not certain. Now, if the forecast is true at birth, then it seems necessary for rain to fall tomorrow—which contradicts the assumption that rain is just possible. And if the forecast is born false, it would seem to be impossible to suffer rain tomorrow, which again contradicts the supposition that rain and not-rain are only possible. To avoid either conclusion we must either assume that forecasts have no truth value or that the facts they refer to are not possible but just impossible or necessary.

Suppose finally the next day is with us and, in fact, it rains. We are justified in saying 'It rains' and even 'It rains as forecast'. But can we say that the statement "Rainy tomorrow" has been verified, i.e. has turned out true? The forecast was not 'Rainy today' but 'Rainy tomorrow', and it would seem that there is nothing today with which to compare the forecast: the true sentence 'It rains' does not seem to cohere with the forecast 'Rainy tomorrow' (apparent failure of the coherence theory of truth); nor does the forecast 'Rainy tomorrow' depict the rain that is now pouring (apparent failure of the correspondence theory of truth).

Let us begin with the latter riddle because it is the simplest. In fact, it is just a bad trick of ordinary language. Replace 'Rainy tomorrow' by 'Rainy on such a such day', i.e. give the exact calendar date, and the paradox dissolves. So far, then, the coherence and the correspondence theories of truth are saved. What about the second paradox, which involved the category of possibility? The simplest way out is to regard projective statements as only possible: instead of insisting that "Rainy tomorrow" should be either true or false today, we may say that it is just possible for "Rainy tomorrow" to be true or untrue. Another solution is to assign possibility to the facts themselves rather than to the propositions expressing them: i.e. to assert "It will probably rain tomorrow" or "It is probable that tomorrow it will rain". But the forecast cannot be reinterpreted at will since it has been deduced from definite premises; such wilful reinterpretations would violate the requisite of semantic closure (see Sec. 7.2). If we hold that the forecaster must mean that it will probably rain tomorrow, in the sense that weather shows a disposition to turn to rainy, then we must show that at least one of his premises was in fact a probabilistic statement, since probability statements can be derived from probability statements alone. And this may not be true, particularly if the forecast was made on considerations of atmospheric physics rather than with the help of low-level statistical generalizations. (Besides, what was initially a definite forecast has become an imprecise and untestable prophecy.)

Let us then try the first reinterpretation, namely, that the proposition "Rainy tomorrow" is to be regarded as possible, or else to be assigned a third truth value (indeterminate). This was, incidentally, the proposal that gave rise to three-valued logic. But then we will have to trace this indeterminateness back to the premises that entail the forecast—again by the principle of semantic closure. Now, if we pronounce either the generalizations or the data possible or indeterminate, we must be consistent and adopt correspondingly a new logic all the way: some system of modal logic in the first case, and a three-valued logical calculus in the latter. But this would be science fiction: in factual science, for the derivation of projections we always employ ordinary (extensional two-valued) logic and mathematical theories that have it built into. Consequently the introduction of modalities and indeterminate truth values at this point is logically illegitimate. The procedure would be legitimate if the whole of mathematics and factual science were reformulated with some such non-ordinary system of logic. But this project, besides being utopian, would have the undesirable consequence that the number of untestable propositions would increase without limit. (See Secs. 5.6 and 10.4 for the untestability of possibility statements.)

The same holds for our first riddle, which is the gravest of all: as a matter of fact when we derive a projection we employ ordinary logic and/or mathematics; this entails that, for inferential purposes, we assume that our premises have a definite truth value, even knowing that they are at most rough approximations and that the truth value assigned to them will eventually have to be modified upon testing the projection. In short, we regard our premises as propositions; if we did not we would be unable to operate the logico-mathematical machin-

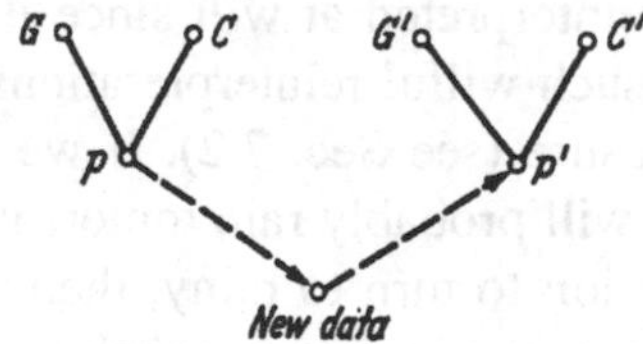

Fig. 10.12. Feeding corrections back into the premises: tests force us to adopt *p′* instead of *p* and *p′* obliges us to search for new premises *G′* and/or *C′*.

ery. Does this commit us to Platonism? Yes, if the usual, monistic theory of truth is retained; no, if a *dualistic theory of truth*, consistent with a fallibilistic epistemology, is adopted. This theory assumes that every factual premise in a reasoning (e.g., in the derivation of a prediction) is asserted as outright true (or false) *for the sake of reasoning* and, at the same time, as at most approximately true as regards its correspondence with facts as tested by experience. In other words, for inferential purposes we treat the projectans premises *as if* they were completely true (or false), even though we may strongly presume that they are just partially true.

The dualistic theory of truth seems to correspond to the actual procedure in research: indeed, in working out the testable consequences of our initial assumptions we make the initial *pretense* that our premises are entirely true, and this is why we call them premises or assumptions. In fact we say or imply 'Suppose the generalizations G and the data *C* are true: assume this if only for the sake of argument; then the projection *p* follows from them, and *p* is true under the same assumption. Notwithstanding, mistrust *p*, and consequently *G* and *C*, subjecting *p* to test. On the basis of such tests assign *p* a new truth value (one which in general will differ from the initial evaluation). Finally, assign the premises a truth value in accordance with the results of the test and correct them if necessary'. (See Fig. 10.12.)

If we assign the premises of our projection two truth values, a theoretical and an empirical one, then the projection derived from them will have a dual status as well: our projective formula will be theoretically true if the projective inference is exact but it will lack an empirical truth value until it is tested. In short, before testing the theorem *t* we do not assign it a total truth value but only a prior verisimilitude, i.e. a degree of truth on the theory T in which it is

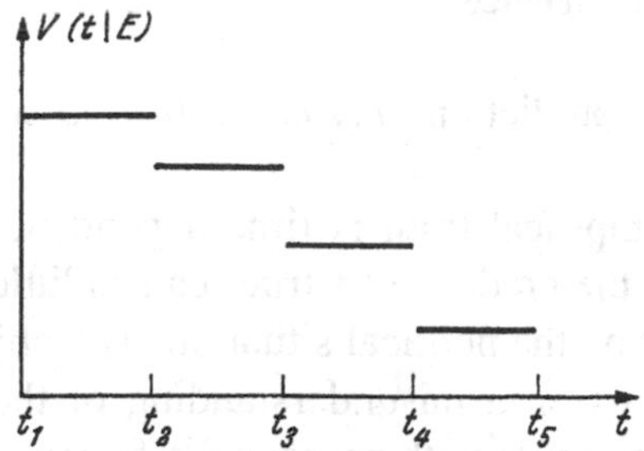

Fig. 10.13. The time-dependence of empirical truth: successive empirical tests, at times t_1, t_2, ..., may yield different values of the empirical truth value $V(t|E)$ of the projection *t*.

derived and the antecedent knowledge A. (This duality has been involved in our treatment of projective performance in Sec. 10.4.) It is not that our projection *t has* all the time an empirical truth value, only we do not *know* it: our projection has only the capacity for acquiring such a property and, whether it acquires it or not, will depend on extrinsic circumstances, such as the technical feasibility and the interest of the empirical test. We might as well say that a projection, while it has not been tested, is potentially true or false as regards experience: it may acquire an empirical truth value but nobody might care to check it, so that it is by no means necessary for a projection to eventually acquire an empirical truth value. Furthermore, whereas the prior truth value or prior verisimilitude of a projection can be a high one (if no additional simplifications are introduced in its derivation from the given premises), its empirical truth value can be anything between falsity and truth.

Now, the empirical tests allowing us to assign a complete (empirical and theoretical) truth value to a projection may not be definitive. A qualitative projection such as "Rainy tomorrow" may in many cases be settled once and for all. But the tests of a quantitative projection may have to be repeated and the various results may not coincide, because different techniques and circumstances are involved. If we can reach a decision at all this decision may be provisional: it may vary in the course of time. In short, empirical truth values are *time-dependent*, (see Fig. 10.13). This statement should scandalize both Platonists and those who have in mind only formal (logical or mathematical) truth in mind. But it is exactly what everyone else means when he says that a prediction *may come true to some extent* in the future.

The thesis that empirical truth is time-dependent may be laughed away by saying that the predicate 'is true' can as little depend on time as on temperature or on the political situation. The objection, however, is analogical and rests on a misunderstanding of the nature of time, which is a physical variable with no causal efficacy. The statement "*p* is true at time *t*" is an abbreviation for this longer statement: "*p* was found true against the background knowledge and the empirical evidence available at time *t*". Or even "*p* was found true against the background knowledge and the empirical evidence available when the event *x* happened." To say that truth is time-dependent is, then, to say that truth values are always assigned on the basis of a body of knowledge, which body changes in time. In other words, to say that truth is time-dependent amounts to say that truth is contextual: that it is not an intrinsic property of propositions but that it is relative to whatever system is actually employed in assigning a truth value. In other words, factual truth does not *inhere* in propositions but is asserted *about* propositions—which requires metastatements. A factual proposition *p* calls for an epistemic metastatement *P* concerning the truth value of *p*, and *P* depends both on *p* and on the knowledge at hand as well as on the truth criteria adopted at the moment the evaluation is performed (recall Sec. 2.1).

In other words, factual truth and factual falsity are not inborn but acquired characteristics. It is we who assign a factual proposition (our own make) now a given truth value, now another, according as our general body of knowledge evolves. This is so to the point that we often do not care to assign a given proposition an empirical truth value: the task may not be rewarding or it may be beyond our present forces. And such a situation will not be paradoxical since, to be regarded as a proposition proper, a formula must have *some* truth value. And the dualistic theory of truth makes provision for this, as it postulates that every proposition, to be examined at all, belongs to some context or other, a context in which it is assigned a theoretical truth value quite apart from what fresh experience may have to say.

The dualistic theory of truth solves then the first riddle of projection. The fourth riddle is this: Suppose we forget about the process whereby a projection is assigned an empirical truth value and relapse into the naive point of view that it is born factually true or factually false. We have then two possibilities: either the projection is true—but then we need not wait for the projected fact to occur or to be found out, or else the projection is false—in which case there is even less reason to wait for its empirical test. In either case, then, the assumption that projections are born with a total truth value leads to the conclusion that experience is irrelevant to them. Whence it would have to be concluded that they are analytic propositions. This absurd consequence follows from having assumed that factual propositions have a single truth value; if this assumption is replaced by the thesis of the double (theoretical and empirical) truth value of every factual proposition, the puzzle is dissolved.

Our fifth problem has kept skeptics and empiricists awake during two millenia. It is the problem "Why should we rely on certain predictions?" Hume (1740) thought that neither logic nor experience warrant such reliance. His reasoning, in a nutshell, was the following. Prediction is "the transferring of past experience to the future". This transference requires the assumption "that the future resembles the past". But this assumption is derived entirely from habit: it is not rooted in the nature of things. A frequent or even a constant observed conjunction or succession need not be conserved in the future; we have no reason to suppose that it will hold beyond the series of observations, and logic alone is equally impotent to warrant it. In conclusion, our belief or assurance in our forecasts is unwarranted.

Hume was certainly right in asserting the *fallibility* of prediction; but this does not establish that approximate prediction is altogether unwarranted. Moreover, the reasons Hume gave to derive his result are invalid. In the first place, Hume did not have *scientific* prediction in mind but the ordinary prognosis that relies on common sense generalizations concerning succession, conjunction, correlation, or empirically found frequencies, such as "One out of twenty ships does not return to its haven" (a favorite example of his). Now, ordinary prognosis is indeed in most cases a mere "transferring of past experience to the future", but scientific prediction is a conceptual operation that transcends experience because it is based on law statements rather than on inductive summaries. In contrast to ordinary prognosis, scientific prediction need not anticipate the repetition of the past: we may predict events of an unexperienced kind and even facts occurring for the first time in the history of the world, such as the trajectory of a rocket aimed at Venus. In the second place, it is true that the "laws" of common knowledge need not be conserved, particularly so the statements about regular conjunction and succession, the underlying mechanism of which is ignored. It is quite different when the mechanism of the emergence of such appearances is found out, i.e. when representational (nonphenomenological) laws and theories are set up. Moreover, there are no phenomenal laws proper but more or less random time series of experiences with no predictive power. This is why, in order to discover laws proper, the scientist hypothesizes diaphenomenal things and properties. Hume was right in denying phenomenal sequences, even the few regular ones, stability and accordingly projective power and reliability; he was mistaken in believing that such are the only regularities man can ever know. If the regular succession of days and nights were just an idiosincrasy of man we would have no reason to expect its continuance; but this has nothing to do with either nomological statements or nomological prediction, such as practiced by scientists.

This new look at the phenomenal regularities employed in common sense prognosis was made possible by the new, mechanistic science that began to flourish one century before Hume wrote, but which he flatly ignored in his discussion of forecast. (Still more strangely, Hume's doctrine that prediction is a kind of induction and is based on empirical generalizations is popular among philosophers after three centuries of theoretical science.) What remains of Hume's analysis of forecast is the thesis that prediction is fallible—but for quite different reasons: a scientific prediction may fail either because the theory which makes it possible is not true enough, or because the information involved is not accurate enough, or because something went wrong in the derivation itself—either plain blunders or too brutal simplifications in the application of the theory. The same holds, of course, for retrodictions. We shall mistrust *every single* projection while at the same time having confidence in the ability of science to make frequent successful projections and, particularly, to improve the present score. We shall be fallibilists but not skeptics, because we know the mechanism whereby projections can be improved—namely, building stronger and deeper theories, producing more accurate information, and performing more exact derivations. Progress along this line will not eliminate uncertainty altogether: we shall continue making incomplete and wrong projections. But we shall learn from some such mistakes, i.e. we shall gradually correct the premises of our wrong projections. In this hope of progress we are sustained by the principle of lawfulness (see Sec. 6.8), since scientific projection requires a lawful universe and a lawful projector. Projection would indeed be impossible in a patternless or chaotic universe, in a Humean or a Jamesian world in which there were no constraints or objective connections, but in which anything whatsoever could happen.

Sixth riddle: How can we predict a radical novelty without at the same time depriving the predicted fact of its newness? For example if we predict that a new, unheard-of chemical compound with such and such properties will come out if such and such operations are performed, what novelty can be left when the substance is actually produced? The puzzle originates in the presupposition that there is a single kind of novelty, while actually three kinds of novelty are involved in this: conceptual, empirical, and ontological novelty. The actual production of the novel compound is not conceptually novel since it had been predicted; but it is empirically novel since we had never experienced with it; and it is ontologically novel as well since it did not exist before—or so we believe.

Our seventh riddle is closely related to the foregoing. Suppose science were to reach eventually a state of perfection in which it made accurate projections only: would this close the future to radical novelty? If we grant that perfect predictability makes radical newness impossible, we fall into the previous riddle. If, on the contrary, we admit that complete predictability is consistent with ontological novelty, we seem to commit ourselves to the view that future novelty is necessary, predetermined, or preordained. In short, predictability would seem to involve the elimination of both chance and freedom, and scientific progress would consequently be dreadful. The source of this puzzle is, of course, the conjecture that perfect prediction is possible, so that we might accurately predict every single mistake we are bound to make and might accordingly prevent its occurrence. To achieve such a state of bliss we would require every possible theory and infinitely many bits of information, which would not be technically feasible. Even so we would be unable to kill chance and freedom altogether because, if the premises necessary for accurate prediction are indeed perfect, then they will take care of the objectively random traits of reality and of the acts of free human volition. Which entails that the perfect premise will make room for unpredictable events. That is, a perfect knowledge will have to count on unpredictables (recall Problem 10.3.6); but, anyhow, a perfect knowledge is predictably unattainable. In any case, the future will always remain open; not only because some chunks of reality will always escape the search of science but also because we are becoming more and more effective masters of reality—which involves purposely loosening some ties and leaving the door of the future ajar. Which solves the riddle.

Eighth puzzle: What science retrodicts is no longer, what it predicts is not yet, and what it describes becomes past. How can we then say that scientific projection refers to reality and, so, that science deals with reality? Does it not mostly deal with unreal events? The source of this riddle is, of course, that on asking the question we have construed the word 'reality' in its narrow sense of actuality, i.e. of the sum total of what is now (relative to a certain frame of reference). But this is not what the scientist usually means by 'reality'. For instance, when the physicist draws a spacetime diagram showing the history of a particle moving in a field of force, he assumes that what is real is the entire history of the given system and not just its instantaneous present. The puzzle is dissolved, then, with a mere redefinition of "reality" as the sum total of existents at any time—or, equivalently, upon redefining "actuality" as the subset of entities now real (in a given reference frame).

Ninth riddle: How is it possible to derive propositions about past or future events from formulas that may not contain the time variable? For example, the laws of psychology that relate the strength of stimuli and responses of certain kinds are mostly static in the sense that they do not refer to processes; yet given a stimulus value at some time they enable us to predict the ensuing reaction value. The puzzle is dissolved upon recalling that static laws silently assume universal quantification of the time variable, at least over (unspecified) time intervals. Thus, when we state that the stimulus S elicits a reaction R that is invariably related to S through a certain function F—i.e. when we write the formula "$R=F(S)$" that exhibits the correspondence between S and R—we imply that this functional relation does not change in time. (In more detail: "For every t and for every x, if t is an instant of time and x is an organism, then $R(x,t)=F[S(x,t)]$". A more exact formulation would make the value of the response at t dependent on the value of the stimulus at a prior time or perhaps on the values of the stimulus over a whole time interval.) A specific value of the time variable will occur in a projective statement if the latter constitutes an adequate answer to the question '*When* did (or will) x happen?'. But if the question is '*What* did (or will) happen?', the answer need not contain a time value. Time will, on the other hand, always occur in the mental process of deriving a projective statement: this operation takes time and so, if the premises are stated at a given time, the projection will be issued at a later time. And this time may be even later than the projected event itself, either because a retrodiction is at stake or because the event happened while the projection was being computed (as is often the case with ballistic and meteorological forecasts).

Our tenth and last riddle is this: Why is projection different from explanation if both have the same logical structure, i.e. both are deductions from generalizations and data? Let us draw a quick list of some of the differences between the two kinds of inference. (i) What is given in the case of explanation is the conclusion of an argument that must be reconstructed, whereas in prediction the premises of the argument are supplied and the conclusion is sought (see Sec. 9.1). (ii) Any system of conjectures, whether scientific or not, can supply a number of explanations, whereas only scientific theories take the risk of issuing projections; thus, e.g., Aristotle's physics and psychoanalysis do not even attempt to make predictions. It is only scientific explanations that can be controlled by projections. (iii) The relation of explanation, "*q* because *p*", can be established among formulas on any levels of a theory and even among formulas belonging to different theories, whereas predictive formulas must be of the lowest level; thus, e.g., we make no nomological prediction of laws although we do sometimes anticipate or prognose in outline their discovery. (iv) There are explanations of varying depth, according to the depth of the theories involved; on the other hand, the term 'projective depth' does not appear to make sense. (v) Some theories can give precise explanations of individual events, such as the emission of light by an atom, or the contagion of an organism by a virus, but they do not yield exact projections of these same individual facts: their projections refer to assemblies of individuals or else they state what the probability of the individual event is: they are stochastic projections (see Sec. 10.2). We have listed five differences between explanation and projection; there might be more. The explanation of the lack of identity of two operations that are logically identical (since both are deductive) is merely that logic does not discern extralogical (e.g., epistemological) differences. What is puzzling about our last puzzle is, then, only this: Why do most philosophers fail to recognize such differences?

The above may be regarded as philosophical jugglery. Yet our puzzles have subjected certain philosophical theories, such as the theory of truth and the theory of determination, to straining tests. And they have shown that the concept of projectibility cannot be taken as a primitive with which to define the concepts of truth, novelty, law, causality, and chance—as is done in the philosophies that regard predictability as the chief desideratum of scientific theories and reject the concepts of objective newness, objective law, and objective causal and random traits. The metascientist values projective performance mainly in its capacity as testability, which in turn is a gateway to factual truth. Only the applied scientist and the man of action might have a right to regard prediction as the sole or even the main aim of science, for being the basis for rational action. Prediction, indeed, bridges the gap between science and action. Over this bridge we shall now go.

Problems

10.5.1. Examine Hume's doctrine of prediction as expounded in his main philosophical work, *A Treatise of Human Nature* (1739/40), Book I, Part III, sec. xii. *Alternate Problem:* Promises and threats, both unconditional (e.g., "I will do *x*") and conditional (e.g., "If *x* does *y* then I will do *z*") are linguistically alike although they are psychologically different. Confront promises and threats with forecasts, both conditional and unconditional, and find out whether or not they differ linguistically from one another. In any case, draw some conclusion concerning the adequacy of linguistic analysis to handle scientific prediction.

10.5.2. Empiricism has traditionally maintained that prediction is an inductive inference and, indeed, the most typical or the most fundamental kind of induction. This applies not only to classical authors like Hume and Mill but also to contemporary authors such as H. Reichenbach, *Experience and Prediction* (1938; Chicago: University of Chicago Press, 1961), sec. 38; H. Jeffreys, *Scientific Inference*, 2nd ed. (Cambridge: Cambridge University Press, 1957), pp. 1 and 13; G. H. von Wright, *The Logical Problem of Induction*, 2nd ed. (Oxford: Blackwell, 1957), p. 1; R. Carnap, *Logical Foundations of Probability* (Chicago: University of Chicago Press, 1950), p. 207; N. Goodman, *Fact, Fiction & Forecast* (London: Athlone Press, 1954), Ch. III; J. J. Katz, *The Problem of Induction and its Solution* (Chicago: University of Chicago Press, 1962), p. 5. Examine this view, in particular the claim that predictive inference cannot be deductive because it transcends the observed facts, being an inference about unknown facts from a sample of past events, i.e. an inductive extrapolation from past uniformities. Hints: (i) see whether the doctrine applies to the prediction of new facts (such as the prediction of the omega-minus particle, actually discovered in 1964); (ii) see whether it applies to ordinary knowledge prognosis.

10.5.3. R. Carnap, in his once influential *Logical Foundations of Probability* (see Problem 10.5.2), p. 575, maintains that "the use of laws is not indispensable for making predictions. Nevertheless it is expedient of course, to state universal laws in books on physics, biology, psychology, etc." If laws are dispensable and if, as traditional positivism maintains, prediction is the gist of science, why do scientists spend so much time inventing and testing law statements: just because these are expedient? If, on the contrary, laws are essential to scientific prediction why should they be belittled? Take into account that the degree of confirmation of a universal law statement is exactly zero on Carnap's theory of confirmation, in which the degree of confirmation is quantified as the ratio of favorable to possible cases.

10.5.4. Consider the following solution to the problem of the truth value of a proposition which is now asserted as true and after a while discovered to be false. "The proposition was false to begin with, only we did not realize it. What did change was not its truth value but our estimate of it; truth values are intrinsic and atemporal properties of propositions." *Alternate Problem:* Report on the relations of truth to time according to A. N. Prior in Time and Modality (Oxford: Clarendon Press, 1957). See the criticisms of R. M. Martin in *Mind*, LXVIII, N.S., 271 (1959).

10.5.5. Examine the problem that originally gave rise to the system of three-valued logic proposed by J. Lukasiewicz and see whether science uses or needs any such logical theory. See W. and M. Kneale, *The Development of Logic* (Oxford: Clarendon Press, 1962), pp. 47ff. and 569. For the thesis that quantum mechanics requires a three-valued logic see, e.g., P. Destouches-Février, *La structure des théories physiques* (Paris: Presses Universitaires, 1951).

10.5.6. According to classical pragmatism, projections are useful or useless rather than true or false. In turn a useful (useless) projection is a successful (unsuccessful) one, i.e. one that does (does not) hit the mark. What, in turn, can 'successful' mean in this context? And how does pragmatism account for the success of certain predictions? *Alternate Problem.* Study the logic of hope. Hints: (i) start by distinguishing pious hopes from nonpious ones and point out what their respective grounds are; (ii) study the relation of founded hope to scientific prediction; (iii) find out what empirical justification, if any, can hope have.

10.5.7. According to pragmatism, conventionalism, and traditional positivism, prediction is all-important whereas explanation is parasitic. Point out the sound and the unsound motivations of this view. Among the former, the fact that speculative theories offer explanations but not predictions; among the latter, the refusal to go beyond experience. Think whether there could be accurate projections without deep explanatory theories; keep in mind Ptolemy's theories of planetary motions. And find out whether real scientists are much more interested in predicting than in explaining.

10.5.8. N. Wiener and A. N. Kolmogoroff have initiated a mathematical *theory of prediction* which supplies the means for extrapolating from the known past of some variable quantity to its future values. (Actually the "data" considered by the theory constitute an *infinite* sequence of numerical values supposedly taken from observations made at equally spaced intervals throughout the past. The actual observational data must be regarded as a representative sample of that infinite imaginary population.) Would this procedure, if perfected, enable us to dispense with substantive scientific theories so far as prediction is concerned? Or does it rather provide a technique for analyzing time series into message and noise components (see Fig. 10.6 in Sec. 10.2), thereby suggesting laws to be eventually theorified? See N. Wiener, "The Theory of Prediction", in E. F. Beckenbach, Ed., *Modern Mathematics for the Engineer*, 1st Series (New York: McGraw-Hill, 1956), and H. Furstenberg, *Stationary Process and Prediction Theory* (Princeton: Princeton University Press, 1960).

10.5.9. In certain chapters of science, such as botanical systematics, no projections are made. Examine the following possible "conclusions". (i) Such nonprojective disciplines are just nonscientific. (ii) Projection is not essential to science: some sciences are nonprojective. (iii) The nonprojective chapters of a science are parts of wider systems having projective power, and their function is chiefly ancillary to such wider systems. The departmentalization of science into projective and nonprojective chapters is justifiable on pragmatic reasons alone. Recall Problem 10.1.10.

10.5.10. Projection is usually held to be possible owing to the uniformity of nature. What is to be understood by 'uniformity of nature'? And granting that this is the ontological basis of projection, what would account for every failure in projection?

Bibliography

Bunge, M.: Causality, 2nd ed., Ch. 12. New York: Dover, 1979.

Dyke, V. van: Political science: a philosophical analysis, Ch. 4. Stanford: Stanford University Press 1960.

Grünbaum, A.: Philosophical problems of space and time, Ch. 9. New York: Knopf 1963.

Hanson, N. R.: On the symmetry between explanation and prediction. Phil. Rev. **68**, 349 (1959).

——— Mere predictability. In: H. E. Kyburg Jr., and E. Nagel (Eds.), Induction: some current issues. Middleton, Conn.: Wesleyan University Press 1963.

Jevons, W. S.: The principles of science (1874), Ch. XXIV. New York: Dover 1958.

Körner, S. (Ed.): Observation and interpretation. London: Butterworths Sci. Publications 1957; esp. W. B. Gallie, The limits of prediction and G. Ryle, Predicting and inferring.

Meehl, P. E.: Clinical vs. statistical prediction. Minneapolis: University of Minnesota Press 1954

Popper, K. R.: The poverty of historicism, 2nd ed., esp. Sec. 28. London: Routledge & Kegan Paul 1960.

——— Conjectures and refutations, Ch. 16 on prediction and prophecy in the social sciences. New York: Basic Books 1963.

Reichenbach, H.: The rise of scientific philosophy, Ch. 14. Berkeley and Los Angeles: University of California Press 1951.

———Experience and prediction (1938), Sec. 38. Chicago: University of Chicago Press 1961.

Rescher, N.: On prediction and explanation. Brit. J. Phil. Sci. **8**, 281 (1958).

Scheffler, I.: Explanation, prediction and abstraction. Brit. J. Phil. Sci. **7**, 293 (1957).

11

Action

In science, whether pure or applied, a theory is both the culmination of a research cycle and a guide to further research. In technology theories are, in addition, the basis of systems of rules prescribing the course of optimal practical action. On the other hand in the arts and crafts theories are either absent or instruments of action alone. Not whole theories, though, but just their peripheral part: since the low level consequences of theories can alone come to grips with action, such end results of theories will concentrate the attention of the practical man. In past epochs a man was regarded as practical if, in acting, he paid little or no attention to theory or if he relied on worn-out theories and common knowledge. Nowadays a practical man is one who acts in obeyance to decisions taken in the light of the best technological knowledge—not scientific knowledge, because most of this is too remote from or even irrelevant to practice. And such a technological knowledge, made up of theories, grounded rules, and data, is in turn an outcome of the application of the method of science to practical problems (see Sec. 1.5).

The application of theory to practical goals poses considerable and largely neglected philosophical problems. Three such problems, the one of the validating force of action, the relation between rule and law, and the effects of technological forecast on human behavior, will be dealt with in this chapter. They are just samples of a system of problems that should eventually give rise to a philosophy of technology.

11.1. Truth and Action

An act may be regarded as *rational* if (i) it is maximally adequate to a preset goal and (ii) both the goal and the means to implement it have been chosen or made by deliberately employing the best available relevant knowledge. (This presupposes that no rational act is a goal in itself but is always instrumental.) The knowledge underlying rational action may lie anywhere in the broad spectrum between common knowledge and scientific knowledge; in any case it must be knowledge proper, not habit or superstition. At this point we are interested in a special kind of rational action, that which, at least in part, is guided by scientific or technological theory. Acts of this kind may be regarded as *maximally rational* because they rely on founded and tested hypotheses and on reasonably accurate data rather than on practical knowledge or uncritical tradition. Such a foundation does not secure perfectly successful action but it does provide the means for a gradual improvement of action. Indeed, it is the only means so far known to get nearer the given goals and to improve the latter as well as the means to attain them.

A theory may have a bearing on action either because it provides knowledge regarding the objects of action, e.g., machines, or because it is concerned with action itself, e.g., with the decisions that precede and steer the manufacture or use of machines. A theory of flight is of the former kind, whereas a theory concerning the optimal decisions regarding the distribution of aircraft over a territory is of the latter kind. Both are *technological theories* but, whereas the theories of the first kind are *substantive,* those of the second kind are, in a sense, *operative.* Substantive technological theories are essentially applications, to nearly real situations, of scientific theories; thus, a theory of flight is essentially an application of fluid dynamics. Operative technological theories, on the other hand, are from the start concerned with the operations of person and man-machine complexes in nearly real situations; thus, a theory of airways management does not deal with planes but with certain operations of the personnel. Substantive technological theories are always preceded by scientific theories, whereas operative theories are born in applied research and may have little if anything to do with substantive theories—this being why mathematicians with no previous scientific training can make important contributions to them. A few examples will make the substantive-operative distinction clearer.

The relativistic theory of gravitation might be applied to the design of generators of antigravity fields (i.e., local fields counteracting the terrestrial gravitational field), that might in turn be used to facilitate the launching of spaceships. But, of course, relativity theory is not particularly concerned with either field generators or astronautics: it just provides some of the knowledge relevant to the design and manufacture of antigravity generators. Palaeontology is used by the applied geologist engaged in oil prospecting, and the latter's findings are a basis for making decisions concerning drillings; but neither palaeontology nor geology are particularly concerned with the oil industry. Psychology can be used by the industrial psychologist in the interests of production but it is not basically concerned with it. All three are examples of the application of scientific (or semiscientific, as the case may be) theories to problems that arise in action.

On the other hand the theories of value, decision, games, and operations research deal directly with valuation, decision making, planning, and doing; they may even be applied to scientific research regarded as a kind of action, with the optimistic hope of optimizing its output. (These theories could not tell how to replace talent but how best to exploit it.) These are operative theories, and they make little if any use of the substantive knowledge provided by the physical, biological, or social sciences: ordinary knowledge, special but nonscientific knowledge (of, e.g., inventory practices), and formal science are usually sufficient for them. Just think of statistical information theory, or of queuing models: they are not applications of pure scientific theories but theories on their own. What these operative or nonsubstantive theories employ is not substantive scientific knowledge but the *method* of science. They may, in fact, be regarded as scientific theories concerning action: in short, as *theories of action*. These theories are technological in respect of aim, which is practical rather than cognitive, but apart from this they do not differ markedly from the theories of science. In fact all good operative theories will have at least the following traits characteristic of scientific theories: (i) they do not refer directly to chunks of reality but to more or less idealized models of them (e.g., entirely rational and perfectly informed contenders, or continuous demands and deliveries); (ii) as a consequence they employ theoretical concepts (e.g., "probability"); (iii) they can absorb empirical information and can in turn enrich experience by providing predictions or retrodictions, and (iv) consequently they are empirically testable, though not as toughly as scientific theories. (See Fig. 11.1.)

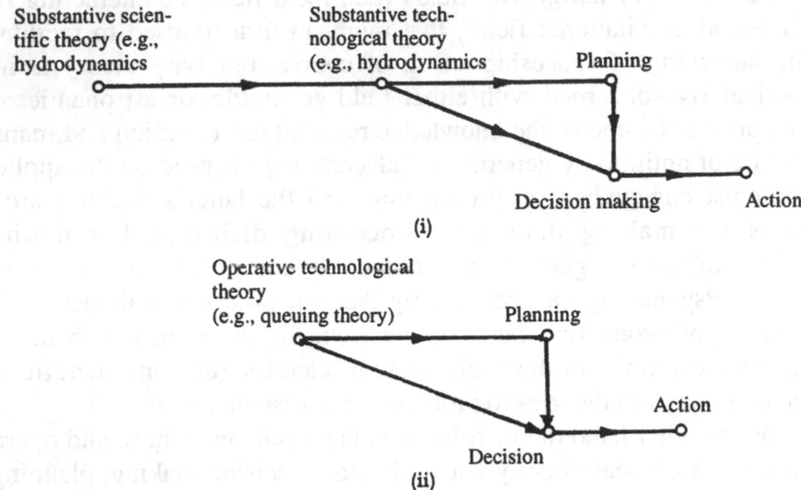

Fig. 11.1. (i) Substantive technological theory is based on scientific theory and provides the decision maker with the necessary tools for planning and doing. (ii) Operative theory is directly concerned with the decision maker's and the agent's actions.

Looked at from a practical angle, technological theories are richer than the theories of science in that, far from being limited to accounting for what may or does, did or will *happen* regardless of what the decision maker does, they are concerned with finding out *what ought to be done* in order to bring about, prevent, or just change the pace of events or their course in a preassigned way. In a conceptual sense, the theories of technology are definitely poorer than those of pure science: they are invariably *less deep,* and this because the practical man, to whom they are devoted, is chiefly interested in net effects that occur and are controllable on the human scale: he wants to know how things within *his* reach can be made to work *for him,* rather than how things of any kind really are. Thus, the electronics expert need not worry about the difficulties that plague the quantum electron theories; and the marketing expert need not burrow into the origins of preference patterns—a problem for psychologists. Consequently, whenever possible the applied researcher will attempt to schematize his system as a *black box:* he will deal preferably with external variables (input and output), will regard all others as at best handy intervening variables with no ontological import, and will ignore the adjoining levels. This is why his oversimplifications and mistakes are not more often harmful: because his hypo-theses are superficial. (Only the exportation of this externalist approach to science can be harmful: see Sec. 8.5.) Occasionally, though, the technologist will be forced to take up a deeper, representational viewpoint. Thus, the molecular engineer who designs new materials to order i.e. substances with prescribed macroproperties, will have to use certain fragments of atomic and molecular theory. But he will neglect all those microproperties that do not appreciably show up at the macroscopic level: after all, he uses atomic and molecular theories as tools—which has misled some philosophers into thinking that scientific theories are *nothing but* tools.

The conceptual impoverishment undergone by scientific theory when used as a means for practical ends can be frightful. Thus, an applied physicist engaged in designing an optical instrument will use almost only ray optics, i.e. essentially what was known about light towards the middle of the 17th century. He will take wave optics into account for the explanation in outline, not in detail, of some effects, mostly undesirable, such as the appearance of colors near the edge of a lens; but he will seldom, if ever, apply any of the various wave theories of light (recall Sec. 9.6) to the computation of such effects. He can afford to ignore these theories in most of his professional practice because of two reasons. First, because the chief traits of the optical facts relevant to the manufacture of most optical instruments are adequately accounted for by ray optics; those few facts that are not so explainable require only the hypotheses (but not the whole theory) that light is made up of waves and that these waves can superpose. Second, because it is extremely difficult to solve the wave equations of the deeper theories save in elementary cases, which are mostly of a purely academic interest (i.e., which serve essentially the purpose of illustrating and testing the theory). Just think of the enterprise of solving the wave equation with time-dependent boundary conditions, such as those representing the moving shutter of a camera. Wave optics is scientifically important because it is nearly true; but for most present day technology it is less important than ray optics, and its detailed application to practical problems in optical industry would be quixotic. The same argument can be carried over to the rest of pure science in relation to technology. The moral is that, if scientific research had sheepishly plied itself to the immediate needs of production, we would have no science.

In the domain of action, deep or sophisticated theories are inefficient because they require too much labor to produce results that can as well be obtained with poorer means, i.e. with less true but simpler theories. Deep and accurate truth, a desideratum of pure scientific research, is uneconomical. What the applied scientist is supposed to handle is theories with a high *efficiency,* i.e. with a high Output/Input ratio: theories yielding much with little. Low cost will compensate for low quality. Since the expenditure demanded by the truer and more complex theories is greater than the input required by the less true theories—which are usually simpler—the technological efficiency of a theory will be proportional to its output and to its operation simplicity. (If we had reasonable measures of either we would be able to postulate the equation: *Efficiency of* T = *Output of* T × *Operation simplicity of* T.) If the technically utilizable output of two rival theories is the same, the relative simplicity of their application (i.e., their pragmatic simplicity) will be decisive for the choice by the technologists; the adoption of the same criterion by the pure scientist would quickly kill fundamental research. This should be enough to refute Bacon's dictum—the device of pragmatism—that the most useful is the truest, and to keep the independence of truth criteria with respect to practical success.

A theory, if true, can successfully be employed in applied research (technological investigation) and in practice itself—as long as the theory is relevant to either. (Fundamental theories are not so applicable because they deal with problems much too remote from practical matters. Just think of applying the quantum theory of scattering to car collisions.) But the converse is not true, i.e. the practical success or failure of a scientific theory is no objective index of its truth value. In fact, a theory can be both successful and false; conversely, it can be a practical failure and nearly true. The efficiency of a false theory may be due to either of the following reasons. First, a theory may contain just a grain of truth and this grain alone is employed in the theory's applications. In fact, a theory is a system of hypotheses and it is enough for a few of them to be true or nearly so in order to be able to entail adequate consequences if the false ingredients are not used in the deduction or if they are practically innocuous (see Fig. 11.2). Thus, it is possible to manufacture excellent steel by combining magical exorcisms with the operations prescribed by the craft—as was done until the beginning of the 19th century; and it is possible to

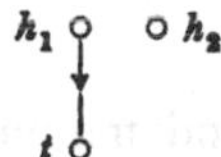

Fig. 11.2. A true theorem *t*, underlying an efficient rule, can sometimes be derived from a verisimilar hypothesis h_1 without using the false (or untestable) hypothesis h_2 occurring in the same theory.

improve the condition of neurotics by means of shamanism, psychoanalysis, and other practices as long as effective means, such as suggestion, conditioning, tranquilizers, and above all time are combined with them.

A second reason for the possible practical success of a false theory may be that the accuracy requirements in applied science and in practice are far below those prevailing in pure research, so that a rough and simple theory supplying quick correct estimates of orders of magnitude will very often suffice in practice. Safety coefficients will anyway mask the finer details predicted by an accurate and deep theory, and such coefficients are characteristic of technological theory because this must adapt itself to conditions that can vary within ample bounds. Think of the variable loads a bridge can be subjected to, or of the varying individuals that may consume a drug. The engineer and the physician are interested in safe and wide intervals centered in typical values, rather than in exact values. A greater accuracy would be pointless since it is not a question of testing. Moreover, such a greater accuracy could be confusing because it would complicate things to such an extent that the target—on which action is to be focused—would be lost in a mass of detail. Accuracy, a goal of scientific research, is not only pointless or even encumbering in practice but can be an obstacle to research itself in its early stages. For the two reasons given above—use of only a part of the premises and low accuracy requirements—infinitely many possible rival theories can yield "practically the same results". The technologist, and particularly the technician, are justified in preferring the simplest of them: after all they are primarily interested in efficiency rather than in truth: in getting things done rather than in gaining a deep understanding of them. For the same reason, deep and accurate theories may be impractical: to use them would be like killing bugs with nuclear bombs. It would be as preposterous—though not nearly as dangerous—as advocating simplicity and efficiency in pure science.

A third reason why most fundamental scientific theories are of no practical avail is not related to the handiness and sturdiness required by practice but has a deep ontological root. The practical transactions of man occur mostly on his own level; and this level, like others, is rooted to the lower levels but enjoys a certain autonomy with respect to them, in the sense that not every change occurring in the lower levels has appreciable effects on the higher ones. This is what enables us to deal with most things on their own level, resorting at most to the immediately adjacent levels. In short, levels are to some extent stable: there is a certain amount of play between level and level, and this is a root of both chance (randomness due to independence) and freedom (self-motion in certain respects). One-level theories will therefore suffice for many practical purposes. It is only when a knowledge of the relations among the various levels is required in order to implement a "remote-control" treatment, that many-level theories must be tried. The most exciting achievements in this respect are those of psychochemistry, the goal of which is, precisely, the control of behavior by manipulating variables on the underlying biochemical level.

A fourth reason for the irrelevance of practice to the validation of theories—even to operative theories dealing with action—is that, in real situations, the relevant variables are seldom adequately known and precisely controlled. Real situations are much too complex for this, and effective action is much too strongly urged to permit a detailed study—a study that would begin by isolating variables and tying up some of them into a theoretical model. The desideratum being maximal efficiency, and not at all truth, a number of practical measures will usually be attempted at the same time: the strategist will counsel the simultaneous use of weapons of several kinds, the physician will prescribe a number of supposedly concurrent treatments, and the politician may combine promises and threats. If the outcome is satisfactory, how will the practicioner know which of the rules was efficient, hence which of the underlying hypotheses was true? If unsatisfactory, how will he be able to weed out the inefficient rules and the false underlying hypotheses? A careful discrimination and control of the relevant variables and a critical evaluation of the hypotheses concerning the relations among such variables is not done while killing, curing, or persuading people, not even while making things, but in leisurely, planned, and critically alert scientific theorizing and experimenting. Only while theorizing or experimenting do we *discriminate* among variables and *weigh* their relative importance, do we *control* them either by manipulation or by measurement, and do we *check* our hypotheses and inferences. This is why factual theories, whether scientific or technological, substantive or operative, are empirically tested in the laboratory and not in the battlefield, the consulting office or the market place. ('Laboratory' is here understood in a wide sense, to include any situation which, like the military manoeuvre, permits a reasonable control of the relevant variables.) This is, also, why the efficiency of the rules employed in the factory, the hospital, or the social institution, can only be determined in artificially controlled circumstances.

In short, practice has no validating force: pure and applied research alone can estimate the truth value of theories and the efficiency of technological rules What the technician and the practical man do, by contrast to the scientist, is not to *test* theories but to *use* them with noncognitive aims. (The practicioner does not event test *things,* such as tools or drugs, save in extreme cases: he just uses them, and their properties and their efficiency must again be determined in the laboratory by the applied scientist.) The doctrine that practice is the touchstone of theory relies on a misunderstanding of both practice and theory: on a confusion between practice and experiment and an associated confusion between rule and theory. The question "Does it work?", pertinent as it is with regard to things and rules, is impertinent in respect to theories.

Yet it might be argued that a man who knows how to do something is thereby showing that he knows that something. Let us consider the three possible versions of this idea. The first can be summed up in the schema "If x knows how to do (or make) y, then x knows y". To ruin this thesis it is enough to recall that, for nearly one million years, man has known how to make children without having the remotest idea about the reproduction process. The second thesis is the converse conditional, namely "If x knows y, then x knows how to do (or make) y". Counter-examples: we know something about stars, yet we are unable to make them, and we know a part of the past but we cannot even spoil it. The two conditionals being false, the biconditional "x knows y if and only if x knows how to do (or make) y" is false, too. In short, it is false that knowledge is identical with knowing how to do, or know-how. What is true is rather this: knowledge considerably *improves* the chances of correct doing, and doing *may* lead to knowing more (now that we have learned that knowledge pays), not because action is knowledge but because, in inquisitive minds, action may trigger fruitful questioning.

It is only by distinguishing scientific knowledge from instrumental knowledge, or know-how, that we can hope to account for the coexistence of practical knowledge with theoretical ignorance, and the coexistence of theoretical knowledge with practical ignorance. Were it not for this the following combinations would hardly have occurred in history: (i) science without the corresponding technology (e.g., Greek physics); (ii) arts and crafts without an underlying science (e.g., Roman engineering and contemporary intelligence testing). The distinction must be kept, also, in order to explain the cross fertilizations of science, technology, and the arts and crafts, as well as to explain the gradual character of the cognitive process. If, in order to exhaust the knowledge of a thing, it were sufficient to produce or reproduce it, then certain technological achievements would put an end to the respective chapters of applied research: the production of synthetic rubber, plastic materials, and synthetic fibres would exhaust polymer chemistry; the experimental induction of cancer should have stopped cancer research; and the experimental production of neuroses and psychoses should have brought psychiatry to a halt. As a matter of fact we continue doing many things without understanding how and we know many processes (such as the fusion of helium out of hydrogen) which we are not yet able to control for useful purposes (partly because we are too eager to attain the goal without a further development of the means). At the same time it is true that the barriers between scientific and practical knowledge, pure and applied research, are melting. But this does not eliminate their differences, and the process is but the outcome of an increasingly scientific approach to practical problems, i.e. of a diffusion of the scientific method.

The identification of knowledge and practice stems not only from a failure to analyze either but also from a legitimate wish to avoid the two extremes of speculative theory and blind action. But the testability of theories and the possibility of improving the rationality of action are not best defended by blurring the differences between theorizing and doing, or by asserting that action is the test of theory, because both theses are false and no programme is defensible if it rests on plain falsity. The interaction between theory and practice and the integration of the arts and crafts with technology and science are not achieved by proclaiming their unity but by multiplying their contacts and by helping the process whereby the crafts are given a technological basis, and technology is entirely converted into applied science. This involves the conversion of the rules of thumb peculiar to the crafts into grounded rules, i.e. rules based on laws. Let us approach this problem next.

Problems

11.1.1. Report on either of the following works dealing with planned action. G. Myrdal, *An American Dilemma,* 2nd ed. (New York: Harper and Row, 1962), especially Appendices 1 and 2.; H. Hart, "Social Theory and Social Change", in L. Gross, Ed., *Symposium on Sociological Theory* (Evanston, Ill.: Row, Peterson, 1959), especially pp. 229ff.; A. R. von Hippel, "Molecular Designing of Materials," *Science*, **138**, 91 (1962).

11.1.2. Comment on the poem in which the Confucian Hsün Tzu (3rd century B.C.) praises the mastering of nature and deprecates the pure knowledge sought in vain by the Taoists. See J. Needham, *Science and Civilization in China* (Cambridge: University Press, 1956), II p. 28: "You glorify Nature and meditate on her; / Why not domesticate her and regulate her? / You obey Nature and sing her praises; / Why not control her course and use it? / You look on the seasons with reverence and await them; / Why not respond to them by seasonal activities? / You depend on things and marvel at them; / Why not unfold your own abilities and transform them? / You meditate on what makes a thing a thing; / Why not so order things that you do not waste them? / You vainly seek into the causes of things, / Why not appropriate and enjoy what they produce? / Therefore I say—To neglect man and speculate on Nature / Is to misunderstand the facts of the universe." Needhams's comment: Hsün Tzu "struck a blow at science by emphasizing its social context too much and too soon" (p. 26). Find similar attitudes in other times and places, and speculate on the possible mechanism of the emergence of pragmatist philosophies.

11.1.3. Pragmatism has often been commended as an antidote to wild speculation, on the assumption that praxis is the truth test of theories. On the other hand medical practitioners, business managers, politicians and other practical men do not cease to invent hypotheses which they have no time—and sometimes no wish—to either ground or test. Clear up this conundrum. *Alternate Problem:* Comment on the paradoxical saying attributed to the theoretical physicist L. Boltzmann: "Nothing is more practical than theory."

11.1.4. Analyze the saying "The proof of the pudding is in the eating". Does it hold for scientific knowledge or rather for pudding recipes? In general: what does the success of a rule of thumb *prove?* Take into account contemporary successful techniques such as electric shock, for the treatment of depression, which at the time of writing is not known why it should work. *Alternate Problem:* Discuss B. Russell's proof that W. James' pragmatist conception of truth leads to an infinite regress. See *A History of Western Philosophy* (New York: Simon and Schuster, 1945), Ch. xxix.

11.1.5. Show the characteristics of "approximate thinking" (including guestimating), i.e. of concentrating on the chief variables and operating with orders of magnitude when setting up preliminary theoretical models both in pure and applied science. See P. E. Morse and G. E. Kimball, *Methods of Operations Research* (New York: Technology Press of M. I. T. and Wiley; London: Chapman & Hall, 1951), p. 38. *Alternate Problem:* Mention some laws and theories of technology and sketch how they are tried out.

11.1.6. Regard social and behavioral engineering as applied sociology and psychology. Analyze the use of the former disciplines in social planning and social reform. *Alternate Problem:* Regard pedagogy as a branch of behavioral engineering and study whether educational theories are tested and, if so, how.

11.1.7. Analyze the relation of psychiatry, as a branch of medicine, to biology, biochemistry, and psychology. Comment on the revolution in psychiatry that is currently being operated by research on psychotropic drugs. *Alternate Problem:* Does Scotland Yard use any theories? If so, are they substantive crime theories? In any case, is crime detection a science or does it just use scientific techniques?

11.1.8. Land surveying is one of the oldest and most exact technologies. Examine the means (techniques) and ends (mapping) of land surveying and show what its theoretical bases are. *Alternate Problem:* The neuroticism criteria currently used (e.g., the Willoughby Schedule) are based on experience with case studies. They have no theoretical foundation and apparently they have not been applied to normal subjects in order to ascertain whether the deviations from normalcy are significant. Moreover, there is no independent normalcy criterion, as would be required for the mere formation of control groups. What do such criteria best measure: the acuteness of a neurotic condition or the present state of psychiatry?

11.1.9. Perform a logical and epistemological study of any of the following disciplines: operations research, game theory, and information theory. Investigate also whether they are related to substantive theories, and if so how.

11.1.10. Study the design of new things, such as synthetic fibres. Does it always consist in combining existing objects (i.e., in putting together pieces of known material) in new ways? Is it possible for science and technology to create new substances, hence to produce new natural laws? See Sec. 6.8. *Alternate Problem:* Study the possibility that the instrumentalist view of science was fostered by a confusion between pure and applied science.

11.2. Technological Rule

Just as pure science focuses on objective patterns or laws, action oriented research aims at establishing stable norms of successful human behavior, i.e., rules. The study of rules—the grounded rules of applied science—is therefore central to the philosophy of technology.

A rule *prescribes* a course of action: it indicates how one should proceed in order to achieve a predetermined goal. More explicitly: a rule is an instruction to perform a finite number of acts in a given order and with a given aim. The skeleton of a rule can be symbolized as an ordered string of signs, e.g., $<1, 2, \ldots, n>$, where every numeral stands for a corresponding action, the last action, n, is the only thing that separates the operator who has executed every operation, save n, from the goal. In contrast to law formulas, which say what the shape of possible events is, rules are norms. The field of law is assumed to be the whole of reality including rule-makers; the field of rule is but humankind: people, not stars; can obey rules and violate them, invent and perfect them. Law statements are descriptive and interpretive, whereas rules are normative. Consequently, while law statements can be more or less true, rules can only be more or less effective.

We may distinguish the following genera of rules: (i) *rules of conduct* (social, moral and legal rules); (ii) *rules of prescientific work* (rules of thumb in the arts and crafts and in production); (iii) *semiotic rules* (syntactical and semantical rules); (iv) *rules of science and technology:* grounded rules of research and action. Rules of conduct make social life possible (and hard). The rules of prescientific work dominate the region of practical knowledge which is not yet under technological control. The rules of sign direct us how to handle symbols: how to generate, transform, and interpret signs. And the rules of science and technology are those norms that summarize the special techniques of research in pure and applied science (e.g., random sampling techniques) and the special techniques of advanced modern production (e.g., the technique of melting with infrared rays).

Many rules of conduct, work, and sign, are *conventional,* in the sense that they are adopted with no definite reasons and might be exchanged for alternative rules with little or no concomitant change in the desired result. They are not altogether arbitrary, since their formation and adoption should be explainable in terms of psychological and sociological laws, but they are not necessary either; the differences among cultures are largely differences among systems of rules of that kind. We are not interested in such groundless or conventional rules but rather in founded rules, i.e. in norms satisfying the following *Definition:* A rule is *grounded* if and only if it is based on a set of law formulas capable of accounting for its effectiveness. The rule that commands taking off the hat when greeting a lady is groundless in the sense that it is based on no scientific law but is conventionally adopted. On the other hand, the rule that commands greasing cars periodically is based on the law that lubricators decrease the wearing out of parts by friction: this is neither a convention nor a rule of thumb like those of cooking and politicking: it is a well-grounded rule. We shall elucidate later on the concept of basing a rule on a law.

To decide that a rule is effective it is necessary, though insufficient, to show that it has been successful in a high percentage of cases. But these cases might be just coincidences, such as those that may have consecrated the magic rituals that accompanied the huntings of primitive man. Before adopting an empirically effective rule we ought to know *why* it is effective: we ought to take it apart and reach an understanding of its *modus operandi.* This requirement of rule foundation marks the transition between the prescientific arts and crafts and contemporary technology. Now, the sole valid foundation of a rule is a system of law formulas, because these alone can be expected to correctly explain facts—e.g., the fact that a given rule works. This is not to say that the effectiveness of a rule depends on whether it is founded or groundless but only that, in order to be able to *judge* whether a rule has any chance of being effective, as well as in order to *improve* the rule and eventually *replace* it by a more effective one, we must disclose the underlying law statements, if any. We may take a step ahead and claim that the blind application of rules of thumb has never paid in the long run: the best policy is, first, to try to ground our rules, and second, to try to transform some law formulas into effective technological rules. The birth and development of modern technology is the result of these two movements.

But it is easier to preach the foundation of rules than to say exactly what the foundation of rules consists in. Let us try to make an inroad into this unexplored territory—the core of the philosophy of technology. As usual when approaching a new subject, it will be convenient to begin by analyzing a typical case. Take the law statement "Magnetism disappears above the Curie temperature [770°C for iron]". For purposes of analysis it will be convenient to restate our law as an explicit conditional: "If the temperature of a magnetized body exceeds its Curie point, then it becomes demagnetized." (This is, of course, an oversimplification, as every other ordinary language rendering of a scientific law: the Curie point is not the temperature at which all magnetism disappears but rather the point of conversion of ferromagnetism into paramagnetism or conversely. But this is a refinement irrelevant to most technological purposes.) Our nomological statement provides the basis for the nomopragmatic statement "If a magnetized body is heated above its Curie point, then it is demagnetized". (The pragmatic predicate of course, 'is heated'. For the concept of nomopragmatic statement, see Sec. 6.5.) This nomopragmatic statement is in turn the ground for two different rules, namely *R*1: "In order to demagnetize a body heat it above its Curie point", and *R*2: "To prevent demagnetizing a body do not heat it above its Curie point." Both rules have the same foundation, i.e. the same underlying nomopragmatic statement, which in turn is supported by a law statement assumed to represent an objective pattern. Moreover, the two rules are equiefficient, though not under the same circumstances (changed goals, changed means). So far, the situation may be characterized by means of the relation of presupposition ⊣, elucidated in Sec. 5.1:

Nomological statement ⊣ *Nomopragmatic statement* ⊣ {*Rule 1*, *Rule 2*}.

*At the propositional level, the structure of both the nomological statement and the nomopragmatic statement is "$A \rightarrow B$". One of the differences between the two lies in the meaning of the antecedent symbol 'A', which in the case of the nomological statement refers to an objective fact whereas in the case of the nomopragmatic statement it refers to a human operation. Rule 1 may be symbolized '*B per A*', which we read '*B* through *A*' or 'To get *B* do *A*' or 'To the end *B* use the means *A*'. The structure of Rule 2 is, on the other hand, '*–B per –A*', which can be read 'To prevent *B* do not do *A*'. The consequent of the law formula "$A \rightarrow B$" has become the "antecedent" of the rule *R*1 and the antecedent of the former the "consequent" of the latter. Or, rather, the logical antecedent of the law formula and its negate are now the means whereas the logical consequent and its negate are the end of one rule each. (But whereas the antecedent of a law statement is sufficient for the occurrence of the fact referred to by the consequent, the "consequent" of the rule may be only necessary for attaining the goal expressed by the "antecedent".) We shall sum up the foregoing results in the following formulas expressed in the metalanguage and valid for elementary laws and rules:

"$A \rightarrow B$" *fund* ("*B per A*" *vel* "*–B per –A*") [11.1]

"*B per A*" *aeq* "*–B per –A*" [11.2]

where '*fund*' means "is the foundation of", '*vel*' stands for "*or*", and '*aeq*' for "equiefficient". These, like '*per*', are rule connectives.

*Notice the deep differences between law formulas and rules. In the first place, the functors '*fund*' and '*aeq*' have no syntactical equivalents. In the second place, "*B per A*" has no truth value. Rules have, on the other hand, effectiveness values. More exactly, we may say that a rule of the form "*B per A*" has one of at least three effectiveness values: it may be effective (symbolized '1'), ineffective (symbolized '0'), or indeterminate (symbolized '?'). This difference is best grasped upon comparing the truth table of "*A*→*B*" with the efficiency table of the associated rule "*B per A*":

Truth table of law "A→B"			*Effectiveness table of rule "B per A"*		
A	*B*	*A*→*B*	*A*	*B*	*B per A*
1	1	1	1	1	1
1	0	0	1	0	0
0	1	1	0	1	?
0	0	1	0	0	?

Whereas the conditional is false in the single case in which its antecedent is true and its consequent false, the only case in which the rule is effective is when the means *A* is applied and the goal *B* is achieved. We can decide that "*B per A*" is ineffective only when the stipulated means *A* is enforced and the desired result *B* is not attained. But if we do not apply the means (last two rows of the table) then we can pass no judgment over the rule, whether or not the goal is achieved: in fact, not to apply the means stipulated by the rule is not to apply the rule at all. The "logic" of rules, then, is at least three-valued.

*We said before that "*B per A*" and "–*B per* –*A*" are equiefficient though not under the same circumstances. This means that in either case there is at least one combination of means and ends that comes under the rule although the combination is not the same in both cases. In fact, the effectiveness tables of the two rules are different, as shown in the following table, in which the four possible combinations of means and goals are exhibited.

A	–*A*	*B*	–*B*	*B per A*	–*B per* –*A*	–*B per A*	*B per* –*A*
1	0	1	0	1	?	0	?
1	0	0	1	0	?	1	?
0	1	1	0	?	0	?	1
0	1	0	1	?	1	?	0

An obvious generalization of the above tables is easily obtained by letting *A* and *B* take any of the three values 1,0 and?. A generalization in a different direction is achieved upon replacing '1' by the relative frequency *f* of success and '0' by its complement 1–*f*.*

The relation between a law formula such as "*A*→*B*" and the rules

"*B per A*" and "*–B per –A*" is not logical but pragmatic. We stipulate the relation by laying down the following *Metarule:* If "*A*→*B*" is a law formula, try the rules "*B per A*" or "*–B per –A*". Our metarule says 'try', not 'adopt', for two reasons. First, every law formula is corrigible, whence the corresponding rule may suffer changes. Second a law formula may refer to a much too idealized model of a concrete system, in which case the corresponding rule will be inefficient or nearly so. Take again the demagnetization rule. In stating the corresponding law statements (the nomological and the nomopragmatic ones) we presupposed that only two variables are relevant, namely magnetization and temperature: we disregarded pressure and other variables that might make a difference. Moreover, we did not even pose the technological problem of building an efficient, rapid and low-cost furnace for heating the material and such that its chemical composition would not alter by its contact with the air during the operation. Now, the neglect of some of these "details" may ruin the efficience of the rule. To take account of them we need additional law statements, even entire additional theories or fragments of such. Even so it might turn out that, for certain purposes, an alternative procedure—one based on different law formulas (e.g., applying a decreasing alternating magnetic field)—may be more effective than heating. We conclude that the truth of a law formula does not *warrant* the effectiveness of the rules based on it. This is why our metarule recommends rather than commands using the rule "*B per A*" once the formula "*A*→*B*" has been established as a law formula.

*If we cannot infer the effectiveness of a rule from the truth of the corresponding law formula, what about the converse procedure? It is even less warranted. In fact, since the rule "*B per A*" is effective if and only if both *A* and *B* are the case, we may satisfy this condition by alternatively adopting infinitely many hypotheses, such as "*A* & *B*", "*A* ∨ *B*", "*A*→*B*", "*B*→*A*", "(*A* & *B*) & *C*", "(*A* & *B*) ∨ *C*", "(*A* ∨ *B*) & *C*", "(*A* ∨ *B*) ∨ *C*", and so on, where '*C*' designates an arbitrary formula. From these infinitely many hypotheses only the third coincides with our law statement "*A* →*B*". In short, given a law formula we may *try* the corresponding rule, as counseled by our metarule, but given a rule *nothing* can be inferred about the underlying law formula. All a successful rule does—and this is a great deal—is to point to possible relevant *variables* and to pose the *problem* of discovering the lawful relation among them.*

The above has important consequences for the methodology of rules and the interrelations between pure and applied science. We see there is no single road from practice to knowledge, from success to truth: success warrants no inference from rule to law but poses the problem of explaining the apparent efficiency of the rule. In other words, the roads from success to truth are infinitely many and consequently theoretically useless or nearly so: i.e. no bunch of effective rules suggests a true theory. On the other hand the roads from truth to success are limited in number, hence feasible. This is one of the reasons why practical success, whether of a medical treatment or of a government measure, is not a truth criterion for the underlying hypotheses. This is also why technology—in contrast to the prescientific arts and crafts—does not start with rules and ends up with theories but proceeds the other way around. This is, in brief, why technology is applied science whereas science is not purified technology.

Scientists and technologists work out rules on the basis of theories containing law statements and auxiliary assumptions, and technicians apply such rules jointly with ungrounded (prescientific) rules. In either case specific hypotheses accompany the application of rules: namely, hypotheses to the effect that the case under consideration is one where the rule is in point because such and such variables—related by the rule—are in fact present. In science such hypotheses can be tested: this is true of both pure and applied research. But in the practice of technology there may not be time to test them in any way other than by applying the rules around which such hypotheses cluster—and this is a poor test indeed, because the failure may be blamed either on the hypotheses or on the rule or on the uncertain conditions of application.

In view of the profound differences between law formulas and rules, the persistent confusion between the two and, what is worse, the characterization of laws as instructions, is hardly justifiable. It can be explained, though, on two counts. Firstly, every law statement can be made the ground of one or more rules; thus, given a law "$L(x, y)$" relating the variables x and y, we can prescribe "In order to measure or compute y in terms of x, use '$L(x, y)$' ". Secondly, most philosophers do not have in mind law statements proper when they deal with laws but rather empirical generalizations, on top of which they formulate such generalizations in pragmatical terms, i.e. in statements containing pragmatical predicates: in short, they start from nomopragmatic statements belonging to ordinary knowledge, from which there is only a short way to rules. Paradoxical though it may seem, an adequate treatment of the pragmatic aspects of knowledge requires a nonpragmatist philosophical approach.

Let us finally inquire into the peculiarities of technological forecast.

Problems

11.2.1. Illustrate the concepts of rule of conduct, rule of work, rule of sign, and grounded rule. *Alternate Problem:* Study the problem of discovering, formulating and learning rules. See F. v. Hayek, "Rules, Perception and Intelligibility", *Proc. Brit. Acad.* XLVIII, 321 (1965) and the bibliography cited therein.

11.2.2. Draw a parallel between a cooking recipe and a law formula *Alternate Problem:* Find out whether the following rules are grounded or not: (i) "If you do not understand it, oppose it"; (ii) "In order to cure mental patients flog them"; (iii) "In order to prevent crime punish it severely".

11.2.3. Among the founded rules occurring in pure and applied science we may distinguish *rules of decision* and *rules of doing,* i.e. rules enabling us to make well-grounded decisions and rules directing us to implement such decisions. Exemplify and analyze these two kinds of rule.

11.2.4. Discuss whether the following uses of the term 'rule' are justified. (i) The parallelogram rule in mechanics; (ii) the right-hand rule and the left-hand rule in electromagnetic theory; (iii) the phase rule in thermodynamics; (iv) the quantization rules in Bohr's quantum theory and the commutation and selection rules in the modern quantum theory.

11.2.5. Examine the characterization of laws of nature as directions and rules of behavior or procedure. See (i) E. Mach, *History and Root of the Principle of Conservation of Energy* (1872; Chicago: Open Court, 1911), p. 55: Galilei's law, "$s=gt^2/2$", is "the rule of derivation by means of which we find, from a given *t,* the corresponding *s,* and this replaces the table [of empirically found values] just mentioned in a very complete, convenient, and compendious manner. This rule of derivation, this formula, this 'law', has, now, not in the least more value than the aggregate of the individual facts. Its value for us lies merely in the convenience of its use: it has an economical value". (ii) M. Schlick, "Causality in Contemporary Physics" (1931), *British Journal for the Philosophy of Science,* XII, 177 and 281 (1961 and 1962). (iii) H. Dingler, *Die Methode der Physik* (Munchen: Reinhardt, 1938).

11.2.6. Are groundless rules set up arbitrarily, i.e. by caprice? That is, from the statement that they are not theoretically justified, does it follow that they are pragmatically unjustified as well? And what kind of logic do groundless rules obey? *Alternate Problem:* The effectiveness of a rule can be measured by its performance. Would such a measure save us the work of explaining why the rule works to the extent that it does?

11.2.7. Is the theoretical foundation of a rule a necessary, a sufficient, or neither a necessary nor a sufficient condition for its adoption? *Alternate Problem:* A rule may be regarded as a means-end relation whereby the means are necessary, in some sense, for goal attainment. Discuss the following formalizations of a rule. (i) "$G \rightarrow M$"; (ii) "$G \rightarrow \Box M$", where 'G' and 'M' designate goal and means respectively, and the square stands for the modal operator "it is necessary that". Notice that in either case (*a*) a proposition not a prescription proper results, and (*b*) the concept of necessity involved is not necessarily either physical or pragmatic.

11.2.8. Psychoanalysts claim that the couch is the probing ground of their hypotheses, thereby meaning that psychoanalytic theory must be true if psychoanalytic therapy proves effective (which, incidentally, it does not). Examine this argument. *Alternate Problem:* Religionists and politicians claim that their following proves the truth of their views: is this claim right?

11.2.9. Both pure and applied science apply and work out rules. What is the difference between a scientific rule of procedure (e.g., a rule for coloring tissues to be observed through the microscope) and a technological rule of procedure (e.g., a rule for dyeing a fabric of a certain kind)? *Alternate problem:* Discuss the thesis that the root of the ethical difference between science and technology is that a single law can be the ground for two contrary rules. See M. Bunge, *Ethics* (Dordrecht-Boston: Reidel, 1989), pp. 258-263.

11.2.10. Consider the principle "Every rule should be either grounded or rejected on the basis of scientific knowledge". Is it a generalization of our experience with rules, a groundless hope, or a programmatic metarule justified by history? *Alternate Problem:* Establish a relation between the logic of rules and three-valued logic. Notice that, if the efficiency table of "*B per A*" is completed by assigning *A* and *B* indeterminate values, it becomes identical with the truth table for the quasi-implication in three-valued logic. For the latter see, e.g., H. Reichenbach, *Philosophic Foundations of Quantum Mechanics* (Berkeley and Los Angeles: University of California Press, 1946), p. 151.

11.3. Technological Forecast

For technology knowledge is chiefly a means to be applied to the achievement of certain practical ends. The goal of technology is successful action rather than pure knowledge. Accordingly the whole attitude of the technologist while applying his technological knowledge is active in the sense that, far from being an inquisitive onlooker or a diligent burrower, he is an active participant in events. This difference of attitude between the technologist in action and the researcher—whether pure or applied—introduces certain differences between technological forecast and scientific prediction. (For the latter see Ch. 10.)

In the first place, whereas scientific prediction says what will or may happen if certain circumstances obtain, technological forecast suggests how to influence on circumstances so that certain events may be brought about, or prevented, that would not normally happen: it is one thing to predict the orbit of a comet, quite another to plan and foresee the trajectory of an artificial satellite. The latter presupposes a choice among possible goals and such a choice presupposes a certain forecasting of possibilities and their evaluation in the light of a set of desiderata. In fact, the technologist will make his forecast on his (or his employer's) estimate of what the future *should* be like if certain desiderata are to be fulfilled: contrary to the pure scientist, the technologist is hardly interested in what would happen anyway; and what for the scientist is just the final state of a process becomes for the technologist a valuable (or disvaluable) end to be achieved (or to be

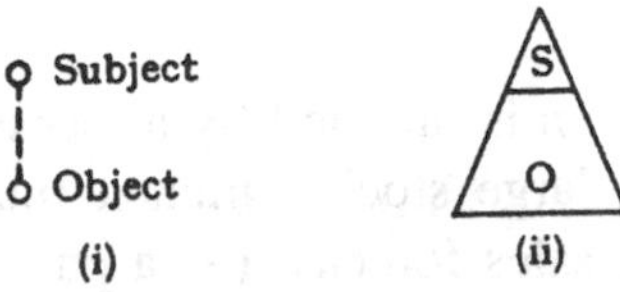

Fig. 11.3. (i) Objectivity: the key to scientific truth. (ii) Partiality: the key to technological control.

avoided). A typical scientific prediction has the form "If x occurs at time t then y will occur at time t' with probability p". By contrast, a typical technological forecast is of the form: "If y is to be achieved at time t' with probability p, then x should be done at time t". Given the goal, the technologist indicates the adequate means and his forecast states a means end relationship rather than a relation between an initial state and a final state. Furthermore, such means are implemented by a specified set of actions, among them the technologist's own actions. Which leads us to a second peculiarity of technological forecast: whereas the scientist's success depends on his ability to separate his object from himself (particularly so when his object happens to be a psychological subject)—i.e. on his capacity of detachment—the technologist's ability consists in placing himself within the system concerned—at the head of it (Fig. 11.3). This does not involve *subjectivity,* since after all the technologist draws on the objective knowledge provided by science; but it does involve partiality, a *parti pris* unknown to the pure researcher. The engineer is part of a man-machine complex, the industrial psychologist is part of an organization, and both are bound to devise and implement the optimal means for achieving desiderata which are not usually chosen by themselves: they are decision makers, not policy makers.

The forecast of an event or process that is beyond our control will not alter the event or process itself. Thus, e.g., no matter how accurately an astronomer predicts the collision of two stars, the event will occur in due course. But if an applied geologist can forecast a landslide, then some of its consequences can be prevented. Moreover, by designing and supervising the appropriate defense works the engineer may prevent the landslide itself: he may devise the sequence of actions that will refute the original forecast. Similarly, an industrial concern may prognose sales for the near future on the (shaky) assumption that a given state of the economy, say prosperity, will continue during that lapse. But if this assumption is falsified by a recession, and the enterprise had accumulated a large stock which it must get rid of, then instead of making a new sales forecast (as a pure scientist would be inclined to do), the management will try to *force* the original forecast to come true by increasing advertisement, lower sale prices, and so on. As in the case of vital processes, a diversity of means will alternatively or jointly be tried to attain a fixed goal. In order to achieve this goal any number of initial hypotheses may have to be sacrificed: in the case of the landslide, the assumption that no external forces would interfere with the process; and in the case of the sales, that prosperity would continue. Consequently, whether the initial forecast is *forcefully falsified* (as in the case of the landslide) or *forcefully confirmed* (as in the case of the sales forecast), this fact cannot count as a *test* of the truth of the hypotheses involved: it will only count as an efficiency test of the rules that have been applied. The pure scientist, on the other hand, need not worry about altering the means for achieving a preset goal, because pure science *has* no goals external to it.

Technological forecast, in sum, cannot be used to test hypotheses and is not meant to be used in such a way: it is used for controlling things or men by changing the course of events perhaps to the point of stopping them altogether, or for forcing the predicted course even if unpredictable events should interfere with it. This is true of the forecasts made in engineering, medicine, economics, applied sociology, political science, and other technologies: the sole formulation of a forecast (prognosis, lax prediction, or prediction proper), if made known to the decision makers, can be seized upon by them to steer the course of events, thus bringing about results different from those originally forecast. This change, triggered by the issuance of the forecast, may contribute either to the latter's confirmation (self-fulfilling forecast) or to its refutation (self-defeating forecast). This trait of technological forecast stems from no logical property of it: it is a pattern of social action involving the knowledge of forecasts, and is consequently conspicuous in modern society. Therefore, rather than analyzing the logic of causally effective forecast, we should start by distinguishing three levels in it: (i) the conceptual level, on which the prediction p stands; (ii) the psychological level: the knowledge of p and the reactions triggered by this knowledge and (iii) the social level: the actions actually performed on the basis of the knowledge of p and in the service of extrascientific goals. This third level is peculiar to technological forecast.

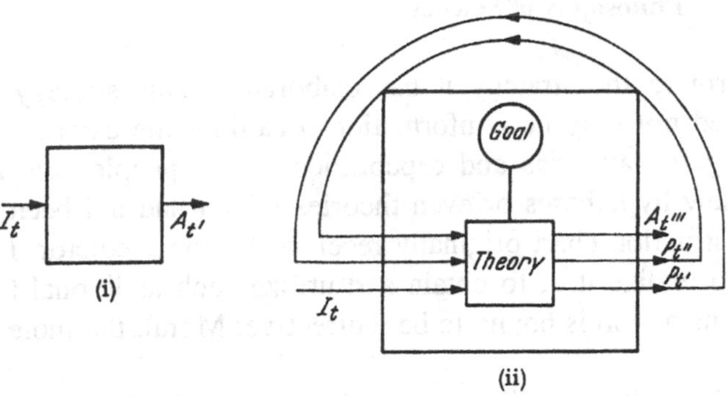

Fig. 11.4. (i) Nonpredicting system (e. g., frog). (ii) Predicting system (e. g., engineer): predictions are fed back and corrected and a new course of action, $A_{t'''}$, is worked out if $P_{t''}$ is reasonably close to G.

This feature of technological forecast sets civilized man apart from every other system. A nonpredicting system, be it a juke-box or a frog when fed with information it can digest will process and convert it into action at some later time. But such a system does not purposely produce most of the information and it does not issue projections capable of altering its own future behavior. [Fig. 11.4 (i).] A predictor—a rational man, a team of technologists, or a sufficiently evolved automaton—can behave in an entirely different way. When fed with relevant information I_t at time t, it can process this information with the help of the knowledge (or the instructions) available to it, eventually issuing a prediction $P_{t'}$ at a later time t'. This prediction is fed back into the system and compared with the preset goal G that controls the whole process (without either causing it or supplying it with energy). If the two are reasonably close the system takes a decision that eventually leads it to act so as to take advantage of the course of events. If, on the other hand, the prediction differs significantly from the goal, this difference will trigger again the theoretical mechanism, which will elaborate a new strategy: a new prediction, $P_{t''}$, will eventually be issued at time t'', a forecast including a reference to the system's own participation in the events. The new prediction is fed back into the system and if it still disagrees with the goal, a new correction cycle is triggered, and so on until the difference between the prediction and the goal becomes negligible, in which case the system's predicting mechanism comes to rest. Henceforth the system will gather new information regarding the present state of affairs and will act so as to conform to the strategy it has elaborated. This strategy may have required not only new information regarding the external world (including the attitudes and capabilities of the people concerned), but also new hypotheses or even theories which had not been present in the instruction chart originally received by the predictor. If the latter fails to realize it or to obtain and utilize such additional knowledge, his or its action is bound to be ineffective. Moral: the more brains the better.

Such a self-correcting process, hinging on the feeding back of predictions into the predictor, need not take place at the conceptual level. Automata could be built to mimic (with purely physical processes) some traits of such behavior. But this imitation can be only partial. In fact, although automata can *store* theories, as well as clear-cut instructions to use them, they lack two abilities: (i) they have no *judgment* or flair to apply them—i.e. to choose the more promising theories or to make additional simplifying assumptions, and (ii) they cannot *invent* new theories to cope with new situations, unpredicted by the designer and to which the stored theories are irrelevant. And automata cannot invent theories because there are no techniques of theory construction out of data and in a psychological and cultural vacuum—if only because no set of data can pose by itself the problems which a theory is supposed to solve. And, if no theory construction technique is available, no set of instructions can be fed into the computer to get it to theorize. (Moreover, a computer's output is a ciphered message such as a punched strip of paper. In order to become a set of ideas this message must first be decoded and then "read" or interpreted. While the decoding can be made automatically by the computer itself, the interpretation requires an expert brain soaked in relevant knowledge. Suppose now a computer did invent a new theory. How can we know it? Being a new theory it will use new concepts, some of them primitive or undefined; these new concepts will be designated by either new symbols or new combinations of old symbols, and in either case no provision for their decoding could have been made: if it had been made the theory would not be genuinely new. If there is no decoding there can be no interpretation: the message remains unintelligible, i e. it is no message at all, and we may as well assume that the machine has run wild.)

The above account of technological forecast is based on the assumption that it relies on some theory, or rather theories, whether substantive or operative. This assumption may be found wanting by anyone knowing that the forecasts issued by experts in medicine, finance, or politics, are often successful and yet involve no great deal of theorizing. True: most often, *expert prognosis* relies on inductive (empirical) generalizations of the form "A and B occur jointly with the observed frequency f", or even just "A and B occur jointly in most cases", or "Usually whenever A then B". The observation that a given individual, say a human subject, or an economic state of affairs, does have the property A is then used to prognose that it also has, or will acquire, the property B. In daily life such prognoses are all we do, and the same applies to most expert prognoses. Occasionally such prognoses made with either ordinary knowledge or specialized but nonscientific knowledge are more successful than predictions made with full fledged but false or rough theories; in many fields, however, the frequency of hits is not better than the one obtained by flipping a coin. The point, though, is that expert forecast using no scientific theory is not a scientific activity—just by definition of "scientific prediction" (Sec. 10.1).

Yet it would be wrong to think that experts make no use of *specialized knowledge* whenever they do not employ scientific theories: they always judge on the basis of some such knowledge. Only, expert knowledge is not always explicit and articulate and, for this reason, it is not readily controllable: it does not readily learn from failures and it is hard to test. For the progress of science, the failure of a scientific prediction is by far preferable to the success of an expert prognosis, because the scientific failure can be fed back into the theory responsible for it thereby giving us a chance to improve it, whereas in the case of expert knowledge there is no theory to feed the failure into. It is only for immediate practical purposes that expert prognoses made with shallow but well-confirmed generalizations are preferable to risky scientific predictions.

Another difference between expert prognosis and technological forecast proper would seem to be this: the former relies more heavily on *intuition* than does scientific prediction. Yet the difference is one of degree rather than of kind. Diagnosis and forecast, whether in pure science, in applied science, or in the arts and crafts, involve intuitions of a number of kinds: the quick identification if a thing, event, or sign; the clear but not necessarily deep grasp of the meaning and/or the mutual relations of a set of signs (text, table, diagram, etc.); the ability to interpret symbols; the ability to form spatial models; skill in realizing analogies; creative imagination; catalythic inference, i.e. quick passage from some premises to other formulas by skipping intermediate steps; power of synthesis or synoptic grasp; common sense (or rather controlled craziness), and sound judgment. These abilities intertwine with specialized knowledge, whether scientific or not, and are reinforced with practice. Without them theories could neither be invented nor applied—but, of course, they are not suprarational powers. Intuition is all right as long as it is controlled by reason and experiment: only the replacement of theorizing and experimenting by intuition must be feared.

A related danger is that of *pseudoscientific projection tools, so* common in applied psychology and sociology. A number of techniques have been devised to forecast the performance of personnel, students, and even psychologists themselves. A few tests, the objective ones, are somewhat reliable: this holds for intelligence and skill tests. But most tests, particularly the subjective ones (the "global evaluation" of personality by means of interviews, the thematic apperception test, the Rorschach, etc.) are in the best of cases inefficient and in the worst of cases misleading. When they have been subjected to the test of prediction—i.e. when their results have been checked with the actual performance of the subjects—they have failed. The failure of most individual psychological tests, and particularly of the subjective ones, is not a failure of psychological testing in general: what is responsible for such failures is either the total absence or the falsity of the underlying psychological theories. Testing for human abilities without first establishing *laws* relating objective indices of abilities or personality traits is as thoughtless as asking a tribesman to test an aircraft. As long as no theoretical foundations of psychological tests are secured, their employment as predictive instruments is not better than crystal gazing or coin flipping: they are practically inefficient and, even if they succeeded, they would not contribute to psychological theory. The limited success of psychological testing has led many to despair of the possibility of finding a scientific approach to human behavior, but the right inference is that such an attempt has only been tried after a large number of alleged tests invaded the market. What is wrong with most of "applied" (educational, industrial, etc.) psychology is that it does *not* consist in the application of scientific psychology at all. The moral is that practical wants—such as personnel training and selection—should not be allowed to force the construction of "technologies" without an underlying science.

Technological forecast should be maximally *reliable*. This condition excludes from technological practice—not however from technological research—insufficiently tested theories. In other words, technology will ultimately prefer the old theory that has rendered distinguished service in a limited domain and with a known inaccuracy, to the bold new theory which promises unheard of forecasts but is probably more complex and partly therefore less well tested. It would be irresponsible for an expert to apply a new idea in practice without having tested it under controlled conditions. (Recall the affair of the mutagenic drugs in the early 1960's.) Practice, and even technology, must be more conservative than science. Consequently, the effects of a close association of pure research with applied research, and of the latter with production, are not all of them beneficial: while it is true that technology challenges science with new problems and supplies it new equipment for data gathering and data processing, it is no less true that technology, by its very insistence on reliability, standardization (routinization) and fastness, at the expense of depth, range, accuracy, and serendipity, can slow down the advancement of science.

Reliability, a desideratum of technological forecast, is of course not always attainable. Frequent sources of uncertainty are: (i) lack of adequate theory and/or accurate information, and (ii) "noise" or random variation of a number of factors that are not under control. These shortcomings are the more acutely felt in the case of technology because of the complexity of the systems it handles and the imperfect control of their variables—a control that can be achieved only in the artificial conditions offered by a laboratory and a few high-precision industries. A third factor of uncertainty in technological forecast is that it often consists in a projection, from a model, to a real system that is a long way from the model: this can be termed a *qualitative extrapolation* to distinguish it from the quantitative extrapolation concerning one and the same system. Examples: the engineer may build a small scale model of a dam and will study its behavior before he builds the large scale system; the aeronautical engineer may make a small scale aircraft and test it in the wind tunnel; and the pharmacologist and the research physician may choose rats or monkeys—preferably rats—as material models of man.

In pure science too such material models and the corresponding extrapolations are made: the biologist will experiment *in vitro* with tissue cultures before doing it *in vivo*, and the psychologist will study how social deprivation affects the behavior of monkeys as a guide to the investigation of human behavior. But the purpose of using such material models is altogether different: the scientist wants to discover and test generalizations that might be extrapolated to the system that ultimately interests him, whereas the technologist uses material models to give his rules and plans a quick and inexpensive preliminary test for effectiveness: if the material model behaves as forecast, the leap to the system of interest (dam, aircraft, patient) may be attempted. In this leap unforeseen events will occur both because a number of new variables—mostly unknown—show up in the real system, and because the control of every variable is now short of impossible. The difference between the actual and the forecasted performance will, of course, lead to alterations in the original plans and eventually in the rules as well, so that new forecasts, with a lesser error, can be made. The process is self-correcting but not foolproof. Hence the philosopher of technology, just as the philosopher of pure science, should be confident in the possibility of progress as well as in the inevitability of error.

This closes our account of the application of scientific ideas. Let us now approach the problem of their test.

Problems

11.3.1. Propose examples of expert forecast and of technological prediction proper. *Alternate Problem:* J. S. Huxley and P. Deane, in *The Future of the Colonies* (London: Pilot Press, 1944), anticipated a number of features of the new Asiatic and African nations. Was this a prophecy, a prognosis, a scientific prediction, or a technological forecast?

11.3.2. Examine the forecasts concerning computer developments made by H. A. Simon and A. Newell, "Heuristic Problem-Solving: The Next Advance in Operations Research", *Operations Research*, **6**, 1 (1958) and R. W. Hamming, "Intellectual Implications of the Computer Revolution", *American Mathematical Monthly*, **70**, 4 (1963).

11.3.3. The production costs of new equipment are usually underestimated by a factor of 2 to 20. Why? See C. Hitch, "Uncertainties in Operations Research". *Operations Research*, **8**, 437 (1960). *Alternate Problem:* How is the feasibility of a new thing or procedure estimated? In particular, how are high level technological devices (for, e.g., the control of thermonuclear power) evaluated?

11.3.4. On the basis of laboratory and field experiments, radio-biologists have made the following forecasts concerning the damages produced by the radiations emitted in nuclear explosions. (i) "Above 5 roentgen, radiation kills any tree" (ii) "Above 500 roentgen, radiation kills any man." (iii) "Above 50,000 roentgen, most bacteria and insects are spared but everything else dies". Make the corresponding forecasts concerning the possible effects of nuclear war.

11.3.5. Study the role of forecast in the tactics and strategy of a political party which proposes itself to seize power within a definite lapse. *Alternate Problem:* Study a problem of technological forecast dealt with in either of the journals *Operations Research* or *Management Science*.

11.3.6. Make a detailed study of the process leading from the research chemical laboratory to the pilot plant and thence to the industrial plant. In particular, study the sources of uncertainty of the forecast from laboratory performance to pilot plant performance and from the latter to production performance. *Alternate Problem:* Technological forecast is made with the help of theories. What about the forecast of the sucess of technological forecast?

11.3.7. Psychoanalysts claim that they need perform no laboratory experiments nor gather and process statistics because psychoanalytic theory—or what passes for such—has been amply proved in the couch. On the other hand statistics and experiments with control groups have established that psychoanalytic therapy is, in the best of cases, innocuous. Analyze the situation. See H. J. Eysenck, Ed., *Handbook of Abnormal Psychology* (London: Pitman Medical Co., 1960), Ch. 18 and bibliography cited therein. *Alternate Problem:* I. Pavlov's work on conditioned reflexes and C. L. Hull's related learning theory have become the basis for behavior therapy, centered on the hypothesis that neurotic symptoms are maladaptive conditioned responses, whence the cure of neuroses consists in unlearning such undesirable habits. See J. Wolpe, *Behavior Therapy Techniques* (Oxford: Pergamon, 1966). Compare this situation, of an applied science, with that of a psychiatry with either no underlying theory or an underlying pseudoscientific doctrine.

11.3.8. Projective techniques such as the Rorschach test claim to be able to predict overt behavior through the disclosure of hidden personality traits, on the basis of the conjecture "Behavior is dynamically motivated" (whatever 'dynamical motivation' may mean). At the very best the correlation between needs and overt behavior, as obtained with the help of projective techniques, is 0.4. Study this situation and, in particular, find out whether the failure of projective techniques is due (i) to the falsity of the underlying hypothesis (which is both programmatic and exceedingly vague), or (ii) to the lack of a theory embodying the hypothesis. For an evaluation of projective techniques, see K. B. Little, "Problems in the Validation of Projective Techniques", *Journal of Projective Techniques*, **23**, 287 (1959), and scan the *Psychological Abstracts* in search for more recent work. *Alternate Problem:* Discuss the ethical problem raised by the use of groundless procedures (such as the projective techniques) for the assessment of personality traits and the forecast of behavior, and examine the ethical implications of technology in general.

11.3.9. Study the forecast of individual technological invention as currently practised by decision makers in applied research institutions. In particular, examine the phrase 'The proposed line of research is scientifically [or technically] possible'. Is it refutable? See A. W. Marshall and W. H. Meckling, "Predictability of the Costs, Time, and Success of Development", in National Bureau of Economic Research, *The Rate and Direction of Inventive Activity. Economic and Social Factors* (Princeton: Princeton University Press, 1962). *Alternate Problem:* Contrast individual invention with total inventive output in the long run and their corresponding forecasts. For the latter, see H. C. Lehman, "The Exponential Increase of Man's Cultural Output", *Social Forces*, **25**, 281 (1947). For extrapolations on the basis of such a trend line (exponential curve), see H. Hart, "Acceleration in Social Change" and "Predicting Future Trends", in F. R. Allen et al., *Technology and Social Change* (New York: Appleton-Century-Crofts, 1957). See also Problem 10.3.9.

11.3.10. Study the problem of the self-fulfilling and the self-defeating forecasts, which the economists call the Morgenstern effect and Popper the Oedipus effect. See R. K. Merton, *Social Theory and Social Structure,* 2nd ed. (Glencoe, Cill.: The Free Press, 1957), pp. 128ff. and 4 Ch. xi, and R. C. Buck, "Reflexive Predictions", *Philosophy of Science*, **30**, 359 (1963). *Alternate Problem:* Could we predict the self-fulfilment (or the self-defeat) of a forecast? Would the prediction of the outcome leading to the forceful refutation of a forecast contradict the forecast itself? If so, what? And would it be possible to perform such predictions with the help of (as yet undiscovered) laws of social psychology?

Bibliography

Ackoff, R. L.: Scientific method: Optimizing applied research decisions. New York and London: John Wiley & Sons 1962.

Agassi, J. : Technology: Philosophical and social aspects. Dordrecht-Boston: Reidel 1985.

Boirel, R.: Science et technique. Neuchâtel: Ed. du Griffon 1955.

Brown, G. S.: New horizons in engineering education. Daedalus **92**, 341 (1962).

Bunge, M.: "Philosophical inputs and outputs of technology." In G. Bugliarello and D. B. Donner, eds., The history and philosophy of technology. Urbana: University of Illinois Press 1979.

——— Philosophy of science and technology, Part II, Ch. 5. Dordrecht-Boston: Reidel 1987.

Child, A.: Making and knowing in Hobbes, Vico and Dewey. Los Angeles: University of California Press 1953.

Churchman, C. W., R. L. Ackoff, and E. L. Arnoff: Introduction to operations research New York and London: John Wiley & Sons 1957.

Le Chatelier, H.: De la méthode dans les sciences expérimentales, chs. 10–13 Paris: Dunod 1936.

Mitcham, C.: Thinking through technology. Chicago: University of Chicago Press 1994.

——— and R. Mackey, Eds. : Philosophy and technology. New York: Free Press, 1983.

Morse, P. M., and G. E. Kimball: Methods of operations research. New York: Technology Press of the M.I.T. and John Wiley & Sons; London: Chapman & Hall 1951.

Simon, H. A.: The new science of management decision. New York: Harper & Row 1960.

Suppes, P.: The philosophical relevance of decision theory. J. Phil. **58**, 605 (1961).

Susskind, C. : Understanding technology. Baltimore: Johns Hopkins University Press 1973.

Technology and culture, II, No 4 (1961), issue devoted to Science and engineering, and VII, No 3 (1966), titled "Toward a philosophy of technology".

Part IV

Testing Scientific Ideas

12. Observation
13. Measurement
14. Experiment
15. Concluding

So far we have dealt with scientific ideas and some of their applications. Henceforth we shall tackle the problem of finding out the extent to which those ideas fit facts. Since this fitness is estimated through experience, we shall study the empirical procedures whereby ideas concerning facts are checked.

Empirical operations are made in science with cognitive aims or with practical ones. Cognitively directed empirical operations are those performed in order to produce data, generate modest conjectures, or test hypotheses and theories. Practically directed empirical operations are those aiming at testing rules of procedure or concrete systems—materials, instruments, persons, organizations. In brief, empirical operations can be classed into cognitively directed or knowledge-increasing (data-gathering, hypotheses-generating, or hypotheses-testing), and practically directed or power-increasing (rules-testing or systems-testing). In the following, emphasis will be laid on cognitively directed empirical procedures.

Now the results of empirical operations aiming at advancing our knowledge are insignificant by themselves: they have to be interpreted and evaluated—i.e. some "conclusions" must be drawn from them. In other words, if such empirical results are to become relevant to scientific ideas then certain inferences must be performed. This is why the present and last part of the book closes with a chapter devoted to the evaluation of hypotheses and theories in the light of scientific experience. In this way we close the loop that starts with facts prompting the questioning that elicit ideas requiring empirical tests.

12

Observation

The basic empirical procedure is observation. Both measurement and experiment involve observation, whereas the latter is often done without quantitative precision (i.e. without measuring) and without deliberately changing the values of certain variables (i.e. without experimenting). The object of observation is, of course, an actual fact; the outcome of an act of observation is a datum—a singular or an existential proposition expressing some traits of the result of observing. A natural order to follow is, then: fact, observation, and datum. Our discussion will close with an examination of the function of observation in science.

12.1. Fact

Factual science is, by definition, concerned with ascertaining and understanding facts. But what is a fact or, better, what does the word 'fact' mean? We shall adopt the linguistic convention of calling *fact* whatever is the case, i.e., anything that is known or assumed—with some ground—to belong to reality. Accordingly this book and the act of reading it are facts; on the other hand the, ideas expressed in it are not facts when regarded in themselves rather than as brain processor.

Among facts we usually distinguish the following kinds: state, event, process, phenomenon, and system. A *state* of a thing at a given instant is the list of the properties of the thing at that time. An *event,* happening, or occurrence, is any change of state over a time interval. (A point event, unextended in time, is a useful construct without a real counterpart.) A flash of light, and the flashing of an idea, are events. From an

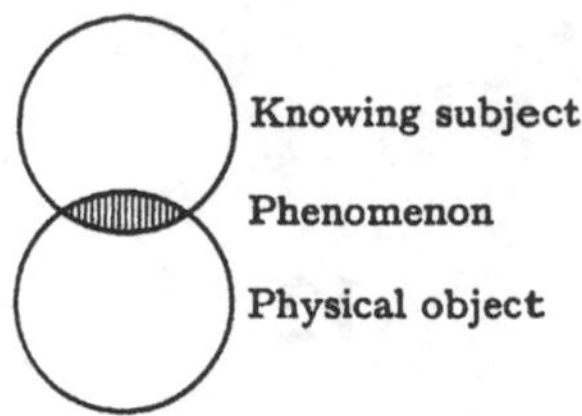

Fig. 12.1. Phenomena as facts occurring in the transactions of the knower with his environment.

epistemological point of view, events may be regarded either as the elements in terms of which we account for processes, or as the complexes which we analyze as the confluence of processes. In science, events play both roles: taken as units on a given level, they become the objects of analysis on a deeper level.

Are there unanalyzable events, i.e. ultimate facts in terms of which all others must be accounted for and which could not conceivably become the object of further analysis? According to one school of thought events such as the quantum jump taking place when an atom emits a light quantum are not further analyzable. We cannot here discuss this question in detail, but as cautious philosophers we should surmise that, from our failure to analyze an event either empirically or theoretically, we are not warranted to conclude that the event itself is atomic or irreducible. The fault may lie with our tools of analysis, either empirical or conceptual—as has so often been the case along the history of science. There might be elementary, i.e. unanalyzable events, but this we may never know: we should always try to analyze them and it should always be too soon to admit our defeat.

A *process* is a sequence, ordered in time, of events and such that every member of the sequence takes part in the determination of the succeeding member. Thus, usually the sequence of phone calls we receive during the week is not a process proper, but the sequence of events beginning with a phone call to our doctor and ending with the payment of his bill includes a process. If analyzed deeply enough, most events are seen to be processes. Thus, a light flash consists in the emission, by a large collection of atoms at slightly different times and at random, of wave groups that are propagated with a finite velocity. It is no easy task to trace processes in the tangle of events. A process is rarely given in experience: in science, at least, processes are mostly hypothesized. Thus, stars are not seen to evolve, but star evolution models must be imagined and then tested by recording and interpreting events such as the traces left by starlight on photographic plates.

A *phenomenon* is an event or a process such as it appears to some human subject: it is a perceptible fact, a sensible occurrence or chain of occurrences. Rage is not a phenomenon except for the subject suffering a fit of rage; but some of the bodily events accompanying a fit of rage—some acts of behavior—are phenomena. Facts can be in the external world but phenomena are always, so to speak, in the intersection of the external world with a cognitive subject (Fig. 12.1.) There can be no phenomena or appearances without a sensible subject placing itself in an adequate observation position. One and the same event (objective fact) may appear differently to different observers, and this even if they are equipped with the same observation devices (see Sec. 6.5). This is one reason why the fundamental laws of science do not refer to phenomena but to nets of objective facts. The usage of 'phenomenon', though, is not consistent: in the scientific literature 'phenomenon' is often taken as a synonym of 'fact'—just as in ordinary language 'fact' is often mistaken for 'truth'.

(An old philosophical question is, of course, whether we have access to anything non phenomenal, i.e. that does not present itself to our sensibility. If only the strictly empirical approach is allowed, then obviously phenomena alone are pronounced knowable: this is the thesis of *phenomenalism.* If, on the other hand, thinking is allowed to play a role in knowledge in addition to feeling, seeing, smelling, touching, and so on, then a more ambitious epistemology can be tried, one assuming reality—including experiential reality—to be knowable if only in part and gradually: this is the thesis of the various kinds of *realism.* According to phenomenalism, the aim of science is to collect, describe, and economically systematize phenomena without inventing diaphenomenal or transobservational objects. Realism, by contrast, holds that experience is not an ultimate but should be accounted for in terms of a much wider but indirectly knowable world: the set of all existents. For realism, experience is a kind of fact: every single experience is an event happening in the knowing subject, which is in turn regarded as a concrete system—but one having expectations and a bag of knowledge that both distort and enrich his experiences. Accordingly, realism will stimulate the invention of theories going beyond the systematization of experiential items and requiring consequently ingenious test procedures. We have seen at various points, especially in Secs. 5.9, 8.4, and 8.5, that science both presupposes a realistic epistemology and gradually fulfils its program.)

Finally we shall call entities, or physical things, *concrete systems*—to distinguish them from conceptual systems such as theories. A light wave is a concrete thing and so is a human community, but a theory of either is a conceptual system. The word 'system' is more neutral than 'thing', which in most cases denotes a system endowed with mass and perhaps tactually perceptible; we find it natural to speak of a force field as a system, but we would be reluctant to call it a thing. By calling all existents "concrete systems" we tacitly commit ourselves—in tune with a growing suspicion in all scientific quarters—that there are no simple, structureless entities. This is a programmatic hypothesis found fertile in the past, because it has stimulated the search for complexities hidden under simple appearances. Let us be clear, then, that by adopting the convention that the protagonists of events be called concrete systems, we make an ontological hypothesis that transcends the scope of the special sciences.

Events and processes are what happens to, in, or among concrete systems. (We disregard the metaphysical doctrine that things are just sets of events, for having no foot on science.) Events, processes, phenomena, and concrete systems are, then, facts; or rather, we shall count them in the extension of the concept of fact. Facts are, in turn, a kind of object. *Object* is, indeed, whatever is or may become a subject of thought or action. Things and their properties are objects; again, concepts and their combinations (e.g., propositions) are objects, but of a different kind: they are often called *ideal objects*. Facts, the concern of factual science, are objects of a different kind: they may be termed *concrete objects*. In brief,

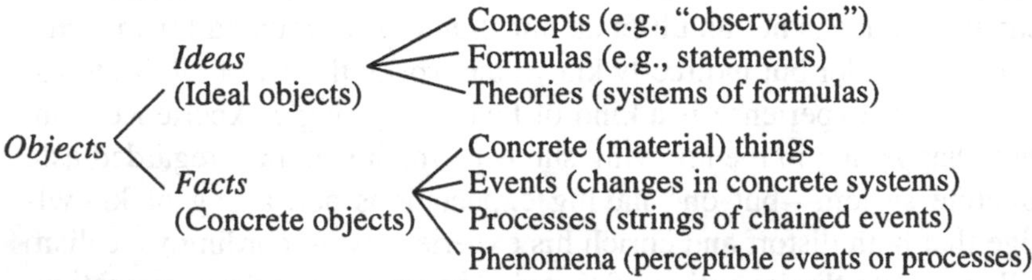

What about physical *properties*, such as weight, and *relations*, such as subordination? Shall they be counted as material objects or as ideal

objects? If we do the former we shall be forced to conclude that properties and relations can exist by themselves, apart from concrete things and their changes (events and processes); also, that there could exist concrete things destitute of every property. Either view is at variance with science, which is concerned with finding out the properties *of* concrete things and the relations *among* them—and, on a higher level of abstraction, with investigating the relations among properties and relations as well. If, on the other hand, we count properties and relations as ideal objects, then we are led to hypothesize that concrete objects have ideal components—the "forms" of ancient hylemorphism. And, being ideal, such properties and relations would not bear empirical examination, which would render factual science unempirical.

What are then physical properties and relations if they can be classed neither as material nor as ideal objects? The simplest solution is to declare them nonexistent. But then we would have again things without properties and events without mutual relationships—and consequently we would be at a loss to account for our success in finding laws. There seems to be no way out but this: there are no properties and relations apart from things and their changes. What do exist are certain things with properties and relations. Property-less entities would be unknowable, hence the hypothesis of their existence is untestable; and disembodied properties and relations are unknown: moreover, every factual theory refers to concrete things and their properties and relations (recall Sec. 7.2). In short, the properties and relations of concrete things have as little autonomous existence as ideas. But the latter, being our own make, can be thought apart from the corresponding brain processes, and this is why we may count them as a special kind of object: after all, ideas exist by being thought by someone, and there is nothing to prevent us from thinking ideas detached from material systems—even though such a thought be a brain process.

And what about *possible* but not yet actual facts: where shall we place them? This is a captious question: from the moment we speak of 'possible facts' we admit them as a subclass of facts—that is, we tacitly assume that the set of facts is the union of actual and possible facts. So far the question is verbal. It becomes an epistemological problem if we ask whether science limits itself to actuals or deals with possibles as well. A look at any special science will show the latter to be the case. Thus, the geneticist may be said to study the class of all the "informations" that *can* be transmitted (both undistorted and with mistakes) by organisms to their offspring, and to estimate the probability that any such encoded "information" will in fact be transmitted and displayed—i.e. the probability that any given possibility be actualized. Likewise, when the applied scientist weighs alternative decisions, he reckons with both knowledge concerning actual facts and with assumptions about possible facts—e.g., the possible moves of his opponent (nature, competitor, enemy). Whether an event is possible or not depends on the laws of nature: these may metaphorically be said to put constraints on possibility. But whether a possibility allowed by the laws of nature will be actualized or not depends on the special state of the thing concerned (e.g., on its initial and/or boundary conditions). Thus, it is possible for a two-headed child to be conceived by a healthy couple: the laws of genetics do not prohibit such an event although they assign it a very low probability. But whether a given couple will in fact engender such a freak will depend on particular circumstances during impregnation.

So far we have dealt with various kinds of facts; let us now cast a glance at facts in relation to the knowing subject. We may deliberately or unwillingly produce facts or we may hypothesize them. Daily occurrences are partly given to us and partly made by us; the experimental scientist finds out further facts, the theoretician hypothesizes and explains them, and the technologist devises recipes for the economical production and avoidance of facts having a practical (positive or negative) value. Science is interested in all of them, but the more so in nonordinary facts, i.e. in facts that are not within the reach of the layman but require special tools, either empirical or conceptual. Thus, the falling of hail on our roof will be of no particular interest to the meteorologists who may on the other hand study the formation of hail clouds in general—a process the details of which are not directly observable but must be hypothesized (or inferred, as is also said). The finding and making of nonordinary facts is particularly interesting to the scientist because it is not a purely empirical matter but involves hypotheses, theories, and instruments designed with the help of such ideas.

What about *the given*? What the scientist is given is rarely the fact itself: usually he produces some evidence in favor or against the conjecture that a given set of facts did occur or may in fact happen. Thus, the meteorologist does not watch the formation of ice crystals in the cloud above his head any more than the historian watches a social process. What can be observed is usually a small fraction of the facts constituting the object of an investigation; the observable facts or phenomena are just documents suggesting or confirming the existence of more interesting facts behind. Facts are iceberg-like: they are mostly submerged under the surface of immediate experience—which experience is very often of a very different kind from the facts it points to. (That the perceptible is only a part of the existent, and that many phenomena originate in unperceptible events, was realized long ago. In the early days of hypothetical thinking the invisible was often equated with the supernatural and inscrutable. It was a merit of the Ionian physicist-philosophers to suggest that the invisible is natural and is scrutable through its effects. What the Greek atomists were not able to do is to control the unseen by acting on it and thereby effectively testing the hypotheses on the unperceptible: this is an achievement of modern man.) The submerged portion of facts must be hypothesized and, in order to test such hypotheses, definite relations between the unobserved and the observed must be added, by which the observed can count as evidence for or against the existence of the hypothetical unseen, and the unseen can explain what can be seen. These relations are represented by hypotheses and theories. We cannot escape them if we wish to get closer to the facts; our sole concern must be to shun ungrounded and untestable hypotheses (recall Ch. 5). In factual science, theory and experience are interpenetrating rather than separate, and theory alone can lead us beyond appearances, to the core of reality.

Let us finally deal with certain misunderstandings and puzzles centered on the term 'fact'. In the first place, note the ambiguity of the expression 'it is a fact'. If we say 'It is a fact that the Earth rotates around the Sun', we may have in mind either (i) the *hypothesis* "The Earth rotates around the Sun" or (ii) the *fact* referred to by this hypothesis. The ambiguity is harmless as long as the expression 'it is a fact' is not employed as a rhetorical device aimed at hiding a hypothesis under the guise of a datum. We should abstain from using 'it is a fact' instead of 'we assert' or 'we assume', as the case may be.

Secondly, some philosophers have purposely called facts all singular factual *propositions*. With this verbal manoeuvre the problem of factual truth, i.e. the problem of evaluating factual propositions with the help of truth criteria, is evaded rather than solved. Moreover, a number of puzzles arise. For instance, negative facts, alternative facts and general facts have to be admitted: if something does not happen "it" is called a negative fact; if two possibilities are open we are led to talk about an alternative fact; and if a set of facts of the same kind occur they are collectively called a general fact. But this is misleading: it leads to confusing facts with ideas about facts. This distinction is entrenched in sound rules of linguistic usage, according to which (i) the name *fact is* given to "positive" (but not necessarily actual) existents and changes alone, (ii) negation is assigned to formulas, not to reality, and likewise (iii) generality is predicated of certain formulas. (To speak of negative, alternative, or general facts may be more than a verbal muddle: it may involve a philosophical doctrine. Thus, certain systems of Indian metaphysics seem to postulate the existence of negative facts. The naive reflection theory of knowledge and the doctrine of the isomorphism between language and reality require them too. Indeed, if every proposition reflects some aspect of the world or other, then a negative proposition must reflect a negative fact, a general proposition a general fact, a contradiction conflicting forces, and so on. Materialists, Hegelians, the early Wittgenstein and at one time Russell have held similar views. We shall abstain from reifying logical operations such as negation, alternation and generalization.)

Thirdly, a signification or meaning is often assigned to facts and indeed both metaphorically (which is harmless) and literally (which can be dangerous). Thus, we might say 'Upon expansion a gas cools down, which *means* that the gas loses thermal energy in surmounting whatever resistances, either external or internal, oppose its free expansion'. The word 'means' can and should be avoided in the preceding sentence: it is advantageously replaced by expressions like 'is due to the circumstance' or 'is explained by the hypothesis'. When a physicist carelessly writes 'the fact x means that y' he does not intend to mean that facts can mean anything in a literal sense. On the other hand traditional philosophers have often asked what the meaning of human existence, or of human history, could be. This again has been a lapse, but a dreadful one: the question was actually what the *purpose* of certain events and processes were, what the hidden master plan was. In this case the question 'What is the meaning of x?' took for granted that x had a "meaning" in the sense of purpose, so that the occurrence of x "made sense" in a certain anthropocentric or theistic picture. We shall abide by the linguistic rule according to which only artificial signs can signify: facts cannot mean anything. (For the concept of meaning see Sec. 2.2.)

A fourth confusion is generated by the expression 'demonstration of facts', not uncommon among scientists. A fact, such as hypnosis, can be shown or made apparent, and a hypothesis concerning a certain fact can be rendered plausible, but only theorems can be *demonstrated,* i.e. proved conclusively. If facts could be demonstrated in a literal sense they would be true, and their absence (the corresponding "negative facts") would be false, which would be a fifth muddle. There are no true facts, of course, but true (or false) accounts of facts. If all that is meant by the phrase 'demonstration of facts' is their monstration, exhibition or display, why not say it correctly?

A fifth misleading expression is *scientific fact.* What is meant by this locution is a fact the occurrence of which is ascertained, certified or controlled with scientific means. In this sense hypnosis is a "scientific fact" but telepathy is not. The expression, though common, should be avoided because facts are susceptible to scientific treatment but, by themselves, they are illiterate. Facts are neither scientific nor unscientific: they just are. What can be scientific or not are ideas and procedures, not their objects.

This should suffice as a linguistic introduction to the study of the most elementary source of knowledge: the observation of facts.

Problems

12.1.1. The expressions 'psychological fact' and 'sociological fact' are frequently used. Are they unambiguous? *Alternate Problem:* Examine the way W. Sellars, in *Science, Perception, and Reality* (New York: Humanities Press, 1963), criticizes the doctrine that sense experience is the ultimate court of appeal.

12.1.2. H. Spencer, in his famous *First Principles* (New York: Appleton, 1896), Part II, Ch. viii, mentions some alleged facts that should be granted before interpreting biological evolution, such as that matter (meaning "mass") is indestructible (meaning "is constant"), that motion is continuous, and that force (meaning "energy") persists (meaning "is conserved"). Are these facts? And are they as hard as Spencer believed?

12.1.3. According to A. Lalande's celebrated *Vocabulaire technique et critique de la philosophie* (see Données) scientific data are identical with "the given": in fact, the dictionary calls *data* the facts presented to the subject and serving as starting points for research. Examine this linguistic proposal and, in particular, the underlying epistemological doctrine that (i) all data are given, and (ii) what is given precedes anything that is conjectured.

12.1.4. Report on Plato' s doctrine concerning the object, its name, description, image, and knowledge. See his Epistle vii, 342A–343B, e.g. in T. L. Heath, *Greek Mathematical Works* (Loeb Classical Library), I, 391 ff.

12.1.5. Examine B. Russell's doctrine of negative and general facts, in "The Philosophy of Logical Atomism", *The Monist*, **28**, 502ff. (1918) and **29**, 42ff. (1919). *Alternate Problem:* Study and illustrate the concept of factual possibility, as involved in the chemist's calculations concerning the possibility of certain reactions or in the biologist's considerations about the possibility of certain lines of descent.

12.1.6. When we face a situation with which we are familiar we may be tempted to say that 'Facts are speaking for themselves'. And when someone makes an incredible forecast or a seemingly unfeasible proposal we are tempted to say 'Let facts speak for themselves'. Analyze these expressions. *Alternate Problem:* Analyze the phrases: 'A fact is a true sentence' and 'That book tells the basic facts about numbers'.

12.1.7. The locution 'scientific object' was popularized by the physicist A. S. Eddington who, in a famous passage in *The Nature of the Physical World* (Cambridge: University Press, 1928), wrote about two different tables: the smooth vulgar table or perceptual one and what he called the scientific table, which is a collection of molecules. Are there two different objects, a vulgar and a scientific one, or rather two different concepts and pictures of one and the same object? And is it possible to have a vulgar notion of every object accessible to science? See L. S. Stebbing, *Philosophy and the Physicists* (London: Methuen, 1937), Ch. III.

12.1.8. Any description of facts involves universal concepts, such as "thing" and "hard", and it may well presuppose law statements, such as the physical and chemical laws of conservation which "guarantee" that our desk will not disappear or turn into an Egyptian mummy while we go out of our study. Hence every description of a fact is, at the same time and to some extent, an interpretation of it in the light of some body of knowledge. Does this efface the distinction between fact and fact interpretation?

12.1.9. Examine the thesis according to which the sciences of nature deal with human experience—with phenomena—rather than with nature—with objective facts. *Alternate Problem:* Analyze the once popular doctrine, expounded by A. N. Whitehead, that physical objects are just systems of point events.

12.1.10. Discuss the doctrine of G. Ryle in *Dilemmas* (Cambridge: University Press, 1954), Ch. VII, that there is no phenomenon of perceiving (e.g., seeing) but only a wrong use of the verb 'to perceive' which deludes us into believing that the word denotes a process that can be correlated with, say, physiological processes. *Alternate Problem:* Is linguistic analysis a substitute for or rather propaedeutical to epistemology?

12.2. Observability

Observation proper can be characterized as *purposeful* and *enlightened* perception: purposeful or deliberate because it is made with a given definite aim; enlightened because it is somehow guided by a body of knowledge. The object of observation is a fact in either the external or the inner world of the observer. In particular, the external object may consist of marks on a paper: reading, which involves both seeing and interpreting, is in fact a typical case of observation. If the fact occurs in the observer, it may have to be introspected: an onlooker could only observe behavior acts accompanying the internal fact. Whether external or internal, the object must be perceptible if we are to speak of direct observation. We shall agree to use the term 'observation' in its strict sense of *direct observation.* The "observation" of molecular collisions or of other people's feelings is indirect: it is a hypothetical *inference* employing both observational data and hypotheses. At most it can be named *indirect observation.*

There is no observation in itself, i.e. observation of no object whatever, just as there is no observation without an observer, whether human or not. If someone says '*x* is observing', he means that there is at least one (actual or possible) object *y* that is being observed by *x*, i.e. "$(\exists y)$ (x is a subject & y is a concrete system & x observes y)". This could be generalized to all experience: experience is not a self-subsistent object but a certain transaction between two or more concrete systems at least one of which is the experient organism. Experience is always of somebody and of something. Accordingly experience sentences are of the form '*x* experiences *y*', and not just '*x* experiences'. This elementary point had to be made in view of the doctrine that modern physics does not deal with physical objects but with experiences in themselves.

The observation of an event, such as the passage of a girl or the click of a Geiger counter, is not an isolated and raw percept but a bundle of momentary sensations, memories, and perhaps expectations of similar events. Habits and beliefs tie the given percept to both recalls of past experiences and expectations of coming events. No percept occurs in a vacuum: if it did we would be unable to discriminate among observations, let alone interpret them. Unlike rough sensation, observation is selective and interpretive: it is selective because it is purposeful and it is interpretive because it is enlightened. If we sit lazily at a café table on a sunny sidewalk, we experience a sequence of impressions which we may care neither to order nor to interpret—having no aim other than enjoying such experiences. If, on the other hand, we sit at a control panel of an electron microscope with the aim of studying the structure of certain cell vessicles, we shall select just those percepts we assume relevant to our aim and shall soak them in our specialized knowledge. (Our aim is, of course, to win abundant and accurate relevant information. The emphasis is on relevancy: excess of detail, if irrelevant, can be as encumbering in science as it is in everyday life. The amateur naturalist will record heaps of minute facts which the field biologist will neglect; the latter will on the other hand record a few essential observations, many of which will have escaped the untrained watcher for want of a definite aim marked off by a conceptual system.) Moreover, we shall at the same time interpret what we see or part of it: what to the layman appears as just a dark patch, the morphologist may interpret as a vessicle. Such an interpretation may, of course, be wrong, i.e. the result of our theory-laden observation may be false. At any rate, interpretation is more than unavoidable in observation: it is welcome since we have to interpret what we perceive if we wish to make a selection of relevant perceptual objects, of perceptual objects relevant to our ideas and aims. If we have neither we may record nothing or everything that falls within our perceptual horizon: it will make no difference so far as science is concerned.

In observation we may distinguish the act from its outcome. But the outcome of an act of observation is just the last link of a complex process. This process may be analyzed into various steps: (i) becoming aware of the object, (ii) recognizing it in broad outline, and (iii) describing it. The first stage is that of the *presentation* of the object, i.e. its perception by the observer. For example, an astronomer sees a light spot when he aims the telescope at a given direction (expecting to find something interesting). In general: The subject *w* has the perception *x* of an object y under the circumstances *z*. The nature of the object is still unspecified. The second stage is that of a *preliminary interpretation* of what the subject has perceived: the nature of the perceived object is thereby determined in outline. For example, the astronomer interprets the spot he has seen as a flare emerging from the solar corona (he may even feel it as a threat, which it is). In general: The subject *w* proposes an interpretation *v* of the perception *x* as caused by the object *y* under the circumstances *z*. The same object is now identified to some extent.

The third stage is that of *description.* every proposition in it is a *datum*—a raw datum for the time being. For example, the astronomer describes in more or less technical language what he has seen and interpreted—i.e. what he has "read" with his telescope. He refrains from describing his own experiences (unless he has a fit of either poetry or craze); he issues what he understands is an objective account of his observation. In giving this account he will screen out his feelings and he will employ certain technical terms each designating a special theoretical or empirical concept. In general: The subject w builds the description (image, representation) *u* of the object *y*, that causes the perception *x* in *w* under the circumstances *z*, and *w* builds *u* on the basis of his preliminary interpretation *v* and with the help of the explicit background knowledge *t*. This final description or set of data may, in turn, be the raw material for the theoretical astronomer, who may supply a *theoretical interpretation* of that raw material, i.e. a deeper and more detailed interpretation than the one contrived by the observer: for example, he may interpret the solar flare as the outcome of a thermonuclear reaction.

The layman may miss all three stages: he may not even *see* what the trained eyes of the specialist perceive. Even if he does see the object he may not *recognize* it, because recognition or preliminary interpretation is the fitting of something into a preëxistent framework and if this framework is missing the observer will be at a loss. Still, the layman may correctly recognize the object but then he may fail to *describe* it accurately and a fortiori to propose a theoretical interpretation of it. Conversely, the specialist may "read" too much in what he perceives. (Even Leonardo da Vinci, the best observer of his age, depicted not only what he saw but also what he had learned in books: for example he drew a second urethral canal along which the soul was to flow into the embryo during conception.) In short, scientific observation is a refined way of apprehending the perceptible world and checking our ideas about it; it is influenced by scientific knowledge and therefore it can be indirect as well as direct, and accurate as well as dead wrong; but, the results of it being public, it can be controlled and eventually corrected by any qualified worker.

Essentially five items can be recognized in the observation process before its results are reported: the *object* of observation, the *subject* or observer (including of course his perceptions), the *circumstances* of observation (or environment of object and subject), the observation *media* (senses, auxiliary instruments, and procedures), and the body of *knowledge* in which the former items are related together. Both the observation media and the body of relevant knowledge can be grouped under the name of observation *tools* (concrete and conceptual), which leaves us with four items. (They are means for the observer, not however for the instrument designer or for the theoretician.) Observation statements have accordingly the following form: "*w* observes *x* under *y* with the help of *z*").

So far we have dealt with actual observation; let us now elucidate the concept of *observability*. To a first approximation we may stipulate that a fact is observable only if there exist at least one subject, one set of circumstances, and one set of observation tools, such that the fact can appear to the subject armed with those tools under those circumstances. This elucidation depends on the psychological concept of appearance and is therefore unsatisfactory; we should exclude the possibility of declaring ghosts to be observable just because someone claims in good faith that they appear to him. In other words, we should specify what is *objectively observable. A* first step in this direction is taken by replacing the human observer by a physical recording device, such as a camera or a thermometer. We accordingly stipulate that x is *observable* only if there exist at least one recording instrument w, one set of circumstances y, and one set of observation tools z, such that w can register x under y helped by z. We have not hereby eliminated the subject but rather the possibility of some of the subject's perceptual delusions: someone has to make the interpretation of the signals registered by the physical recorder. The latter might have a built-in code enabling it to make the interpretation, but this would not eliminate the flesh and bone observer: he would act as the apparatus designer or builder. Physical recorders, whether automatic or not, do not replace the human operator but make his job lighter and the results public and less subject to unwanted fluctuations, hence more accurate and reliable. All we have done so far is to add a public control device to the human observer. At any rate, the existence of the observation apparatus, the additional tools and the circumstances are necessary for the given fact to be observed: not being sufficient they do not ensure that it will be observed—this being why we wrote 'x is observable only if...' rather than the other way around. (A physical condition for the observability of a some trait of a concrete system is that the latter interact with an observation instrument. Now, the recording system may not be sensitive enough: in other words, the interaction may not be strong enough to overcome the threshold peculiar to the apparatus, above which it will respond to the stimulus. In this case the object under observation would be unobservable with the tools concerned, and the technical problem of devising and building a more sensitive apparatus would be posed.)

We have just insisted on the public or intersubjective character of the results of scientific observation, a necessary condition for science to be objective. (It is not sufficient, though: collective delusions have been known to occur and, what is more common, unanimous mistakes due to the use of wrong theories must be reckoned with. In any case, intersubjectivity does not coincide with objectivity but is a condition of it.) The public character of scientific observation is frequently expressed in the form of a rule to the effect that observation results must be *reproducible* by other workers in similar conditions. But this rule is too strong, since it excludes all nonrepetitive facts. The outburst of a nova, the eruption of a vulcano, a political event, or an emotional crisis may not be reproducible at will and consequently their observation may not be replicated. All we should require is that observation results of *the same kind* (e.g., about volcanic eruptions) should be reproducible by qualified observers, otherwise kept in suspense. Exact duplication is desirable but not always attainable. And in any case it does not guarantee objectivity: independent observers may make the same wrong observations because they employ the same inadequate equipment or the same false hypotheses. There are no guarantees of accurate observation: all we can do is to multiply the controls, i.e. the empirical and theoretical procedures for the detection and correction of errors of observation.

How are observation and physical existence (reality) related? A number of doctrines have been or could be proposed. In a nutshell these are:

Doctrine 1: "Observability is necessary for physical existence—i.e. if anything exists it is observable. Equivalently: if anything is unobservable then it does not exist." This doctrine is too narrow, as it prohibits the existence of what we have not been able to conceive as observable. After all, observability is not a property inherent in things but is a complex relational property involving our cognitive equipment—one which certainly is historically conditioned.

Doctrine 2: "Physical existence is necessary for observability—i.e. if anything is observable then it exists. Equivalently: if anything is nonexistent then it is unobservable." This doctrine is too liberal, as it warrants the existence of whatever can conceivably be observed. It only takes some imagination to devise tools and conditions under which the ghostliest thing might be observed.

Doctrine 3: "Physical existence is necessary and sufficient for observability—i.e. anything is observable if and only if it exists." This doctrine is false since it is the joint assertion of Doctrines 1 and 2 which were independently found to be false.

Doctrine 4: "Actual observation is necessary for physical existence—i.e. if anything exists then it is observed." This doctrine is even narrower than Doctrine 1 and a fortiori it cannot be accepted either.

Doctrine 5: "Physical existence is necessary for actual observation—i.e. if anything is observed then it exists." If taken literally this doctrine warrants the inference of physical existence from observation, without making room for the possibility of wrong observation. All that can be concluded from the actual observation of x is the *likelihood* of the hypothesis that x in fact exists: from favorable observation reports we "conclude" a proposition concerning the truth value of another proposition that has the rank of hypothesis if x happens not to be directly observable. But this qualification can be taken care of without modifying the core of Doctrine 5, by just regarding both observation reports and existence assumptions as *corrigible*. We shall then adopt the following version of Doctrine 5 regarding the relation of observation to reality:

Doctrine 6: "If anything is in fact observed with the help of competent empirical and conceptual tools, then the hypothesis of its physical existence may provisionally be entertained." Should anyone feel that this doctrine is overcautious, we might remind him that demons, ghosts, flying saucers and so forth have in fact been observed many times, which has not established the hypothesis that they exist. We may call them wrong perceptions or wrong inferences from right perceptions if we wish, but at any rate they were observations. Should we feel reluctant to call them by this name, we had better recall that observation, and particularly so scientific observation, is far from being a purely physical relation between two systems, the object and the observing system: observation is a process in which the habits, expectations, know-how and scientific knowledge of the observer play a decisive role, if not during the actual recording of impressions, which can be automatized, during the design of the observation and during the interpretation of its outcome.

Moreover observation reports, far from being accepted at their face value, are confronted with other reports of the same kind and are analyzed in the light of theories in order to see whether they are possible at all. As was said before, scientific observation is purposeful and enlightned perception, a selective and interpretive operation in which ideas have at least as much weight as sense impressions: this makes it relevant to conceptual knowledge and, at the same time, constitutes a source of error. As a compensation, errors of observation can in principle be detected and subsequently corrected to some extent. In short scientific observation, far from being the solid rock upon which scientific knowledge is built, is but a changing component of it: it is as fallible and corrigible as scientific theory, with which it is constantly interacting.

The role of ideas in observation is apparent in the case of indirectly observable objects, i.e. in the case of inference from symptom to cause by way of hypothesis. The rotation of an anemometer informs us about the velocity of wind, in itself unobservable; the click of a counter, about the passage of a cosmic ray; the anomaly in a star's motion, about possible planets perturbing its path; blushing, about emotions of a certain kind—and so on. In every such case an event or physical property is regarded as a symptom or natural sign of another object which is occult to direct perception. To find indicators or objectifiers of occult things, properties and events, is a major task of science, one which precedes observation—which in turn checks the conjecture that such directly observable objects are in fact manifestations of others. An observable property may be said to be a *reliable indicator* of another property only if there exists a definite constant relation between the two, such that any observable change in the former can be interpreted as, or attributed to a definite change in the latter. The property the existence or value of which is inferred in this way is called *indirectly observable.* Indirect observation, then, is possible through theory alone. The more precise the theory, the more accurate such an inference will be. This is one reason for valuing mathematical theories: because in this case the relations take precise forms. (More on this in the next section.)

The dependence of observability upon theory raises interesting questions. One of them is: How certain can we be in concluding that a certain transempirical property is observable or unobservable in principle? If a property is declared observable—indirectly observable, that is—and efforts to observe it fail, something may be learnt. On the other hand, if it is declared to be unobservable, whole lines of research are doomed beforehand. For example, until a few years ago one taught that atoms would forever remain invisible, even with the assistance of the most powerful electron microscopes. Fortunately E.W. Müller (1951) doubted this and, by operating an ion microscope at low temperature, took photographs of individual atoms; he was even able to see one atom after another breaking away from the crystal lattice. This is, to be sure, indirect observation, but so is most scientific observation.

In the preceding discussion the so-called *causal theory of perception* was taken for granted. According to it, our perceptions are never either spontaneous or at random but are lawfully elicited by extra-perceptual objects, whether within our bodies or outside them. We shall regard this hypothesis as satisfactorily corroborated by psychology, although some room should be made for a small random component ("noise"), particularly since we know that even the most perfect recording device is affected by it. The causal theory of perception presupposes in turn the philosophical (ontological) hypothesis that there are objective facts, that these are ontologically prior to experiential facts, and that every phenomenon is a fact appearing to a subject. Accordingly, phenomena are primary as regards knowledge, not however ontologically: we must start with phenomena and ideas if we want to know anything about the external world, but reality is self-sufficient and is the object of our research. This thesis we regard as both presupposed and corroborated by scientific research. Suffice it to mention two scientific supports of the hypothesis of the primacy of objective fact over phenomenon. First, one of the epistemological teachings of relativity theory is that the basic laws of facts are absolute (i.e. dependent on no reference frame and, in particular, on no observer), whereas observable facts (phenomena) are relative to the reference frame adopted in each case (see Sec. 6.5). Second, we do not look for laws of phenomena because there are none: appearance has no structure. The stream of experience constitutes a random time series, whence patterns are sought and found beyond appearance, in a reality that is supposed to be there, that must be hypothesized since it cannot be directly perceived. In other words, science itself has debunked phenomena and enthroned in their place objective facts—mostly unperceivable—and their invisible patterns. Phenomena are ontologically derivative, however much epistemologically primary they can be as purveyors of problems and evidence. Observation, then, is a gate to fact and law, the objects of scientific inquiry. Accordingly, the concept of observation cannot be regarded as a primitive in terms of which everything else is to be elucidated: far from this, observation is to be analyzed as a complex process taking place on all the levels crossed by man.

Our next task will be to analyze indirect observation somewhat closer.

Problems

12.2.1. Do scientists observe and describe the totality of the given?

And do historians care for every document and try to reconstruct every minute in the past? *Alternate Problem:* Comment on Macaulay's following theses concerning the historians' work. (i) The historian "must be a profound and ingenious reasoner. Yet he must possess sufficient self-command to abstain from casting his facts in the mould of his hypothesis" (ii) History must be true, but selectively true rather than wholly true: "Perfectly and absolutely true it cannot be; for, to be perfectly and absolutely true, it ought to record *all* the slightest particulars of the slightest transactions—all the things done and all the words uttered during the time of which it treats"—which is practically impossible. (iii) "He who is deficient in the art of selection may, by showing nothing but the truth, produce all the effects of the grossest falsehood". See *Miscellaneous Works of Lord Macaulay,* Ed. by Lady Macaulay (New York: Harper & Brothers, s.d.), pp. 154, 161 and 162 respectively.

12.2.2. In the early 1960's observations of sight through the face skin ("dermooptical phenomena") were reported: blindfolded subjects were said to read texts without using their eyes. Observations were accumulated although no possible mechanism was in sight, until fraud was discovered. Discuss this episode in the light of the official philosophy that observation is the basis of science. *Alternate Problem:* Examine the views of R. Avenarius and E. Mach that (i) perception is not *of* something and (ii) things are more or less stable complexes of sensations (The latter characterization of "thing" was first suggested by Berkeley and adopted by J. S. Mill and R. Carnap.)

12.2.3. The only directly measurable magnitudes in astronomy are star brightness and the angle subtended at the telescope by two celestial objects. All other magnitudes, in particular distances and masses, are inferred with the help of theories and the above-mentioned angle data. Can astronomy accordingly be regarded as an *observational science?*

12.2.4. Look at the fascinating photographs of atoms taken with the field ion microscope: E. W. Müller, J. *Appl. Phys.* **28**, 1 (1957), and reflect on the "proofs" that atoms are in principle unobservable. *Alternate Problem:* Analyze the camera attached to a telescope as an observer by proxy.

12.2.5. Consider the following cases of unobservability in principle. (i) Very distant stars and even whole galaxies are presumably forming right now, which do not yet interact with our planet: the light and the gravitational field they are emitting will reach us, if at all, in the distant future. (ii) It is logically possible that other universes exist which are self-contained, i.e. in no interaction with our own. Examine the two conjectures as to testability and ground.

12.2.6. Is there a foolproof observational procedure whereby empirical information can be secured with complete certainty? In case no such error-free observation procedure exists, would it be possible to design one? Or would it be more reasonable to improve the current procedures for spotting and correcting observation mistakes? See G. Ryle, *The Concept of Mind* (London: Hutchinson, 1949) p. 238. *Alternate Problem:* Study the problem of the alteration of the observed object by the very observation process, as in the case of the mitochondria, which were destroyed by the more common fixatives, rich in acetic acid. Is it possible to prevent such an alteration?

12.2.7. Around 1950 certain astronomical observations seemed to refute the predictions of some relativistic models of the universe, which were given up in favor of alternative theories, particularly the steady-state or continual creation theory. It was subsequently shown that the observations had been wrong, and relativistic cosmology was favored again. Draw some moral—as the Duchess would do.

12.2.8. Speculating on extraterrestrial life is often regarded as science fiction since there are no *direct* observational data concerning planetary systems other than our own. The sole indications are so far certain theories of star formation and the anomalies in the motions of certain stars. Does that condemn "exobiology"? See G. G. Simpson, *This View of Life: The World of an Evolutionist* (New York: Harcourt, Brace & World, 1964), Ch. 13.

12.2.9. Examine the argument that the attempt to show the reliability of observation is circular, because a decision concerning observation reports can only be grounded on further observation reports. See H. Mehlberg, *The Reach of Science* (Toronto: University of Toronto Press, 1958), p. 92.

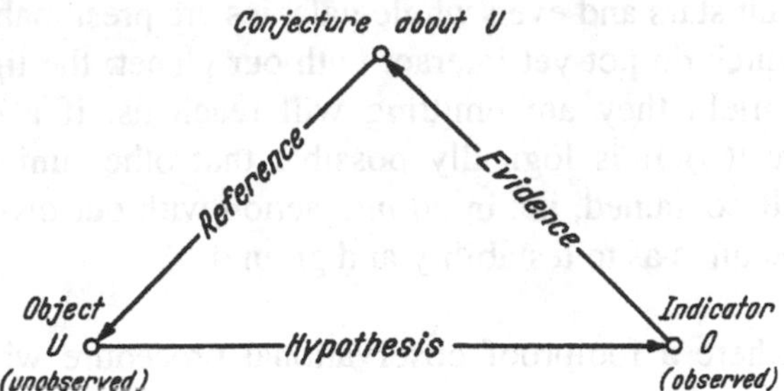

Fig. 12.2. The object-indicator relation, a physical relation, is expressed by a hypothesis enabling us to infer the object from observations made on its indicator.

12.2.10. Visit a nuclear physics laboratory and ask someone in the staff what he is observing. Then try to fit his reply into a phenomenalistic epistemology. Hint: do not waste your time. *Alternate Problem:* Most microscopic observation involves producing changes in the observed object, so that after all something different was being observed. How could anything be inferred about the original observandum? And what changes have been brought about by the invention of the phase contrast microscope (F. Zernike, 1935)?

12.3. Indicator

Most of the facts we know something about are vicariously observable, i.e. they can only be inferred through the intermediary of perceptible facts by way of hypotheses. Thus, wind is not directly observable but rather inferred from watching bodies supposedly moved by it, or feeling its pressure on our body: in either way we conjecture the existence of a gust of wind and, if we are interested in a more accurate evaluation of our surmise, we have to check it because we might have inferred wrongly (the motions might be due to an earthquake or to the observer's drunkenness). Similarly, if we see someone blushing we attribute it to shame, embarrassment or rage, according to the circumstances accompanying the fact (additional information) and our knowledge of psychology. And we hypothesize that a political event is taking place if we see tanks surrounding public buildings and people running in the streets.

In all the preceding cases we *objectify* an unobservable fact by positing its lawful relation to some perceptible fact (or set of facts) serving as an *indicator* of the former. In other words: we make hypotheses concerning unperceived facts and test them by means of evidence consisting of data about directly observable facts on the assumption that the latter are concomitant with or effects of the former. That such a lawful relation between observables and unobservables should hold is, of course, a *hypothesis*—although it is often called an *operational definition.* Fig. 12.2 summarizes what has so far been said.

The above can be made somewhat more precise with the help of the function concept. Let U be a member of a set of possible values of an unobservable property such as atmospheric pressure, and O a member of a set of values of a directly observable property such as the height of a barometer column. Suppose further there is a theory stating a functional relation between the O's and the U's: $O=F(U)$. Then by inverting this relation and measuring O we can infer the corresponding U values. In the formula "$O= F(U)$" both U and O are theoretical concepts but O stands for a directly observable property whereas U is an interpretive concept. And the relation F itself is a hypothesis that must somehow be both grounded and confirmed. In the case of wind O might be the measurable angular velocity of the anemometer and U the unobservable wind velocity. In the case of biological evolution O might be the degree of anatomical resemblance among two species and U the corresponding degree of species affinity. And in the case of a rat subjected to instrumental conditioning, O might be the number of bar pressings per minute and U the drive strength—e.g., the sugar water appetite. If O is a quantitative concept (a magnitude), it is often called a *measure* of U—not to be confused with a *measurement* of U.

When a scientist claims to have discovered a new "effect" he is expected to exhibit evidence for it, i e. to offer objective information supporting his assertion which, if interesting, will be about unobservables. Now, evidence can be direct or indirect, either in a logical or in an epistemological sense. A *logically direct evidence* regarding a proposition is a datum that can be compared with the latter—or with an instance of it if the proposition is general. Thus any new measurement of the speed of light yields direct evidence for or against the previously found values; and it is also evidence for the general hypothesis that light velocity is an absolute: it is a case of the latter. A *logically indirect evidence* for a proposition is one matching it without being comparable with it. (In other words, indirect evidences are not extensionally related to the corresponding hypothesis unless intermediate statements are added. In particular, their logical extension is not contained in the extension of the hypothesis: they are not substitution instances of the latter.) For example, comparative anatomy, physiology, and embryology, provide logically indirect ("collateral") evidence in favor of the hypothesis that man descends from certain anthropoid primates. Only the hypothetical reconstruction of any such anthropoid primate, on the basis of a few extant bones, provides a logically direct evidence in support of the hypothesis.

Indirect evidence is much more important in science than direct evidence: direct evidence serves only to support or undermine low level empirical generalizations, and these are more typical of ordinary knowledge than of advanced science. In fact, no logically direct evidence can support or undermine a theory's axioms: the latter's low level consequents must be worked out and checked with empirical data. Moreover, whatever evidence is claimed to validate or invalidate a theory proper is not only logically collateral but also *epistemologically indirect,* in the sense that transempirical hypotheses can only be tested through checking their observational consequences. Consequently, the evidence relevant to a theory will be different from the theory's referent: otherwise the theory would not deal with transempirical objects. For example, the evidence relevant to phylogeny is nonphylogenetic, since we cannot observe the mutations, adaptations, and other processes that determine the characteristics of the descent: all our data are poor documents, such as a few old bones, from which hypotheses are "concluded" In short, the axioms of scientific theories do not, unlike the empirical evidence relevant to them, refer to directly observable facts. Therefore the appropriate symptoms or indicators of unobservables must be devised—which raises a cluster of methodological problems.

The clothes a man wears, the car (or donkey) he drives, and his manners, belong to a constellation of objective indicators of social status—in itself unobservable. Therefore, given the hypothesis "*c* belongs to the rank *R* in the social hierarchy prevailing in his society", we may test it by observing certain properties of the individual *c*: such data will be used as evidence relevant to our hypothesis because we accept certain generalizations regarding the relation between the indicators (wrongly called "status symbols") and social status. In science such relations among indicators and the corresponding unobservables are (i) *postulated* by theory and (ii) *independently checked* whenever possible. For example, the identification of an animal as a member of a given species, or the inclusion of a species in a genus, are made on the basis of clusters of indicators of half a dozen different kinds: anatomical physiological, biochemical, ecological, biogeographical, and behavioral. Anatomical features are the most readily observable of all, and practically the sole available ones in the case of fossils, but the others are becoming more and more reliable as more is known about them. In any case they all provide evidence for identification and classification purposes on the basis of certain law statements, such as "Each species has its own ecological niche". (Incidentally, the property of belonging to an animal species may be regarded as a macroproperty even in the case of microorganisms. It is interesting to note that some of the indicators of taxonomy, such as the histological and biochemical ones, are microproperties often unobservable in practice. This goes to refute the belief that only microproperties require indicators and that the latter must always be macroproperties.)

As another illustration take the objectification of dreams. Dreams were not subjected to scientific study until objective dream indicators were devised: before that, from the ancient Egyptians to psychoanalysis, dreams were left to wild interpretation. Two objective indicators of dreaming are the rapid eye movements, detected by means of sensitive electrodes placed in the eyelids, and disturbances in the brain wave patterns as shown by the electroencephalograph. These data-gathering techniques enable the investigator to establish whether the subject is in fact dreaming, but so far they do not reveal the dream's content or story. Yet there is no reason to doubt that the latter will eventually be objectified as well. At any rate, just as in the case of the neutron detector or the lie detector, the future dream detectors will be designed on the basis of certain hypotheses, and eventually theories stating certain definite relations among unobservable and observable facts, relations of the form "$U=F(O)$". Because this is, by the way, one of the functions and achievements of scientific theories: to bridge the gap between unobservables and observables by rendering some of the former indirectly observable or, rather, objectifiable.

When the postulated relation between the unobservable variable U and its observable manifestation O is not somehow *justified by a theory* and *independently testable,* the claim that a certain U is in fact objectified by a certain O is hardly more than either a hunch or a dogma.

This is, unfortunately, the case of most psychological tests used in clinical psychology at the time of writing: they are claimed to objectify certain mental functions and personality traits but they are seldom tested independently, let alone embedded in a theory capable of explaining how they work or why they should work. (See Sec. 11.3.) In the best of cases they are empirically justified. Thus, intelligence tests are justified by follow-up studies on scholastic achievement but, then, one is not sure whether one is measuring native intelligence, memory, tenacity, or some combination of these traits.

The indicator hypotheses "$O=F(U)$" are not arbitrary: they are hypotheses which, upon corroboration and theorification, become law statements. In no case are they conventions—for example definitions, as is claimed by operationalism. The operationalist, anxious to avoid metaphysics, tends to regard every objectification (or object-indicator) relation as a definition. For example, when he postulates that hunger can be measured by the time of food deprivation, he claims that the hypothetical construct "hunger" is thereby *eliminated* in favor of a measurable variable ("duration"). But this is conceptually wrong, for not every relation between variables constitutes a definition (see Sec. 3.6). And it is empirically wrong as well, because the particular (linear) relation between hunger and deprivation time mentioned a while ago is an oversimplification that can be kept as a rough approximation only as long as no more reliable (e.g., physiological) indices of hunger are found. To regard object-indicator hypotheses as conventions, hence as fiats in no need of either theorification or independent empirical checking, is as dangerous as to claim the possession of unique faculties (e.g., clairvoyance) or tools (e.g., crystal balls) for the penetration into the unseen.

Now that we have taken a look at the object and means of observation we may examine the final result of this operation.

Problems

12.3.1. Mention actions that externalize feelings and thoughts and can therefore be employed to "read" the latter. *Alternate Problem:* Some madmen learn to behave normally in order to elude internment. Are there means to detect their hidden abnormalcy? And what is the bearing of this case to behaviorism?

12.3.2. Examine how any of the following instruments are supposed to work. (i) The voltmeter (for objectifying electric field strengths). (ii) The electroencephalograph (for detecting brain injuries). (iii) The polygraph.

12.3.3. Analyze the technique of photogrammetry, in particular for inferring soil conditions (and even large scale scars left by human settlement) from aerial photographs. *Alternate Problem:* Examine the hypotheses underlying the claim that share values are indicators of a country's economy.

12.3.4. Compare alternative hunger indices (measures). See N. E. Miller, "Shortcomings of Food Consumption as a Measure of Hunger", *Annals of the New York Academy of Sciences*, **63**, 141 *(1955)*. *Alternate Problem:* Analyze the technique of photoelasticity for visualizing inner stresses in materials.

12.3.5. Discuss the merits and demerits, as to sensitivity and reliability, of the various indicators of thinking: speech, writing, *EEG* patterns, upper lip action potential, etc. *Alternate Problem:* Study the general problem of justifying (validating) indicators or measures in psychology.

12.3.6. The objectifier or index should not react significantly on the event or process it helps manifest. For instance, if a measuring apparatus is connected to a system it should consume only a small fraction of the latter's energy. Why? And is this precaution observed by psychologists who use indirect introspection, as e.g. when attempting to disclose problem-solving thinking by asking the subject to verbalize the various steps in the strategies he chooses? *Alternate Problem:* Can a change in scientific theories force changes in the list of observables? See A. D'Abro, *The Decline of Mechanism* (New York: Van Nostrand, 1939), p. 812.

12.3.7. Animals under severe stress have their internal organs remarkably enlarged and deformed. Examine such anatomical indicators of strain. See H. Selye, *The Stress of Life* (New York: McGraw-Hill, 1956).

12.3.8. Some states, such as pain, are untransferable, i.e. private. Does it follow that they cannot be objectified and must therefore remain beyond the reach of science? Do physiologists and psychologists pay any attention to this claim of some philosophers of mind or do they rather devise ways of objectifying such inner states and so independently checking introspective reports on them? *Alternate Problem:* Suppose mental states were not just private (untransferable, unshareable) but also occult (inscrutable), hence unknowable from the outside. Then if someone said 'I am in pain' he would not be expressing a proposition, since propositions must be true or false whereas the above is just a complaint—hardly more than a groan. What is wrong with the yelp doctrine of pain and pleasure?

12.3.9. Consider the evidence statements supporting or undermining the hypotheses "It is very humid today", "He was proud of his children", and "That country is stagnant". Do the evidence statements contain the constructs occurring in the hypotheses? What kind of evidence are they? And how could we go about justifying the claim that they are pieces of evidence at all? Or can we afford to ignore this question?

12.3.10. Examine the following statements. (i) "Thermometers record temperature variations". (ii) "Thermometers record temperature variations of the systems to which they are attached". (iii) "Thermometers cause temperature changes in their environment". (iv) "Thermometer readings and environmental temperatures happen to be roughly the same". (v) "Thermometers generate thermal states". (vi) "Thermometers define the temperature concept". *Alternate Problem:* Study the general problem of the relation between transempirical properties and their observable manifestations. Hint: Avoid the usual confusion between "dispositional" and "unobservable".

12.4. Data and Evidence

The ideas expressing the outcome of a run of observations are a set of data. A datum is a singular or an existential factual proposition such as, e.g., "Rat 13 was injected 1 mg nicotine on the first day". Suppose now we keep rat 13 under observation and eventually discover that it has developed lung cancer. We may then issue the following observation report: "Rat 13 developed lung cancer around 20 days after being given a 1 mg nicotine dose". This too we shall call a datum, although it has not been derived from direct observation: it required the previous dissection of the animal and a microscopic examination of its lung tissues. Lung cancer may have been conjectured from certain perceptible symptoms but this hypothesis was subsequently tested with the help of the microscope and a body of knowledge concerning cancer cells. Even the initial datum concerning the nicotine injection involved some knowledge beyond immediate experience: the experimenter did not report that he had injected some of the contents of a flask labeled 'Nicotine'; instead, he took the risk of assuming that the label was correct—which assumption might have been disproved by a chemical analysis. If the experimenter had reported: 'I filled a syringe with a solution of what looked as nicotine to me and then seized what I took to be rat 13 and gave it what I felt was the needle', he would have portrayed a slice of his private flux of experience—but then none save perhaps his psychiatrist would have been interested in his report which would certainly be a datum but not a *scientific datum.*

There are several reasons for not regarding reports on private experiences as scientific data. First, because the experimenter is supposed to report on objective facts *(objective referent).* Second, because his personal experiences are irrelevant to such objective facts. (They may be of interest in psychology as long as they can be related to and so controlled by objective data concerning either behavior or physiological events.) Third, because his reports are supposed to be controllable by his colleagues *(public control),* whereas his personal experiences are not so easily checked. Fourth, because the scientist's data are supposed to be charged with interpretation (see Sec. 12.2) and even couched, at least partly, in a theoretical language (Sec. 8.4). If scientific data were not required an objective referent, a public control, and a minimum of interpretation in terms of accepted theories, they could be invented at will and they would be irrelevant to the ideas they are supposed to support or undermine. Consequently the so-called *sense data* are not scientific data. Pure perceptions, if there are any, are not the building blocks of science but a problem for psychology: science—including psychology—begins with expert questioning and conjecturing rather than with feeling. In short, science is interested in impersonal data concerning objective (though perhaps hypothesized) facts, and moreover in such data as can "make sense" in some body of knowledge: in systematizable data.

Another peculiarity of scientific data is that, contrary to the etymology of the word, they are not given but must be produced, very often by hard work. Physical objects of a sort—perceptible bodies and some of their changes—are given to us as such, i.e. as physical objects. Thus, sound waves of certain characteristics are perceptible, but they are not perceived as waves: the wave nature of sound is a hypothesis, hence data on the wavelength and amplitude of a sound wave do not express experienced facts (phenomena). As *objects of knowledge*, physical objects are reconstructed by us on the basis of (i) their appearance, (ii) our handling of them, and (iii) our inventing and subsequent testing of models of them. (Phenomenalism has stressed the first aspect, pragmatism the second and idealism the first part of the third one: each of these philosophies saw a part of the whole truth.) Cognitive objects are not placed ready-made in our senses or in our intellect but are built by both, and data are at once the result and the raw material of this construction.

This is not to deny the existence of objective facts. They are there by themselves, at least most of them. But in order to catch them, to turn them into cognitive objects, the scientist may have to use devious means, all of which include ideas about which he cannot be (or he should not be) completely sure. Consider the physicist who wants to study the behavior of electrons in a metal or the historian who wants to reconstruct the behavior of the Sumerians. None of these facts and patterns are accessible to the senses: all there is is a set of data (physical information, historical documents) which, if placed against a background of specialized knowledge, can suggest (psychologically, not logically) certain hypotheses that must be tested by searching for further data. These further data, if produced ("gathered"), will serve as evidence in favor or against the hypotheses concerned. (Notice once again that the method of the physicist is the same as that of the historian; they differ in the special techniques for producing data and in the power of their theoretical models.)

Empiricists like Bacon and Comte, and intuitionists like Bergson and Husserl, have demanded the gathering of data without previously making any assumption: in this way "pure" and "hard" data, untainted by ideas and therefore safe, would be gathered. Actually nobody searches for anything—in this case data—without having in mind a fan of possibilities regarding the properties of what he is looking for. Even subhuman animals search against a background of expectations.

Otherwise (i) the searched-for thing would not be recognized when met (i.e. it would never be found) and (ii) we would not know how to search. The more one ignores the likely look of the sought object the more the imagination has to be exerted; and the more one knows about it the more hypotheses are available to guide the search. Take for instance the radioastronomical search for signals from possible extraterrestrial civilizations: people would not know what to look for unless they made definite assumptions concerning extraterrestrial intelligent beings, such as that they would communicate if at all with radio signals (hence: Use radiotelescopes!), that they may gravitate around any star (hence: Scan the whole sky!), and so on. Of course these assumptions may be false, but if they were not made we would have no hope of ever producing empirical evidence concerning extraterrestrial rationals.

Every evidence is a datum but not every datum is *evidence*. The evidential character of a datum is not an intrinsic property of it: evidence are such data as are relevant to some idea. For example, atomic theory leads us to suspect that charged particles traversing a photographic plate may leave a trace in it. If we in fact find such tracks upon microscopic examination, we interpret them as effects produced by charged particles and we thereby obtain evidence of a new kind in favor of the atomic theory. The latter predicts then its own evidence—and furthermore an evidence that cannot support the continuum theories of matter. Similarly, a fossil is interpreted as the remains of a deceased organism and, moreover, of one probably belonging to an extinct species: this interpretation is made in terms of the evolution theory. In turn, we regard every fossil finding as relevant to the theory of evolution.

In short, we believe a datum constitutes an evidence in favor of a theory and we assign the theory some credence because it accounts for or even predicts that evidence. Of course this is circular: we validate a datum by means of a theory and conversely, which is as though an accused were to judge himself. *Empirical confirmation is circular* and therefore has no value by itself: recall Alternate Problem 8.4.10. If a theory is confirmed just by the sort of evidence it suggests itself, this confirmation is worthless. In order to carry some weight, empirical confirmation should be complemented by the satisfaction of additional conditions, chiefly the theory's continuity with the bulk of science. It is possible to contrive any number of theories of ghosts and to confirm

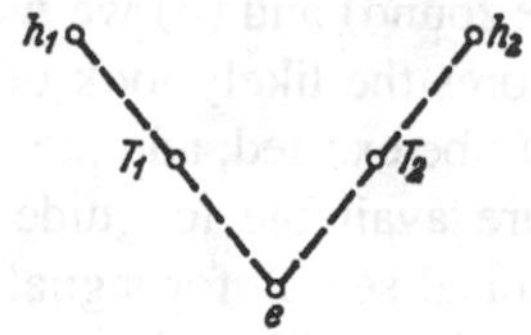

Fig. 12.3. The relativity of data: one and the same datum points ambiguously to two different hypotheses via two different theories.

them empirically with reports from psychotics, charlatans, and even retired scientists: what is impossible is to square such theories with the rest of science.

The concept of evidence is therefore complex: it has to be related to the specific hypothesis to which it is relevant and to the background knowledge—often a set of fragments of theories T—on the strength of which we claim that the datum is an evidence relevant to the hypothesis. Accordingly, the expression 'The datum *e is* an evidence relevant to the hypothesis *h'* is incomplete. It should be replaced by the expression 'On the theory [or the background knowledge] T, the datum *e is* an evidence relevant to the hypothesis *h*'. The logical structure of the evidence concept is then that of a ternary relation. In other words, every evidence is relative to some hypothesis by virtue of a body of theoretical knowledge: no evidence is absolute and no evidence is prior to every theory. For example, the observed deviation of a magnetic needle in the vicinity of an electric circuit (datum *e)* supports the hypòthesis h_1 that electricity is flowing through the circuit, on the theory T_1 that electric currents produce magnetic fields which in turn interact with the fields of magnetic needles. But exactly the same datum *e* might be taken as an evidence in favor of the rival hypothesis h_2 that a big magnet somewhere nearby has been switched on, on the theory T_2 that magnets can interact directly with one another (see Fig. 12.3.). The given datum is then *ambiguous* and only an *independent checking* of h_1 and h_2, i.e. a test independent of *e*, will enable us to reach a decision between the two rivals, at least satisfactorily enough for the moment. That is, new evidence will be needed in order to evaluate the old one. In short, no datum is by itself an evidence but it may *become an evidence* upon being interpreted with the help of some theory, and it can as easily be debunked from its status as the theory itself.

Before an observation is performed to produce evidence, the *kind* of data that will count as such must be stipulated. Otherwise dishonest attempts to save inept theories, as well as fruitless observations and controversies may take place. For example, a scientific psychologist will not accept as evidence for a theory purely introspective, hence uncontrolled reports, save for the sake of asking questions and devising hypotheses to be tested by objective data. Let us then adopt the following *Rule:* The kind of data that are to count as evidence must be agreed upon in advance of observation, and on the basis of theory. Needless to say, both the data and the theory will have to be scientific in order for the evidence to be acceptable. This is why the data that are claimed to constitute evidence for the hypothesis of precognition are unacceptable: neither the data are scientific (according to the criteria laid down at the beginning of this section) nor the hypothesis is scientific (according to the requirements stipulated in Sec. 5.7).

Whenever a scientist obtains data that seem incompatible with a well corroborated theory, far from merrily getting rid of either the data or the theory he tries to *reproduce* the data. If they do not recur he will set the anomalous data aside as suspect of some systematic error. And, whether they fail to recur or not, he will try to *explain* the anomaly—to legalize it—by applying or building hypotheses or theories that can be checked independently, i.e. that can be confronted with data other than the given anomalous data. One of the troubles with the alleged anomalous data of the parapsychologist is that he has no theory to account for them; as a consequence he does not know where to fit such data nor what value they have. The data of parapsychology are not scientific data, much less evidence, because there is nothing definite—no full-fledged hypothesis or theory—to test in parapsychology. No theory, no evidence relevant to it.

Are there *ultimate data*, i.e. singular or existential propositions in no need of support either empirical or theoretical? Several philosophical schools, notably phenomenalism and intuitionism, have held that certain factual statements are self-supporting, either because they are a direct expression of sense data (allegedly incorrigible) or because they are self-evident (evidently true). Thus, e.g., it is claimed that an observation statement such as "This here is green" is such an ultimate datum: it would be self-sufficient (in no need of background knowledge) and definitive (incorrigible). But psychology teaches that color perception is learned and can be wrong. Reports on inner experience are not better off: they are not more certain but just less interesting and less readily testable—hence less subject to actual test—than reports on objective facts. Suppose someone claims having had a certain unusual experience. A second observer tries to repeat the observation but fails to confirm the former report. He may wonder whether the first observer did in fact have the experience in question, or whether he interpreted rightly what he experienced. He may even attempt to falsify that report on theoretical grounds, by showing that it is inconsistent with the best established knowledge. The first observer might still claim that his was an internal experience he did have and moreover one to which available knowledge is irrelevant. In this case his assertion may not be falsified for the time being: worse than this, it will be shelved because it serves no scientific purpose. Data, in order to be scientific, must not be irrefutable but testable, and if they are not somehow justified by some theory or other they may be left in suspense. In conclusion, there are no ultimate data in science: there are only data which, at a given stage of research, are not further analyzed and controlled and which are used as evidence relevant to certain hypotheses or theories. At a later stage those same data may be subjected to critical scrutiny. Many evidences are *practically conclusive* just because we must reach decisions at certain points, but these decisions are not arbitrary, and no evidence is ever *theoretically conclusive*. Scientific data are not harder and longer-lived than the hypotheses and theories that help produce them or than those ideas they in turn support or undermine. To legislate that data of a certain kind are ultimate would be to enforce a dangerous dogma—one which, of course, runs counter scientific practice.

In fact, far from being accepted at their face value, observation reports are trimmed with theoretical shears. In the first place they are *critically scrutinized* in search for errors of observation. There are norms of observation, which every discipline devises with the help of substantive theories, and there are norms for the correction of observational data, which are built both with substantive theories and with the general theory of errors of observation (a methodological theory included in mathematical statistics). Secondly, the data are universalized or *standardized*. Thus, e.g., length measurements are "reduced" to standard or conventional temperature and pressure in order to eliminate contingent local conditions. A datum such as "The brightest red line of the Cadmium spectrum measured in Austin, Texas, by Profes-

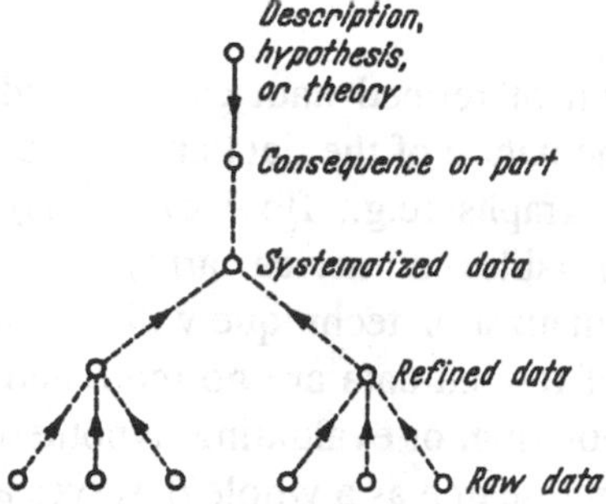

Fig. 12.4. Data are processed (refined) and systematized to compare them to descriptions or theories.

sor Doe while President Kennedy was being assasinated, was so many angstroms long", is no doubt precise but, on the one hand, it contains redundant information and, on the other, it fails to mention the temperature and pressure values.

The *raw* datum may well contain any information, but the *refined* datum must convey only relevant and universally useful information; this is achieved both by pruning the raw datum and by referring it to a standard (conventional) condition, such as 15 °C and 76 cm of mercury. It is not, therefore, a question of maximizing empirical information but rather of optimizing the precision and relevance of universal information, of information that can be checked by anyone competent in the same field, and eventually improved on. Thirdly, whenever statistical data such as averages or spreads around averages are sought—as in the case of quantitative measurement—the raw data are subject to *reduction*: the final information is actually a construction out of the raw data.

Some information is always lost in the process of critical scrutiny, standardization and reduction of raw data. Consequently, given a set of refined data—i.e. of data elaborated by critical scrutiny, universalization, and reduction—the original, much bulkier set of raw data cannot be retrieved: data refinement is an irreversible process. Hardly anybody deplores this loss of raw information, because data are means rather than ends: we are interested in *systematizing* data with the aim of disclosing patterns in them, and this cannot be done unless the "noise" is eliminated by the pruning process we have just described (see Fig. 12.4).

The systematization of refined data can be made in a number of ways, according to the nature of the data and to the end in view. They can be displayed in graphs (e.g., flow diagrams), histograms (frequency distributions), tables, numerical arrays (matrices), and so on. None such data systematization technique will, of course, replace theorizing: ordered sets of refined data are no more and no less than food for problemizing, theorizing, or evaluating hypotheses or theories. But the function of data in research as a whole deserves a new section.

Problems

12.4.1. Report on W. H. Ittelson and F. P. Kilpatrick's "Experiments in Perception" (1952), reproduced in D. C. Beardslee and M. Wertheimer, *Readings in Perception* (Princeton: Van Nostrand 1958). *Alternate Problem:* Examine the influence of points of view and theories on the gathering and use of anthropological data. See M. J. Herskovits, "Some Problems of Method in Ethnography", in R. F. Spencer, Ed., *Method and Perspective in Anthropology* (Minneapolis: University of Minnesota Press, 1954).

12.4.2. Examine the role of theory in the selection of the facts to be observed, as well as in the interpretation of data, in the field of psychology. See D. O. Hebb, "The role of neurological ideas in psychology", *Journal of Personality* 20: 39–55 (1951).

12.4.3. A field geologist finds some fossil bones and calls in two palaeontologists. One of them thinks that the remains belong to an ape, the other that they are evidences for (the hypothesis that there existed) a primitive man. What makes them differ despite their presumably having the same sense impressions? *Alternate Problem:* Photographic plates "showing" nuclear disintegrations, positrons, and the planet Pluto were available before the actual discoveries of these objects were made. Draw a moral.

12.4.4. Discuss the so-called *principle of total evidence,* according to which a hypothesis should be judged in the light of the totality of available evidence. Is this enough to give the hypothesis a fair deal?

12.4.5. Are data justified by tracing their sources in the observers?

See K. R. Popper, *Conjectures and Refutations* (New York: Basic Books, 1963), Introduction. *Alternate Problem:* What guides and motivates scientific observations? See Nobel prize biochemist A. Szent-Györgyi, "Lost in the Twentieth Century", *Annual Review of Biochemistry*, **32**, 7 (1963): "I make the wildest theories [hypotheses?], connecting up the test tube reaction with [the] broadest philosophical ideas, but spend most of my time in the laboratory, playing with living matter, keeping my eyes open, observing and pursuing the smallest detail [...] most of the new observations I made were based on wrong theories".

12.4.6. Subjectivists, intuitionists and empiricists have posited the existence of *pure* human experience, i.e. of experience unspoilt by prior belief or expectation, and have regarded reports on pure experience (protocol sentences) as the untouchable and unanalyzable foundation or starting point of knowledge. Pure experience has been given various names: Maine de Biran called it "primitive fact", Comte "positive fact", Mach "element", Husserl "pure fact", Bergson "immediate datum of consciousness", and so on. Does pure experience (perception without interpretation and expectation) exist? Assuming it does, is it regarded as belonging to science? And are such raw data or protocols accepted as basic by scientists in the sense that they serve as building blocks? *Alternate Problem:* Examine Kant's criticism of the idea of pure experience and see whether his medicine (namely, the idea that the understanding supplies an *a priori* framework to experience) was any better than the ailing he sought to cure.

12.4.7. Check whether any rational choice models in social science, such as game-theoretical models in political science, enjoy any empirical support. See M. Bunge, *Finding Philosophy in Social Science* (New Haven: Yale University Press) and *Social Science under Debate* (Toronto: University of Toronto Press, 1998).

12.4.8. C. Darwin, in his *Descent of Man* (1871), proposed his theory of human evolution on no direct empirical evidence. Quite the contrary, the theory directed palaeontologists to search for evidence. Darwin himself explained the lack of data by saying that (i) the most adequate regions had as yet not been explored by geologists, and (ii) geological processes had probably destroyed a part of the fossil record.

Discuss this case in terms of the doctrine that data must always precede theory. See W. le Gros Clark, "The Crucial Evidence for Human Evolution", *Proceedings of the American Philosophical Society*, **103**, 159 (1959).

12.4.9. Examine the following views concerning the scientist's decision to accept or reject an observation report. (i) The decision is *altogether justified* by the observer's perceptual experiences. (ii) The decision is *arbitrary* and any search for justification is logically unjustified because it leads to an infinite regress. (iii) The decision is *partially justified* by both intersubjective perceptual experiences and scientific theories—but even so only provisionally.

12.4.10. Data consisting of singular propositions contain either proper names or *definite descriptions* such as 'the new ammeter'. This expression is symbolized '$(\iota x)Ax$', read 'the (single) x such that x is an A', or 'the A' for short, and is treated as a proper name in the sense that anything can be truthfully or untruthfully predicated of it. A proposition such as "The new ammeter is burning" can then be symbolized: '$B(\iota x)Ax$'. According to B. Russell the proper analysis of "The new ammeter is burning", i.e. "$B(\iota x)Ax$", is "There exists an individual such that it is a new ammeter and is burning and, if anything is a new ammeter, then it is that particular individual". Briefly,

$$B(\iota x)Ax \leftrightarrow (\exists x)\, [Ax \,\&\, Bx \,\&\, (y)\, (Ay \rightarrow y = x)].$$

See I. Copi, *Symbolic Logic* 2nd ed. (New York: Macmillan, 1965). In this way, given a statement such as "The boss of the Olympus was fond of girls", we need not introduce an ideal object—the one apparently denoted by the description 'the boss of the Olympus'—since no such object exists. On the other hand, a datum in which a definite description occurs does not have the form of an atomic (not further analyzable) statement, as a universal quantifier occurs in it. And this is inconsistent with even the most tenuous empiricism, since it resolves certain singulars into universals rather than the other way around. Examine the following proposals. (i) Rule that data are not to contain definite descriptions but only proper names—and make a count of your following. (ii) Declare Russell's analysis irrelevant, i.e. hold that his logical analysis need not be paralleled in epistemology or that it should not be applied in science—and supply a ground for your decision. (iii) Replace Russell's analysis with the following:

$$(a=(\iota x)Fx)=_{df} Fa \ \& \ (x)(Fx \rightarrow x=a).$$

See M. Bunge, *Interpretation and Truth* (Dordrecht and Boston: Reidel, 1974), pp. 155–61.

12.5. Function

Most metaphysical system-builders, from Plato and Aristotle through Descartes and Leibniz to Hegel and Hartmann, have claimed that they had *a priori* knowledge of reality, that is, factual knowledge independent of controlled experience. They left science the boring task of finding the details of this knowledge but its essence they held, without proof, to be attainable either by a special intuition (intuitionism) or by reason alone (rationalism), in any case without the assistance of extrospection and experiment. Such systems have failed in a definitive way: it is not just that they happened to contribute no factual truth but that, because of their wrong approach (method and goal), they could not possibly replace science, which is both theoretical and empirical.

The failure of apriorism has always stimulated aposteriorism, i.e. the empiricist epistemology according to which experience alone and, in particular, purposeful and controlled experience, could provide factual knowledge. Empiricist philosophers in the past have done much to spread the idea that data gathering and systematization are goals in themselves. We have seen at various places, however, that this is a picture of protoscience rather than of science, so that pure aposteriorism is as false a philosophy of science as pure apriorism.

*Aposteriorism is sometimes expressed by saying that, ultimately, factual knowledge is condensed in *state descriptions,* every state description being a set of singular propositions exhaustively describing the state of a finite part of the cosmos—or rather of the universe of experience. The examination of an example will show the right size of this idea. Imagine a universe consisting of just three structureless things that can be arranged linearly in a nonmetrical space, so that only their mutual positions are of interest. There are 1·2·3=6 ways in which the constituents of our microuniverse can be arranged. Any one of these possible states of our imaginary universe can be described by a statement such as "Thing 1 is at *n*, thing 2 at *p*, and thing 3 at *q*". In turn, every one of these descriptions of a possible state of our microuniverse is made up of three propositions. If we adopt numerals to individualize positions we can write any such atomic proposition in the form "Thing *m* is at *n*", where both *m* and *n* are (ordinal) numbers. Even shorter: let 'P_{mn}' designate the statement "Thing *m is* at *n*". With this notation, state 231 would be described by the conjunction "P_{21} & P_{32} & P_{13}". All in all there are $3^2=9$ such propositions. For *N* different objects there will be N^2 atomic propositions by means of which $1 \cdot 2 \cdot 3 \cdot \ldots \cdot N$ complex propositions descriptive of one state each can be built. These N^2 atomic propositions can be arranged in a square table or matrix embodying all the possible information bits concerning the fictitious universe: $[P_{mn}]$, to which the pompous name of *universe state matrix* can be given.

*But all this is practically and theoretically pointless. Practically, because the task of performing an exhaustive description of any part of the world is hopeless. Suffice it to recall that for only 10 simple (structureless) things that can be arranged linearly the number of possible states is $10!=3{,}628{,}800$. Moreover, no interesting item in either an actual or a possible state description could be obtained without a number of theories, since most facts are only indirectly observable (recall Secs. 12.2 and 12.3). If the objects of interest happen to be atomic or subatomic, then the value of (indirect) observation is even slighter because no amount of empirical information suffices for the unique determination of the state of such a system: in fact, there is only a probability relation between every state and any given set of (indirectly) observable values of the microproperties (e.g., the momentum) of the system. Conversely, if the state of a system is "known" (that is, assumed), and any property *P* of it is in fact "observed" a number of times, then all we can infer is (i) that any of a certain infinite set of values of *P* will appear, and (ii) that every such value will occur with a certain probability which, if the assumptions are true, will be close to the actually observed frequency of occurrence. In short, in the case of atomic systems no set of observations suffices to uniquely determine a state description—which description is couched in theoretical rather than in empirical terms. So much concerning the practical impossibility of making an exhaustive inventory of the universe. As to the theoretical pointlessness of it, it is clear considering that nothing could be inferred from even a complete state matrix, which is just a compendium of loose and separate information. What science aims at is *laws* and *systems of laws* (theories) rather than collections of unrelated bits of information. (A set of relations among the various matrix elements P_{mn} of the universe state matrix, or of statements concerning their evolution in time, would constitute a theory—but this is precisely the monster that was supposed to be eluded with the state description approach.) The emphasis on state descriptions is just a relic from the pretheoretical stage of science, when the aim of research seemed to be the accumulation of observation reports that solved no problems and were accounted for by no theories. This shopkeeper philosophy has become obsolete even in management science, where full-fledged mathematical theories of inventories are built.*

We may call *dataism* the belief that every piece of scientific knowledge is a set of data. Dataism is the companion of *dadaism,* the doctrine that theories are to be the simplest (most economical) systematizations of data: if there is no knowledge other than empirical knowledge and this consists of a set of data, then it is foolish to organize data in complex ways. Dataism and dadaism are the hard core of strict empiricism, a philosophy incongruously held by many theoreticians and underlying much research planning. It encourages the blind accumulation of superficial information that leads nowhere because it comes from nowhere and takes place in a vacuum of ideas. This kind of industrious groping in the dark is exemplified by the professor who fed dogs with everything he could find in the grocery, not excluding detergents of various makes, and carefully measured the volume of gastric juice he collected through fistules. Aimless work of this kind can consume as much manpower and equipment as a genuine piece of research sparked off by an interesting problem occurring in a body of ideas and aiming at probing some of them. It can even become an expensive fraud and, what is worse, it can mislead countless gifted men into believing that scientific research consists in observing for the sake of observing rather than in asking, thinking and checking—both empirically and theoretically—in order to discover general patterns. No idealist philosophy could hamper scientific progress so much as that idealess research policy that passes for scientific just because, according to the dominant philosophy of science, facts accurately reported are all that matters and scientific ideas are nothing but fact summaries or handy tools for establishing links among facts. What is saved by refraining from theorizing and from elaborating a philosophy of science consonant with advanced research is lost many times in the form of blind search.

Dataism and dadaism are also behind the preference for what is called Big Science, consisting of vast research projects employing large masses of scientific workers and equipment and aiming at gathering huge piles of data. No doubt "big" (i.e., expensive) science is necessary and even admirable: who can help but admire those vast and disciplined masses painstakingly and patiently taking and processing myriads of data hoping that these data will eventually be understood by some theorist? "Big" science is necessary as long as it does not overshadow "little" science, which supplies the former with problems and theories, and into which the data must be fed back if they are eventually to make any sense at all. "Big" science is small science compared with "little" science, whence it would be dreadful if "big" science ended up by swallowing its parent and teacher. To prevent this catastrophe from occurring it is necessary to elaborate a philosophy of science avoiding the two extremes of apriorism and aposteriorism and putting observation in its right place—Ch. 12.

A contemporary empiricist would certainly grant that observation is not the sole concern of science, yet he would claim that all science is somehow *based* on sense data or perceptual information collected through observation, measurement and experiment. He might, in addition, hold that such a perceptual basis is unshakeable—whence the name *basis.* It is, in fact, widely held that sense data are basic both psychologically and logically: i.e., that they are in fact prior to ideas and that all constructs (theoretical concepts, hypotheses, and theories) are somehow distilled from observation. This is both psychologically and logically mistaken. First, sense data are often preceded by expectations born from more or less vague beliefs; moreover, they are rendered possible by an expectant, alert attitude—particularly so in the case of scientific experience. Second, the sense data of scientists are, in and by themselves, irrelevant to science: scientific observation and experiment do not collect sensa but objective and controllable data couched in an impersonal language. Third, no datum is sought, let alone used, outside some body of knowledge: there are no such things as spontaneously given scientific data or a primaeval collection of pure data upon which a science can be grounded in isolation from prior knowledge. Fourth, constructs are invented or created rather than distilled (e.g., induced) from sense data, precisely because they go beyond data. Fifth, only theory can convert some data into evidence concerning unobservable objects. In short, sense data are neither psychologically nor logically the foundation of science. Moreover, there is no absolute and incorrigible foundation of science: hence all kinds of fundamentalism are wrong in metascience.

As it often happens with extremes, metaphysical apriorism and positivist aposteriorism miss each one side of the coin: scientific knowledge has always resulted from an interplay of experience and reason. In particular, observation—the simplest kind of scientific experience—is no more and no less than a component of scientific research. In effect, the functions of observation in factual science are: (i) to supply information, (ii) to generate problems concerning such data, and (iii) to test hypotheses conceived to solve such problems. Neither of these is a goal in itself: the final desideratum is the disclosure of patterns, and these are represented by theories which observation can contribute to test. Observation, indeed, is insufficient: if we want deep and accurate factual knowledge we have got to measure and experiment whenever possible. But that will be another chapter.

Problems

12.5.1. The astrophysicist A. Eddington claimed—in the tradition of Kant—to possess a priori knowledge of the physical world, a knowledge derived purely from epistemological principles. Examine this claim and the accompanying programme: see Eddington's *The Philosophy of Physical Science* (Cambridge: University Press, 1939). Next, see the criticisms of L. S. Stebbing, *Philosophy and the Physicists* (London: Methuen, 1937), and M. Born, *Experiment and Theory in Physics* (London: Cambridge University Press, 1944). Finally, discuss whether such a priori attempts as Eddington's could possibly succeed.

12.5.2. Flowers had been observed by man during one million years before it was suspected (1694) that they are the sexual organs of plants; moreover, artificial pollination was extensively practiced by the Sumerians five millenia ago. Why did observation and action fail to generate botany?

12.5.3. Discoveries have often been made by chance. For an enthralling narration see R. Taton, *Reason and Chance in Scientific Discovery* (New York: Philosophical Library, 1957). Now, L. Pasteur said that chance can teach something only to prepared minds. Prepared by what?

12.5.4. Do atomic physicists and psychologists take cognizance of their objects through the senses? Make sure to distinguish the referents of their ideas from the evidence relevant to those ideas. *Alternate Problem:* Descriptive anatomy is a set of accurate observation reports. Is it an independent science?

12.5.5. Drawing room physics became fashionable towards the middle of the 18th century and remained so until late 19th century. Favorite exhibitions were those of electric and magnetic phenomena, which were shown for entertainment rather than for understanding: no science of electricity and magnetism was evolved by that means. Similarly, countless observations were made on drawing room hypnotic phenomena, which added nothing to an understanding of hypnosis. Why this sterility of accurate observation? *Alternate Problem:* In 1776 M. van Marum observed that electric discharges magnetized iron pieces, but failed to "interpret" this as an interaction between electricity and magnetism, and his observation remained buried. In 1887 H. Hertz observed what we now call the photoelectric effect but failed to "interpret" it as the extraction of electrons by light, whence he is not usually credited with the discovery. Draw some moral.

12.5.6. Evaluate (i) the knowledge concerning isolated facts and (ii) the knowledge of isolated generalizations. Regarding the former see H. Helmholtz, *Popular Scientific Lectures* (London: Longmans, Green, and Co., 1873), p. 369. Concerning the latter, see E. Schrödinger, *Science: Theory and Man* (New York: Dover, 1957), p. 93. *Alternate Problem:* By building more and more powerful particle accelerators, the number of exceptional facts can be accumulated without limit. Would there be any point in continuing along this line without the parallel construction of theories explaining and predicting such facts? And is it not exactly this what has been happening at the high energy laboratories over the past few decades?

12.5.7. Comment on the second of the following sentences taken from B. Schultzer, *Observation and Protocol Statement* (Copenhagen-Munksgaard, 1938): (i) "By a protocol statement we propose to understand a statement which, on the basis of a given observation, assigns certain characteristics to what is observed in this observation" (p. 5). (ii) "A protocol statement is said to be valid when its formulation in a given protocol language is unambiguously determined by what is observed (in connection with an accepted reference system and an accepted tolerance of inaccuracy)" (p. 9). Are observation reports justified by observation alone? And is there a protocol (or neutral observation) language, i.e. one not "contaminated" by hypotheses and theories? In case none is in existence, would it be possible to construct it? And if it were possible, would it be worth while, if not for scientific purposes at least in order to make motion pictures about the Australopithecus? *Alternate Problem:* Comment on the following sentence taken from a biological journal: "The high resolving power of the electron microscope together with biochemical techniques will ultimately uncover the precise role(s) of the mitochondria". Can you think of any biological function that has been "uncovered" by sheer observation? Recall that the reproduction mechanism was a mystery until the 18th century, and that the function of large size organs such as the panchreas and the thymus was not discovered until recently.

12.5.8. Comment on the following texts. (i) J. Tyndall, *Fragments of Science*, 5th ed. (New York: Appleton, 1880), pp. 425–426: We may not be able to transcend experience, "but we can, at all events, carry it a long way from its origin [...]. Urged to the attempt by sensible phenomena, we find ourselves gifted with the power of forming mental images of the ultra-sensible; and by this power, when duly chastened and controlled, we can lighten the darkness which surrounds the world of the senses". (ii) H. Hertz, *The Principles of Mechanics* (1894; New York: Dover, 1956), p.25: "If we try to understand the motions of bodies around us, and to refer them to simple and clear rules, paying attention only to what can be directly observed, our attempt will in general fail. We soon become aware that the totality of things visible and tangible do not form a universe conformable to law, in which the same results always follow from the same conditions. We become convinced that the manifold of the actual universe must be greater than the manifold of the universe which is directly revealed to us by our senses. If we wish to obtain an image of the universe which shall be well-rounded, complete, and conformable to law, we have to presuppose, behind the things which we see, other, invisible things—to imagine confederates concealed beyond the limits of our senses".

12.5.9. Is observation the source of all knowledge or rather a source of descriptive knowledge and a test of hypotheses and theories? *Alternate Problem:* Discuss W. V. Quine's "Posits and Reality", in S. Uyeda, Ed., *Basis of the Contemporary Philosophy* (Tokyo: Waseda University, 1960), vol. V.

12.5.10. Thomas Aquinas, following Aristotle, held that sense impressions are necessary, though not sufficient, for acquiring intellectual knowledge of bodies: the intellect provides the universals without which there is no such knowledge. See *Summa theologiae*, Q. LXXXIV, Arts. 6 and 8, and Q. LXXXV, Art. 1, and *Summa contra gentiles,* Bk. I, Ch. xii. Query 1: Why are Aristotelianism and Thomism so often presented as aprioristic? Query 2: Why is it so often stated that only modern philosophy appreciated the importance of sensory experience for knowledge? Query 3: Which doctrine is closer to science: Cartesian apriorism, Baconian aposteriorism, or Aristotelianism? *Alternate Problem:* When did people realize that the attention of the philosopher of science should shift from the psychological problem of the sources of knowledge to the methodological problem of the test of ideas?

Bibliography

Ayer, A. J.: The problem of knowledge, Ch. 3, London: Penguin Books 1956.

Blanshard, B.: The nature of thought, 2 vols., especially Book I, London: Allen & Unwin; New York: MacMillan 1939.

Bunge, M.: Metascientific queries, especially Chs. 5, 6 and 8. Springfield, III.: Charles C. Thomas 1959.

—— Exploring the world. Dordrecht-Boston: Reidel 1983.

Daedalus **87**, No 4 (1958): issue On evidence and inference, particularly M. Deutsch, Evidence and inference in nuclear research and P. F. Lazarsfeld, Evidence and inference in social research.

Frank, P. (ed.):The validation of scientific theories. Boston: The Beacon Press 1956.

Hanson, N. R.: Patterns of discovery: An inquiry into the conceptual foundations of science, Chs. i. and ii. Cambridge: University Press 1958.

Körner, S. (Ed.): Observation and interpretation. London: Butterworth Scientific Publications 1957; New York: Dover 1962.

Lenzen, V. F.: Procedures of empirical science, vol. I, No 5 of the International encyclopedia of unified science. Chicago: University of Chicago Press 1938.

Lewis, C. I.: An analysis of knowledge and valuation. La Salle, Ill.: Open Court 1946.

Russell, B.: An inquiry into meaning and truth, especially Chs. XXI and XXII. London: Allen & Unwin 1940.

Sellars, W.: Science, perception, and reality. New York: Humanities Press 1963.

Waismann, F.: Verifiability. In: A. Flew (Ed.), Logic and language, I. Oxford: Blackwell 1951.

13

Measurement

What can be observed either directly or indirectly is a set of traits of some concrete system. If observation is to be precise it must be quantitative because concrete systems have quantitative properties, if only because they exist in determinate amounts and in spacetime. Quantitative observation is measurement. Whenever numbers are assigned to certain traits on the basis of observation, measurements are being taken. There are as many kinds of measurement as kinds of properties and measurement techniques. Essentially we can count, compare, or measure in a strict sense. In order to decide what kind of measurement is to be done, an analysis of the concept denoting the corresponding property must be performed. Accordingly the nature of quantification must be analyzed before the features of measurement can be understood.

13.1. Quantitation

We shall call *numerical quantification* any procedure whereby certain concepts are associated with numerical variables. Thus, volume is quantified bulkiness, probability can be interpreted as quantified possibility, and probability of success is quantified effectiveness. Since there is a variety of kinds of concept we shall presumably find various sorts of quantification.

Let us recall the classification of concepts given in Sec. 2.2 (see Table 2.6). We distinguished four genera of concepts: individual, class relation, and quantitative. Individual concepts, i.e. concepts the referents of which are individuals, are clearly nonquantitative (qualitative), unless they consist of individual numbers. (But numbers are nowadays regarded as classes rather than as primary individuals.) Individuals can certainly be assigned *numerals* (number names) in place of names, as are prisoners. But this *nominal* use of numbers is purely conventional and has no theoretical significance: in fact, numerals can in this case be freely exchanged for signs of any other kind. Hence mere assignment of numerals is insufficient for either measurement or formalization.

Class concepts, again, are qualitative. Yet they, too, can be assigned numbers in a conventional way. In effect, a class concept such as "man" gives rise to the predicate "is a man", which in turn can be true or false of any given individual. If a given individual named c is in fact a man, then we can symbolize this fact by writing down the numeral '1'; if, on the other hand, c is not a man, then we may write '0'. That is, we may regard class concepts as generating dichotomic predicates, i.e. predicates that can assume either of two values: presence or absence of the corresponding property. This assignment of numbers is, again, conventional: we might as well have chosen the pair (–17,3).

*There is a second way in which class concepts can be assigned numbers. Let G be a genus composed of two well-defined species, S_1 and S_2. We may also think of G as a set of states consisting of two possible states, S_1 and S_2 (for example, open and closed in the case of doors and electric switches). In any case, let S_1 and S_2 be nonintersecting subsets of G. Given an individual x there are two possibilities: either x is in G or x is not in G. We are interested in the first case alone; if x is in G, then it is either in S_1 or in S_2. A useful representation of these two possibilities is the following: (i) If x is in S_1, we describe the state of x by means of the column matrix or vector

$$\psi_1(x)=\begin{bmatrix}1\\0\end{bmatrix}.$$

(ii) If x is in S_2, we describe the state of x by means of

$$\psi_2(x)=\begin{bmatrix}0\\1\end{bmatrix}.$$

(iii) If the only thing we know about x is that it is in G but do not know whether it is in S_1 or in S_2, we write

$$\psi(x)=c_1\psi_1(x)+c_2\psi_2(x)$$

where c_1 and c_2 are definite functions that may, for instance, be related to the respective probabilities of x being in S_1 or in S_2. This association of class concepts with numbers and numerical functions is conspicuous in quantum theories; there is no reason why it cannot be employed in other sciences, particularly in biology.*

The most common way in which numbers are assigned to classes is by determining the number of elements of a class. The numerosity or *cardinality* of a set S is a set function—which we shall symbolize '*Card*'—that maps S into the set N of natural numbers (nonnegative integers). In short, $Card\ (S)=n$, where n is in N. If the elements of the set are factual (e.g., planets), the whole functor is a "concrete" number (e.g., "9 planets"); in such a case cardinality is a *physical* property, and the calculus of cardinality may not coincide with the arithmetic of natural numbers. In other words, in the case of factual sets the calculus of cardinality will depend on the nature of the sets and on the concrete processes or operations corresponding to the arithmetical operations.

Let S_1 and S_2 be two sets of marbles and let '$\dotplus$' designate the physical process (or the empirical operation) of juxtaposing or bringing together the two collections. Then $Card\ (S_1\dotplus S_2)=Card\ (S_1)+Card\ (S_2)=n_1+n_2$, i.e. arithmetical addition mirrors physical addition. But if the same sets are physically added by a high-energy collision, the resulting set of bodies can be more numerous than before, i.e., $Card\ (S_1\dotplus S_2)\geqq n_1+n_2$, so that arithmetical addition will cease to mirror physical addition. The philosopher who ignores these differences between physical addition and mathematical addition is bound to ask the Kantian question 'How are synthetic a priori judgments possible?'

The cardinality of an empirical set is a finite nonnegative number. But science is not limited to considering empirical sets, all of which are supposed to be subsets of factual sets. And a factual set can be infinite, as is the (actual) set of points in a field of force or the (potential) set of energy levels of an atom. (The set of stable states of an atom is denumerably infinite, but an atom can also be in any of a continuous set of unstable states.) From a mathematical viewpoint the important cleavage is not the one between finite and infinite sets but between denumerable or countable, whether finite or not, and nondenumerable or continuous. But from a methodological viewpoint the cleavage is between finite and infinite sets, since the former could in principle be exhaustively scanned (not exhaustively known, though) whereas we can only sample an infinite population and there is no guarantee that the chosen sample will be representative of the whole.

The concept of cardinality of a set opens the door to quantitative mathematics and thereby to the quantification of more complex concepts. Consider the psychological concepts of independence of judgment in the face of group pressure. An objective indicator of independence is the correctness of appraisals of matters of fact (e.g., length estimates) despite contrary opinion prevailing in a group. This indicator can easily be quantitated in the following way. Let the basic set U be the set of appraisals of a certain factual state of affairs made by an individual or a group under social pressure (prejudice, deliberately misleading information, etc.). This set can be partitioned into the set C of correct estimates and its complement $\overline{C}$, the incorrect estimates: $U=C\cup\overline{C}$. We can measure the degree of independence by the fraction of correct estimates in situations where misleading social factors are operating. In other words, we can introduce the quantitative concept of independence of judgment $I(x)$ of a social unit (person or group) x, taking its numerical variable y equal to the ratio of the cardinality of the set of correct estimates to the cardinality of the total set of estimates:

$$I(x)=y=\frac{Card(C)}{Card(C\cup\overline{C})}.$$

(Values as high as .43 have been obtained by S. E. Asch in situations under experimental control.)

We have examined three different ways in which numbers, and even functions, can be assigned to classes (or class concepts), i.e. three different kinds of quantitation of qualities, one of which was nominal or conventional. Let us now analyze the quantitation of relation concepts, which in Table 2.6 were divided into noncomparative and comparative. In turn we subdivided noncomparative concepts into relations proper, such as "belonging to", and operators, such as "+". Clearly, noncomparative concepts cannot be quantitated except in a nominal (conventional) way. In effect, with regard to a relation of the kind of "is the mother of" or "exerts an influence on" we can say that it either holds or does not hold among two given items. If it holds we may agree to write '1', otherwise '0'. Similarly, for operators such as "+" and "/". For any such dyadic relations we can, then, build an array or matrix $R=[r_{ij}]$ the elements r_{ij} of which represent the satisfaction or nonsatisfaction of the given relation R by the pair (i, j). If R holds between i and j we set $r_{ij}=1$, otherwise $r_{ij}=0$. Such matrices are used in sociology. A number assignment of this kind has a pragmatic significance only: it is practically convenient to use the numerals '1' and '0' but these perform exactly the same functions as the checking sign '∨'. Only comparative concepts, such as "higher than", can be subject to non-nominal quantification.

Comparative concepts are those by means of which sets can be ordered or, equivalently, by means of which the elements of a set can be ranked. The relation of being harder than performs such a function with regard to bodies, hence "harder" is a comparative concept in the set of bodies. In more detail: given a set B of bodies we may order them according to their hardness; i.e., the relation of hardness is satisfiable in the set of bodies—not however in the set of waves or in the set of feelings. The relation is, moreover, asymmetrical: in fact, if x is harder than y then it is not the case that y is harder than x. Secondly, the relation is irreflexive, i.e. it does not hold for a single body. Thirdly, the relation is transitive: if x is harder than y and y is harder than z, then x is harder than z. Since a number of equally hard bodies can be found, the relation does not establish a strict linear order; rather, it orders subsets of bodies. In other words, given any two bodies x and y, it is false that either $x>y$ (meaning "x is harder than y") or $x<y$ (meaning "x is softer than y"). It may be the case that $x=y$ (meaning "x is as hard as y"). In pragmatic terms: if from the set of bodies two items are picked out at random, what will be empirically found to hold among them is the weaker relation $\geqq$. This relation holds among individual bodies and generates a weak order.

We may look at this matter in the following way. We start with the class or qualitative concept "hard"—or, if preferred, "bright". Then it will be a matter of convention where we draw the line between hard and non-hard (soft) bodies. In other words, we may nominally associate a numeral, such as '1', to certain bodies, namely those we include in the set of hard bodies, and the numeral '0' to those we exclude from this set. A finer partition of the set of bodies as regards their hardness can be performed if the comparative concept "harder" is built out of the class concept "hard": moreover, the set may be arranged chain-wise. Finally, a numeral may conventionally be assigned to every link in the chain. We may, in fact, agree to associate the numeral '1' to the least hard of all known bodies, '10' to the hardest known, and the digits in between to other kinds of bodies. None of these numerals will, however, be significant in itself: alternative numerals, letters, or ad-hoc signs could have been chosen, and their order might have been reversed: this is why we do not speak of *numbers* measuring hardness but of *numerals* (numerical signs). All they indicate is the relative place of an item in a simple ordering. Consequently, from the fact that body *a* is assigned hardness 6 and body *b* hardness 2, we cannot infer that body *a* is thrice as hard as body *b:* we cannot *divide numerals* any more than we can divide any other tags. In order to avoid misunderstandings it is convenient to write 'n^*' to designate the numeral or mark corresponding to the number *n*. What we have said does not entail that hardness is an *inherently* nonquantitative property, as is so often heard: quantitative concepts of hardness have in fact been introduced in the theory of metals; the comparative concept of hardness is peculiar to the empirical approach.

The assignment of numbers to comparative (order) concepts is purely nominal or conventional unless a quantitative concept is made to underly the comparative one. Take, for instance, height. Psychologically we start with the class concept "high"; on the basis of it we next introduce the comparative concept "higher"; finally we build the quantitative concept "height". Once we have secured the latter we may go back to the comparative concept and quantify it. Indeed, we may compare the numbers (not the numerals) expressing different heights and say, for instance, 'The Himalayan mountains are twice as high as the Alps', which is mathematically correct because we are now comparing numbers. The introduction of quantitative concepts is then logically prior to the quantification of comparative ones.

A *metrical predicate* (numerical functor, quantity, or magnitude) designates a quantitative property. A magnitude, such as length or stimulus strength, is a complex concept that can be analyzed into object variable(s), numerical variable(s), and a function from the former to the latter. (Recall Sec. 2.3.) The object variables designate the entities of which the magnitude concerned predicates a given property, and the numerical variables designate the numbers associated with this property. Take, for instance, the length concept, which can be analyzed thus: $L(x, u)=y$, read 'the length of *x* equals *y* length units'. The object variable *x* of this concept runs over the set of bodies and usually is not mentioned: it is taken for granted that, in factual science, lengths are lengths of things, not lengths of hopes or lengths in themselves. (In a different context we may assign 'length' a different meaning while keeping the above mentioned logical structure. Thus, we may introduce the concept of length of an idea by agreeing to characterize it, e.g., as the number of symbols occurring in its logical symbolization with the help of some calculus.) And the numerical variable can in this case take any real value in the $[0, \infty)$ interval; in other cases the numerical variables can take on negative or even complex values in a variety of ranges. In any case quantification does not eliminate physical objects: we are not replacing things by numbers but associating numbers with concepts that stand for properties of things.

If two or more objects are considered for measurement, more than one object variable may enter the magnitude. For example, if two rods, x and y, are juxtaposed end to end, a third object z is produced which may be said to be the *physical sum* (or joining) of x and y. We have denoted physical addition by '$\dot{+}$' to distinguish it from the corresponding arithmetical operation: the former regards bodies, the latter numbers. If we now ask what the length $L(x\dot{+}y)$ of the composite rod $z=x\dot{+}y$ is, the answer will be: the numerical value of the total length equals the sum of the partial lengths, i.e.

$$L(x\dot{+}y)=L(x)+L(y) \qquad [13.1]$$

regardless of the length scale and unit.

The foregoing is a synthetic (nonlogical) formula: universes are conceivable in which lengths do not add in this simple way—i.e. in which length is not an additive measure. (Moreover, the *measured* values of $L(x)$ and $L(y)$ are not strictly additive: in some cases the two operations may involve a larger error, in other cases a smaller error than the operation of measuring the total length.) After all we are not concerned with mathematical quantities, such as distances in a metrical geometry, but with a physical property. Suppose now we consider two contiguous regions x and y in a space and compute their volumes. In all geometries volumes are additive, just as distances are. If, on the other hand, we perform the physical operation of adding a body of water to a body of alcohol, we shall find that the total liquid volume does not equal the sum of the partial volumes: $V(x\dot{+}y)\neq V(x)+V(y)$. This, again, is a synthetic (nonanalytic) truth. Experience suggests that volumes of matter are additive on condition that they have the same chemical composition. Letting 'S' stand for "chemical species", we have

$$\text{If } x \text{ and } y \text{ are in } S, \text{ then } V(x\dot{+}y)=V(x)+V(y). \qquad [13.2]$$

Even this clause is insufficient in some cases: just think of the joining of two uranium bodies the total mass of which exceeds the critical mass at which a chain reaction starts. All these are cases of conditional additivity. Some magnitudes are not even conditionally additive. Thus, if the two rods x and y of our previous example are at different temperatures, then the composite body's temperature, i.e. the temperature of $z \dot{=} x \dot{+} y$, will not be the sum of the partial temperatures but will have some value in between. I.e., if x and y are different systems or different parts of a system, then their temperatures are not additive:

$$\text{If } x \neq y, \text{ then } T(x \dot{+} y) \neq T(x) + T(y). \qquad [13.3]$$

Magnitudes that are exactly additive with respect to the object variable are called *extensive*. More precisely, magnitudes for which a physical addition operation $\dot{+}$ exists such that a law of the form [13.1] holds for them shall be called *unconditionally extensive*. If a weaker law of the form [13.2] holds for them we shall call them *conditionally extensive*. Those which, like mass, are approximately additive with respect to the object variable (recall the mass defect) will be called *quasiextensive*. All others will be called *intensive* magnitudes. Distance, area, duration, electric charge and many other magnitudes are extensive. Mass, energy, physical volume (bulk) and others are quasi-extensive. Densities, whether of mass, population, or anything else, are essentially intensive magnitudes; thus, when we say that the specific gravity of iron is about 8, this is not equivalent to saying that the physical addition (juxtaposition) of 8 equal unit volumes of water will yield a body with the same specific gravity as iron. Further intensive magnitudes are frequency, wavelength, refractive index, solubility, magnetic permeability, intelligence quotient, and per capita wealth, as well as many other index numbers used in social science.

*Let us note some delicate points. In the first place, the extensive character of a magnitude depends on the existence of a physical addition operation, but not every joining can be regarded as a physical addition. Thus, electrical resistances are additive because, through the operation of series connection, they can be combined in such a way that the total resistance equals the sum of the partial resistances. But if combined in parallel no such result is obtained, whence parallel connection is not a kind of physical addition. (On the other hand conductances, i.e. the reciprocals of resistances, are additive when combined in parallel.) In the second place, it is not true that, as is sometimes held, some extensive magnitudes are not additive: if they are not additive they do not deserve to be called extensive—by definition. The case of the addition theorem for velocities in relativistic kinematics has been offered as a counter-example but wrongly so: in fact, object variables of different kinds happen to occur in the three velocities concerned, namely the generic mass-point x, the laboratory reference frame L, and a second reference system M in motion with respect to L. Calling $U(M, L)$ the velocity of M with respect to L, $V(x, M)$ the velocity of the mass-point in the frame M and $V(x, L)$ the velocity of the same object x relative to L, we have: $V(x, L) \neq V(x, M)+U(M, L)$, which is not a substitution instance of [13.3]. A third mistake we must be guarded against is to believe that intensive quantities are those which can be ordered but not added. Thus, it is often claimed that a formula such as "2 °C+ 3 °C = 5 °C" is false or even meaningless, and that subjective psychological magnitudes such as loudness are ordinal rather than cardinal simply because they are not extensive. This is a confusion: all magnitudes, whether extensive or not, are cardinal or quantitative concepts. What cannot be the subject of mathematical operations are intensive quantities referring to *different systems* or to different parts of one and the same system. Thus, it is wrong to add the numerical values of the population densities of, say, England and Australia. It does make sense, on the other hand, to compute the rate of growth of the population density of a given region, to multiply it by the area of the region and, in general, to subject it to mathematical operations. If intensive magnitudes could not be subject to any calculus they would not enter quantitative scientific theories, which they do.

Whether a given magnitude is additive with respect to the object variable must ultimately be decided by experience but it can be anticipated by theory—not by any theory but by a sufficiently deep theory. The clue is this: intensive magnitudes refer to nondistributive, nonhereditary, or *collective* properties. The population density of a country is not the same as that of all its constituent regions and when it comes to a small plot of ground the concept is no longer applicable. Similarly, it is incorrect to speak of the temperature of a molecule: temperature is a statistical property, a property of any large collection of atomic objects—and so are solubility, refractive index, magnetic permeability, and per capita wealth, all of which are intensive magnitudes. Classical thermodynamics was not in a position to establish that temperature is an intensive quantity: it had to take this for granted as an empirical proposition; but statistical mechanics can show that the

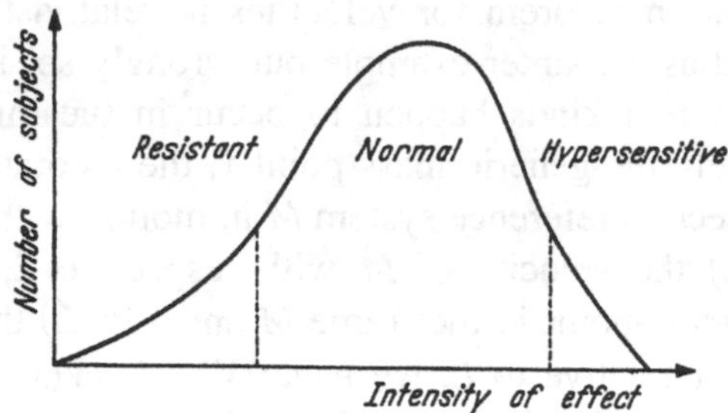

Fig. 13.1. A typical dose-effect curve. It generates the partition of any group of subjects into resistant, normal, and hypersensitive.

object variable of the temperature functor can only be a large assembly of microsystems.

This completes our account of numerical quantification as a conceptual operation. At least three questions of a philosophical interest are raised by it: Why do we value quantification, whether quantification underlies measurement of conversely, and whether there are inherently nonquantifiable properties. The first question is easily answered by listing the advantages of quantification. These are: (i) *concept refinement*—thus, the magnitudes "heat content", "specific heat", and "temperature" are different refinements of the ordinary concept of heat; (ii) *precise description*—think of the description of technological evolution in terms of the concept of velocity of technological innovation; (iii) *precise classification*—think of the grouping of subjects as regards their reaction to a given drug on the basis of a precise effect distribution function (see Fig. 13.1); (iv) formation of *exact hypotheses and theories,* by making a precise interrelation among variables and formulas possible with the help of mathematics; (v) *rigorous test* of hypotheses and theories.

Our second philosophical problem concerns the relation of numerical quantification to measurement. It is a widespread belief that measurement is the foundation of quantitation or, to put it in a different way, that (i) quantitative properties appear upon measurement, and (ii) measurement generates quantitative concepts. The belief is mistaken, as shown by the existence of quantitative theories in mathematics, which is unconcerned with the empirical operation called 'measurement'. Moreover, the concept must be at hand before it is assigned a number with the assistance of an empirical operation; and some logical analysis of the concept is required if the measurement is to be correct. Take, for instance, the velocity concept: it was not until physicists realized that velocity is a three-component vector that accurate velocity measurements could be planned aside from the elementary case of straight line motion. To measure a velocity is, in fact, to assign a numerical value to each of its components on the basis of certain observations. First comes the concept, then the corresponding measurement—the more so if the object of measurement happens to be only indirectly measurable. Just think of measuring the wavelength of light without having made the hypothesis that light comes in waves. Medieval measurements were mostly fruitless because they were not preceded by the corresponding conceptual analysis. Quantification precedes measurement because, by definition, to measure is to assign particular values to the numerical variable(s) of a quantitative *concept* on the basis of observation.

The third philosophical question we wished to tackle is associated with Galilei, who stated the *Rule:* "Measure the measurable and try to render measurable what is not yet". Clearly, the history of science has been in part the story of the quantification of initially qualitative concepts. The questions are: (i) is it possible to quantitate *every* concept? and (ii) is it *desirable* to attempt the quantification of what is still qualitative? The first question is definitely answered in the negative: we have seen a while ago that individual concepts such as "Russell" and noncomparative relations such as "is in" are nonquantitative. They can be assigned numerals (not numbers) in a conventional fashion, but this cannot be regarded as quantitation proper, whence the allotment of such numerals upon observation cannot be regarded as a kind of measurement. The question as to the desirability of quantifying what is still qualitative and is neither an individual nor a noncomparative relation, can be answered affirmatively. Yet it might be rejoined that a property such as maleness cannot be assigned numbers except in a conventional manner: it would appear that any sexually reproducing animal is either a male or a female. Biologists, however, have been forced to go beyond this dichotomy, introducing a concept of degree of maleness: they can experimentally produce a whole gamut of individuals of certain species, such as the fruit fly, with varying degrees of maleness. Moreover, there is no reason why this comparative but still qualitative concept of degree of maleness could not be quantified by equating it with some hormone ratio. Something of the sort has happened with the concept of hardness: it was possible to quantify it when its relation to more fundamental physical properties was found.

Still it might be objected that such a procedure cannot be carried over to all the sciences of man. Thus, it would seem that it is mistaken to say that a person was twice as afraid as another person. The reason for this would be that feelings and urges cannot be divided into parts each of which is a less intense feeling or urge. Yet everyone has experienced varying degrees of fear, so why not try to quantitate this feeling? Actually this can in principle be done on the basis of a well known relation between fear and adrenal level in blood: this biochemical variable may be used as an objective quantitative indicator of fear. It is only when a superficial approach is stuck to that such a quantitation is impossible; but if the physiological substratum of psychical life—ignored both by introspectivism and by behaviorism—is taken into account, a wealth of quantitative indicators becomes available. Similarly with value: if we wish to quantitate and measure value we must attempt to correlate it with some quantitative property or set of properties instead of regarding it as a basic and unanalyzable notion. Suppose we suceeded in quantitating the needs or wants of humans as regards, say, aesthetic stimuli, just as we quantitate and measure our vitamin requirements. We might next agree to define the aesthetic value of an object x for an individual y as the extent to which x satisfies the aesthetic needs of y. This would be a fraction between 0 (no satisfaction) and 1 (total satisfaction). Aesthetic value would then be quantitated and, moreover, on the basis of objective indicators: therefore, it would be measurable. The nonexistence of quantitative theories in aesthetics, or at any rate of satisfactory theories of this kind, does not prove their logical impossibility or their empirical untestability: it may be due just to a long tradition of loose thinking and to the mistaken belief that quantitation and, in general, mathematization, is essentially linked to the subject matter. Once we realize that it is not the subject matter but *our ideas* concerning it that are the subject of numerical quantification, no unsurmountable barriers to quantitation remain.

Let us now turn to the effective assignment of numbers on the basis of empirical procedures, i.e. to measurement.

Problems

13.1.1. Analyze the quantitation of some scientific concept. *Alternate Problem:* Quantitate some concept hitherto treated as nonquantitative.

13.1.2. Are physical magnitudes empirical in the sense that they denote our sense experiences or in the sense that they can be the object of empirical procedures? Compare the heat sensation with the temperature concept. Take into account phenomena such as this: the cup on a wooden table is felt colder than the table although their temperatures are the same. The temperature depends neither on the state of our hands nor on the thermal conductivities of the things concerned, which properties determine on the other hand the sensations of cold and warmth.

13.1.3. Trace the history of the norm that every magnitude is to be defined in terms of measurements.

13.1.4. Take any law statement and show which variables and parameters in it are extensive and which intensive.

13.1.5. Examine the following indicators. (i) The importance of plant domestication in a given site can be estimated on the basis of the number of cultivated species and the number of wild species in the same region, i.e. $D=C/(C+W)$. Similarly for the importance of animal domestication. (ii) The intensity of cultural diffusion, i.e. the strength of intercultural contacts, can be quantitated as the ratio E/C of the number of objects exchanged (imported or exported) to the total number of items consumed in a given region during a unit period. *Alternate Problem:* Discuss the biological similarity coefficient introduced by P. H. A. Sneath and R. R. Sokal, "Numerical Taxonomy", *Nature*, **193**, 855 (1962).

13.1.6. Examine the genetic determination of the affinity of biological species. *Alternate Problem*: Report on any quantitative approach to problems of personality. See e.g. S.E. Asch, "Studies of Independence and Conformity", *Psychological Monographs*, No. 416 (1956).

13.1.7. Study the possibility of quantitating innovation, or any other variable important for the sciences of man. *Alternate Problem:* Examine the possibility that, when psychologists fail to measure certain traits and attitudes, it is because they have failed to analyze them into simpler components—perhaps because they have not realized their complexity or even the need to start with conceptual analysis rather than with empirical procedures.

13.1.8. Discuss the introduction of pseudoquantitative terms such as 'quantity of psychical energy', 'excitation volume', 'excitation level', 'unconscious force', 'motivational interaction' and others as a device for giving pseudoscientific doctrines a scientific flavor. Begin by laying down the conditions(s) for a term to designate a *genuine* quantitative concept. Establish next the condition(s) for applying quantitative concepts. See finally whether certain terms abundantly used in the behavioral sciences and pseudosciences comply with those conditions.

13.1.9. Examine the opinion that a discipline may employ quantitative *procedures,* with eminently practical aims, without thereby using quantitative *concepts.* See A. C. Crombie, "Quantification in Medieval Science", *Isis*, 52, Pt. 2, 143 (1961) and H. Guerlac, "Quantification in Chemistry", *ibid.*, 198 (1961). *Alternate Problem:* Examine the view that it is possible to set up *a priori* a general and basic theory of magnitudes of a given kind—e.g., extensive magnitudes. A number of such theories (called 'theories of measurement') have been advanced by mathematicians, logicians and psychologists on the presupposition that the laws of the combinations of magnitudes (such as associativity, commutativity and additivity) are purely mathematical, hence independent of the substantive (factual) laws occurring in factual science. Are such theories realistic, i.e. applicable to existing scientific theories? And can a theory of measurement proper be built without the help of fundamental substantive theories? Take the case of physics, which contains as many theories of measurement as theories of physical systems.

13.1.10. Analyze the operationalist doctrine that measurement engenders quantitative concepts and is the basis of every theory—a belief behind the common practice of beginning science textbooks with measurement. *Alternate Problem:* According to certain philosophies

Table 13.1. The mensurandum and its estimates

Item	Symbol	Example	Level
Mensurandum (degree of property)	$\dot{r}$	*rod*	reality
		←length→	
Measured value of property	$m(\dot{r})$	(100±.1) cm	experience
Numerical value of property	r	100 cm	theory

of mathematics (extreme formalism in the sense of Hilbert, nominalism, and pragmatism) numbers are just inscriptions or marks, i.e. physical objects of a kind. What is the impact of this doctrine on the theory of quantitation? What becomes of the distinctions (i) between nominal (entirely conventional) and nonnominal quantitation, and (ii) between a numerical value occurring in a factual theory and the numeral pointed at by the pointer of a meter—which numeral need not designate the number representing the real value that is being measured?

13.2. Measured Value

Quantitation is a process whereby a qualitative formula, such as "$P(x)$" (e.g., "x is long"), is replaced by a quantitative formula, say "$P(x, \text{cm})=y$" (e.g., "the length of x in cm equals y") where x is a generic object (individual or class) and y a generic number. When an individual value of y is assigned to a certain property of a definite object with the help of observation, we perform the empirical operation called measurement. For example, comparing the weight of a certain body with the weight of a standard body is measuring the value of the weight $W(c)=w$ of that particular object c. What is measured—to be called the *mensurandum*—is always a particular value of a numerical variable of a quantitative concept (magnitude) representing a property of some concrete system.

Three concepts have been distinguished in the above characterization of measurement: (i) a *property,* such as height, that is supposed to be objectively graded; let '$\dot{r}$' designate any of the degrees, amounts or intensities of a given property, and $\dot{R}$ the set of particular values of $\dot{r}$. (ii) The *measured values* of the degrees $\dot{r}$; we shall call them $m(\dot{r})$, and the corresponding set of measured values will be called $\{m(\dot{r})\}$. (iii) The *numerical values* of the magnitude representing the property; let 'r' designate these numbers and R the set of r's. See Table 13.1.

While the amount $\dot{r}$ is usually assumed to be a trait of a piece of reality, its empirical estimate $m(\dot{r})$ is an outcome of a measurement act and r a component of a theoretical concept. The $\dot{r}$'s can at best be perceived: they are objective traits, independent of conceptualization and measurement. On the other hand, both the $m(\dot{r})$'s and the r's are numbers. For example, $\dot{r}_1$ and $\dot{r}_2$ may be two perceptibly different heights, $m(\dot{r}_1)$ and $m(\dot{r}_2)$ their corresponding (approximate) values as obtained with the help of certain measurements, and r_1 and r_2 the corresponding values of the numerical variable occurring in the quantitative concept "height". In scientific theories r-values rather than $m(\dot{r})$ values will be found to occur, but both of them will represent the same set of $\dot{r}$-values (if the measurement process does not appreciably modify the mensurandum). The above distinctions may look artificial but they are often made in science and have a profound philosophical import. The discussion of a few examples should bear out this contention and prepare the ground for a systematic investigation.

Example 1. Consider a material flat square made of steel or any other substance. A simple conceptual representative of this thing is a perfect geometrical square of, say, unit side. This ideal model neglects, of course, all the irregularities of the material square's edges and surfaces; anyhow it is the conceptual model, not the thing itself, that will be able to enter a scientific theory (for example, statics). The Pythagorean theorem yields $\sqrt{2}$ as the length of any of the square's diagonals. The two numbers, 1 and $\sqrt{2}$, are r-numbers representing each an $\dot{r}$-amount, namely, the side and the diagonal of the real square. Suppose now a length scale s is adopted and measurements of both mensuranda are taken with a device giving an error of 2 in one thousand (e.g., two millimeters per metre). The measurement outcomes will then be the following data:

$$m(\text{side of square}, s)=1\pm.002$$
$$m(\text{diagonal of square}, s)=1.414\pm.003.$$

where 's' designates the length scale and unit adopted for the purpose. The first statement may be read in a number of equivalent ways: 'The measured value of the side of the square lies between .998 and 1.002u's', 'The measured value, in units u, of the side of the square is an unknown number in the interval [.998, 1.002]', 'The most probable value of the measured side of the square is 1 u with a 2 per thou. error', etc. Similarly for the second statement. Notice, in the first place, the uncertainty or latitude peculiar to measurement values $m(\dot{r})$ but absent from theoretical values r. Second, note that the mean value 1.414 obtained for the diagonal can be expressed as a fraction, namely 1414/1000, which is close to 43/3. Every measurement yields a fractionary number, whereas theoretical values can be irrational (non-fractionary), as is the case with the r-value $\sqrt{2}$ for the diagonal of the square.

Example 2. It is a plausible assumption of electron theory that the electric charge of every electron in the universe is the same irrespective of its interaction with the rest of the world. Yet if any two electron charges are measured, or even one and the same electron's charge is measured twice, two slightly different values will most likely be obtained. In other words, a number of slightly different $m(\dot{r})$-values may be obtained for what can be assumed to be one and the same r-value mirroring a single $\dot{r}$-amount.

Example 3. Suppose we want to estimate the importance of animal domestication in a given prehistoric community. A first thing to do is not to rush to the site but to sit down and invent a quantitative concept of degree of animal domestication. Such an index may be the ratio of the number of domesticated animals to the total number of animals both domesticated and wild, that any given community consumes over a given period. In obvious symbols, $I=D/(D+W)$. (Recall Problem 13.1.5.) In order to measure the value of this index for a given community, say the one that dwelt in a particular cave at some time, we shall count the number of bones of domestic and wild animals found in the cave and shall finally perform the arithmetical operations indicated in the formula for the domestication index. The value found in this way will not be an entirely reliable measure of the degree of domestication attained by our ancestors: it might well have happenened that the domestic species had consistently more (or less) bones than the wild species, or that their bones were harder (or softer) and consequently better (or worse) preserved than the others, or that most domestic animals were not killed but milked or sheared, and so on. At any rate, it is conceivable that the count of bones does not provide an entirely faithful estimate of the actual value of the degree of domestication. An alternative index or a different measurement technique may eventually be constructed, but meanwhile we may have no better means for estimating the importance of domestication in prehistoric times.

Measured values, then, may differ from actual values. (Consequently statements concerning measured values may differ from statements concerning actual values. One possible source of discrepancy is the *error* involved in most measurement operations, from which only the simplest counting operations can be free; this error is peculiar to the technique employed. Accordingly the typical form of measurement data, i.e. of propositions reporting on results of measurement, is not "$m(\dot{r}, s)=r$" but rather

$$m(\dot{r}, s)=\frac{p}{q}\pm\sigma \qquad [13.4]$$

where p and q are integers and $\sigma\geq 0$ is the error. A second possible source of discrepancy between actual and measured values is that, whereas the former can be irrational numbers—i.e. real numbers r such that $r\neq p/q$—every measured value is a *rational* (fractionary) number. This limitation cannot be avoided because it arises from the very act of measurement, which involves comparing an $\dot{r}$-amount with a unit of the same property. Irrational values, which are theoretically all-important, are empirically inaccessible: measurement yields only more or less coarse *fractionary approximations* to actual values. A strictly empiricist science ought to use only rational numbers because these alone can be obtained in measurement, all others being hypothesized. But such a science, avoiding continuous variables, would be prevented from using mathematical analysis, which involves the continuum of real numbers. Accordingly a strictly empiricist science would not have advanced beyond the 16th century.

The discrepancy between actual values and their empirical estimates is in most cases unknown and *unknowable:* the most we can do is to estimate bounds beyond which it is extremely unlikely that the discrepancy will lie. In short, the difference $\Delta=r-m(\dot{r})$ is in most cases unknowable (Only the error ε_i of an individual measurement $m_i(\dot{r})$ relative to an "exact" value $\overline{m}(\dot{r})$ can be computed by means of the definition: $\varepsilon_i=m_i(\dot{r})-\overline{m}(\dot{r})$. In this expression, $\overline{m}(\dot{r})$ is either an average of previous measurement values or a theoretical value.) That unknowability is one reason why empiricists avoid the expression 'actual value'; but monsters are not avoided by refraining to talk about them. The discrepancy must be admitted or suspected if we wish to understand (i) why different measurement techniques may yield different estimates of what theory assumes to be one and the same value, and (ii) why new particular procedures (techniques) for measuring the values of a given quantity may constitute significant scientific advances.

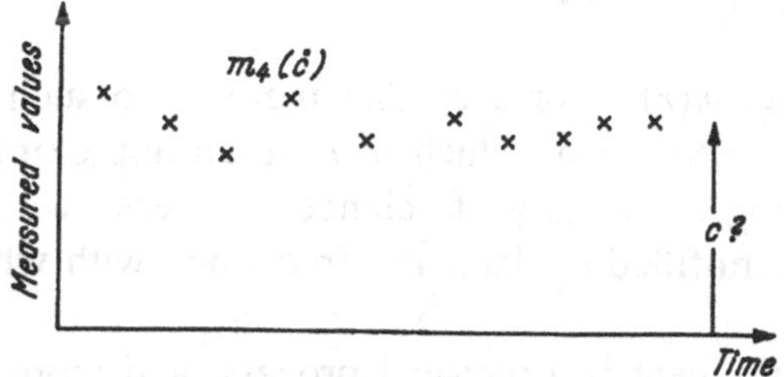

Fig. 13.2. An imaginary temporal sequence of the measured values $m_n(\dot{c})$ of a constant $\dot{c}$ as obtained by trying successively improved techniques.

Take, for example, the various existing techniques for measuring the velocity of light in vacuum, $\dot{c}$: they all yield slightly different values. As new measurement techniques are invented, new c-values are obtained not only further decimal figures but occasionally new results affecting significant decimals of older results. Shall we infer that the velocity of light has changed in time, so that the different measurements gave different actual values of c? This conjecture—that the speed of light changes in the course of time—has in fact been advanced. But it is an altogether unfounded hypothesis: there is no indication whatsoever either in the theory of light or in cosmology, that such a change in the measured values reflects a real change. Consequently it is more reasonable: (i) to reject the philosophical assumption that measured values are all there is, and (ii) to assume that what has changed in the course of time have been the measured values of $\dot{c}$, a universal constant that is approximately known through those measurements. In short, $\dot{c}$ is unknowable but, with various empirical and theoretical tools, we get so many estimates. In general: every measurement technique T_n yields a peculiar value $m_n(\dot{c})$ (actually an average $m_n(\dot{c})$) of any given constant, or a peculiar set $\{m(\dot{R})\}$ of values of a given set R of values of a set $\dot{R}$ of degrees of a given property. We hope, but cannot guarantee, that the sequence of measured values will in the long run *converge to the actual* but unknown value (see Fig. 13.2). These various values obey no law, and our sole ground for hoping that they will converge to the actual value is that we try hard to fulfil this hope. Our ignorance of the actual value c and our hope of getting closer and closer to it are in turn the only motives for trying new measurement techniques. To state that the value of a given property *is* the value given by a run of measurements—i.e. to assert that in every case $m(\dot{r})=r$, or even that there is no such thing as an objective mensurandum $\dot{r}$ of which $m(\dot{r})$ is an approximate value—requires neglecting the history of science. The ease with which this condition can be fulfilled explains the frequency with which that belief is held.

Since a measurement is a physical process, and moreover one involving at least one macroscopic system (the measurement set-up), and since there are no two identical things on the macroscopic level (see Sec. 6.1), it is unlikely to obtain two identical measurement results. Slight changes will occur in the mensurandum, in the instrument, and in the environment of the two, and consequently the numerical outcomes will differ even if slightly. In other words, the outcomes of any run of accurate measurement are different because all the systems involved in these processes are subject to perturbation: just as Heraclitus could not step twice into *the same* river, so we never measure twice *the same* thing with *the same* apparatus. Therefore we need collective representatives of runs of measurement, such that they absorb the individual differences and have some stability. One such stable representatives of a run of measurements of a given mensurandum is the average. Let $m_i(\dot{r})$ be the i-th measured value of the mensurandum $\dot{r}$. A set of numbers like that will be the raw outcome of a run of N measurements. None of them is likely to be adopted: what is finally adopted is a logical construction out of such numbers, namely the *average*

$$\overline{m}(\dot{r})=\frac{m_1(\dot{r})+m_2(\dot{r})+\ldots+m_N(\dot{r})}{N}. \qquad [13.5]$$

In fact, experimentalists usually adopt the following *Rule:* As an estimate of the actual value of a mensurandum adopt the average of a long sequence of accurate measurements of it. This is not an arbitrary but a grounded rule: in fact, it can be shown that random errors of measurement do distribute symmetrically about the average, i.e. they tend to *compensate* each other in the long run and to bunch around the average, which is a comparatively stable value. The average value is then the most probable value; it is also called the exact value, but actually it is only an *estimate* of the exact value, which is itself unknown in most cases.

In sum, we must distinguish the following three sets of numbers associated with one and the same mensurandum $\dot{r}$: (i) its *real* but unknown value r, i.e. the numerical value that the amount $\dot{r}$ has independently of our measurements; (ii) the various individual results $m_i(\dot{r})$ of accurate (but still imperfect) measurements of $\dot{r}$, i.e. the various *individual measured values* of $\dot{r}$; and (iii) the *most probable value* $\overline{m}(\dot{r})$, equal to the average of the individual values obtained in a given run of measurements made with optimal means (instruments and techniques) and under optimally controlled circumstances. The average is the value usually quoted as *the* measured value of the mensurandum. Actually (i) it is improbable to find it on measurement: it is the outcome of a conceptual operation done on measurement results, and (ii) it is not a single number but a pair of numbers: the average proper, $\overline{m}$, and the error or scatter σ around the average. This pair of numbers affords a collective or statistical characterization of a set of measurements. What is fed to a theory is the central value $\overline{m}$.

The average $\overline{m}$ is often termed the *true value*, but this is a misnomer: we do not know the true value, and if we knew it for a given mensurandum we would stop trying to improve the corresponding measurement techniques. All we know is that the so-called true (or exact) value provides *the best available estimate* of the actual value r. The actual value r is probably close to $\overline{m}(\dot{r})$, it is likely to lie between $\overline{m}(\dot{r})-\sigma$ and $\overline{m}(\dot{r})+\sigma$ and unlikely to fall outside this interval, but there is no more certainty and no more comfort than this and the hope that future, undreamt-of efforts, may shrink the margin of error.

Now that we have gained some clarity about the difference between a magnitude and its measurement, and of the complexities involved in the latter, we may take a look at the simplest kind of measurement: counting.

Problems

13.2.1. Is "The weight of this body is 2" an acceptable measurement report?

13.2.2. Suppose the electric charge Q of a corpuscle is measured and found to be 10.1 electron charges. Compute the error knowing that, according to our present knowledge, electric charges come in units, namely, electron charges.

13.2.3. A large wheel has random numerals inscribed on its border. We spin the wheel and, whenever it comes to rest, one of those numerals falls under a fixed pointer; we read the numeral and record it. Does the whole constitute a measuring apparatus, and is this assignment of numerals to physical events a measurement?

13.2.4. Comment on the following statement by E. Schrödinger in *Science. Theory and Man* (New York: Dover, 1957), p. 73: "however far we go in the pursuit of accuracy we shall never get anything other than a finite series of discrete results which are *a priori* settled by the nature of the instrument. [. . .] every measurement is an interrogation of nature and it is we who have arranged in advance a finite number of replies, while nature is always in the position of a voter in a ballot".

13.2.5. Compare the following statements of the Pythagorean theorem for a right-angle triangle of legs a, b and diagonal c:

$c=\sqrt{(a^2+b^2)}$	(*Pure Euclidean geometry*)
$c=\sqrt{(a^2+b^2)}$, with $I(a, b, c)$ = sides of a triangle made of light rays	(*Theoretical Euclidean physical geometry*)
$m(c)\cong\sqrt{\{[m(a)]^2+[m(b)]^2\}}$	(*Experimental Euclidean geometry*).

13.2.6. Analyze the much quoted and copied definition proposed by N. R. Campbell in *Foundations of Science* (New York: Dover, 1957), p. 267: "Measurement is the process of assigning numbers to represent qualities". Does this sentence describe measurement proper or rather quantification? Recall Sec. 13.1.

13.2.7. Discuss the claim of M. Born and others that all science is at bottom statistical because exact measurement involves a random error subject to a stochastic law. Hint: keep reference and evidence distinct. *Alternate Problem:* Analyze the opinion that the choice of the mean value as an estimate of the true value is conventional (groundless).

13.2.8. Examine the contemporary version of Berkeley's philosophy condensed in the unspoken but widely held maxim "To be is to be measured". See M. Bunge, *Metascientific Queries* (Springfield, Ill.: Charles C. Thomas, 1959), Ch. 8.

13.2.9. Discuss Kant's thesis that it is pointless to assume the existence of things-in-themselves, such as our $\dot{r}$-degrees and the corresponding real but unknowable r-values.

13.2.10. Every single macroscopic measurement result is actually an average over a volume and over a time interval. Why then do we have to average such averages?

13.3. Counting

Counting is the basic measurement operation since, in the last analysis, even the measurement of values of a continuous magnitude like time consists in counting a number of units of the magnitude. In some research fields, particularly in the social sciences, counting is still the major measurement method.

To count is to establish a one-to-one correspondence between the set of objects to be counted and a subset of the positive integers (see Fig. 13.3). The order in which this correspondence is established is immaterial: all we are interested in when counting is the largest member of the set of natural numbers coordinated with the given set. This presupposes that the process of counting does not affect the numerosity of the original set; or, if it does affect it because in order to discriminate among the members of a given set we have to exert some influence on them, counting presupposes that there are means for estimating the changes introduced in the number by the counting process.

The set to be counted may be nonempirical, as is the case of the set of primitive concepts of a theory. In this case we shall not regard counting as a kind of measurement, because it will involve no empirical operation proper. Counting becomes an empirical operation when it bears on a material collection, such as a set of marks or a set of shudders. It may, of course, be argued that even the counting of ideal objects is an empirical operation since it takes a thinking person going through the experience of coordinating two sets of ideas. But we are not here concerned with this broad notion of experience: we shall disregard inner experience and shall restrict our attention to *empirical counting*. This is an operation requiring the assistance of the senses and eventually of special instruments, such as scintillation counters,

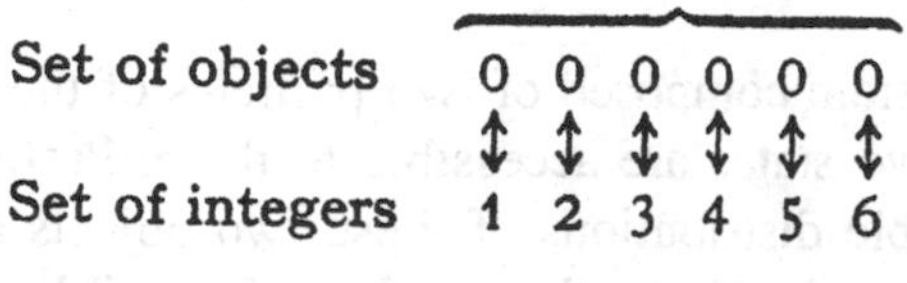

Set of objects	0	0	0	0	0	0
	↕	↕	↕	↕	↕	↕
Set of integers	1	2	3	4	5	6

Fig. 13.3. Counting: a 1:1 correspondence between the counted set and a subset of the natural numbers.

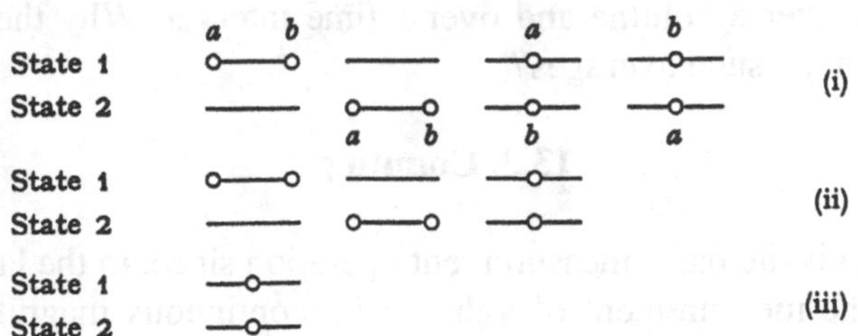

Fig. 13.4. Result of counting dependent on nature of objects concerned. (i) Non-identical particles: 4 possibilities. (ii) Identical particles: 3 possibilities. (iii) Identical and mutually exclusive particles: 1 possibility.

and special theories such as the theory of the ionization of matter by the passage of electrically charged corpuscles through it.

For a collection of things or events to be empirically *countable* it must consist of empirically distinguishable members. A continuous physical system such as the gravitational field filling our office, is not discrete, hence it has no natural (nonarbitrarily cut off) parts that can be counted: it constitutes a nondenumerable set. But the distinctness of the members of a material set is not enough to become the subject of counting: their separateness must be empirically accessible, i.e. it must be possible to discriminate among its elements either by the senses or with the help of special tools.

Discrimination, a simple affair with regard to medium sized bodies becomes a thorny problem on both extremes of the size scale. As regards the counting of astronomical objects, suffice it to recall that until a couple of centuries ago only stars were countable: binary stars had not been analyzed into their components and nebulae had not been "observed", partly owing to poor instrumental resolution, partly because astronomers were not prepared to perceive them: they had not imagined that anything like that could exist, they thought only in terms of separate simple stars. The situation is even more difficult in the microscopic extreme of the scale, as will be seen from the following example.

Consider a system composed of two particles of the same kind and such that only two states are accessible to them. Problem: Count the number of possible distributions of these two objects among the two states. In other words: Count the number of possible states in which this miniature universe can be. This is no game with state descriptions (Sec. 12.5) but a genuine and tough theoretical problem that, surprisingly enough, admits three different solutions (see Fig. 13.4). First solution: if the two particles are different, hence distinguishable in principle, such as two marbles, there are four possibilities: both marbles are in state 1, both in state 2, one in state 1 and the other in state 2, and conversely. Second solution: if the particles are identical, i.e. have no individuality, they cannot be labeled, and there will be a single situation in which each particle occupies one state; this is the case of helium atoms and the other "bosons." Third solution: if the particles are not merely identical, but every state has at most a single occupancy, then there is a single possibility: one particle per state; this is the case of electrons and the other "fermions." Our original question, then, has been given three answers depending not on arithmetics or on the resolution of microscopes but on the nature of the objects counted, i.e. on physical laws. First moral: Before counting physical alternatives—e.g., physically possible states, or possible moves—decide on how they are to be counted, and base this decision on a consideration of the nature of the objects concerned. Otherwise the same alternative may be counted twice or even more times. This rule, which now sounds obvious, was not learnt until the late 1920's. Second moral: There can be no universal rules or axioms for counting, i.e. independent of subject matter: what is counted counts as much as how it is counted.

Counting may be *direct* or *indirect,* according as the set to be counted is, or is not, made up of elements that are perceptible and countable in a reasonable lapse. Scientifically, the most important case of direct counting is that of time intervals. Time measurements are, in effect, nothing but counting periodic (recurrent) events of a given kind. In principle nothing else is needed, save a theory justifying the assumption that the events in question are in fact periodic. It is only for practical reasons that dials, hence length measurements, are introduced to measure time. (The philosophical question of the nature of time is most important but it is hardly enlightened by time measurement, which raises only technical problems.)

Philosophically, the most interesting cases are of course those of indirect counting, i.e. counting performed with the help of law formulas and definitions. Blood corpuscles are counted indirectly for practical reasons: it would be unnecessarily time- and work-consuming to count them all individually, though this is in principle possible. On the other hand atoms are counted indirectly both because of their number

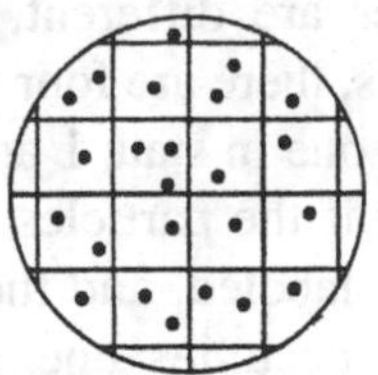

Fig. 13.5. The blood corpuscles in a known volume consisting of a few dozens of square cavities are counted.

and because as yet they are not easily made manifest. In the case of blood corpuscles, stars, and many other visible things, statistical formulas are employed in connection with the sampling techniques. In the case of atoms, atomic collisions, and other invisible things and events, statistics plays a role, too, but the essential formulas are substantive rather than methodological hypotheses. Let us examine briefly these two procedures of indirect counting.

In order to count the total number of red corpuscles belonging to a given individual, a small blood sample is extracted on the assumption that it will be a *random sample,* i.e. that the extraction procedure has not influenced the concentration, so that the sample is a fair representative of the whole population. A number of techniques can then be applied to this sample: from actually counting the red corpuscles in the sample with the help of a microscope, to employing some automatic device built on the basis of certain laws regarding the migration of red corpuscles in an electric field. An intermediate technique, which may be regarded as a typical sampling procedure, is the following. The total volume V of the subject's blood is first estimated on the basis of some empirical generalization (such as the weight-blood content correlation) and of data such as the subject's weight. Then the volume v of the blood sample is measured in order to find the total number N of red corpuscles by means of the formula $N=n\cdot V/v$, where 'n' designates the number of corpuscles in the sample. (The hypothesis "$n/N=v/V$", true for a droplet, will be false for a millionth part of it.) Since the sample is still too large, it is diluted in a known amount of water and a single drop of this solution is actually observed through the microscope—on the assumption that the dilution process has not produced local accumulations or rarefactions, so that this second sample is representative of the first. The number of corpuscles in the drop is estimated by counting the number of corpuscles in a certain known area and assuming again a uniform distribution (see Fig. 13.5). The total number of corpuscles in the sample is computed from this result and from the volume ratio. Finally the total number of corpuscles is inferred along the same line. The hypotheses and inferences at work in sampling are usually simple but uncertain, particularly if the sample is small; in any case they are as important as the actual count.

The results of the above technique or of any other indirect counting procedure are not accepted blindly but are subjected to criticism of two kinds: *statistical checking,* and *comparison with alternative techniques.* The statistical checking of a sampling like the one described above consists in comparing the number of corpuscles in every square (see Fig. 13.5) with the probable values predicted by a certain statistical formula (Poisson's distribution). If the discrepancy between the actual counts per square and the expected values is large, it can be suspected that the number of observed squares has been insufficient or that some other trait of the sampling has been defective. The second kind of control alluded to consists in comparing the results obtained with the given technique with those produced by another technique involving different "principles" (laws): in this way the original count is confirmed or disconfirmed by an independent witness, so to say. The objectivity of any counting technique is controlled rather than accepted on faith; otherwise it is not called a scientific technique (recall Sec. 1.3). Imagine a "haemoanalyst" claiming to be able to discover the number of red corpuscles in any subject just by asking him a few questions and refusing to let either his technique or his conclusions to be checked and doublechecked by independent means.

Counting molecules illustrates the second kind of indirect counting, namely the one involving a sizable amount of substantive scientific knowledge. We may call it *atomic-theoretical counting* for want of a better name. One of the earliest operations of this kind was the determination of the Avogadro number, i.e. the number of molecules per mole of a gas at normal pressure and temperature. This number, equal to 6.023×10^{23}, was introduced as a component of the atomic theory, i.e. as a theoretical concept. Various theories suggest so many ways of "counting" this number, by deriving relations between it and other quantities. (For example, $N=F/e$ and $N=R/k$, where 'F' stands for the faraday—an electrochemical constant—'e' for the electron charge, 'R' for the gas constant, and 'k' for the Boltzmann constant.) In other words, a number of mutually independent checkings of any single "counting" of N is suggested by physical theories—which would not be possible if N were just the end result of blind empirical manipulations.

The numbers most important for science are those determined by indirect counting of the atomic-theoretical kind, i.e. the counting of which requires sophisticated theory and sophisticated measurement. The reason is this: easily accessible numbers, such as the number of parts of a leaf or the population of a given region, can occur in descriptions but seldom in theories, if only because (i) no theories are usually necessary to obtain them, and (ii) there are few theories, if any, into which they can be fed. However important they may be for practical purposes, directly measurable numbers have little cognitive significance as compared with the "hidden" numbers, those accessible through theory alone. For all we know, directly countable numbers (i) derive from more basic traits of reality, and (ii) they might change within bounds without the world changing into some qualitatively different thing. Change by one million the population of the world: no deep transformation will probably occur; but add or subtract just one particle in an atom and a qualitatively new thing will emerge.

To sum up, the following kinds of counting may be distinguished:

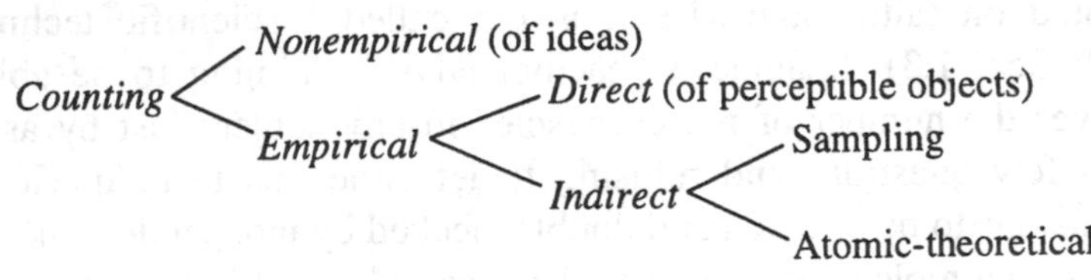

Problems

13.3.1. Expound a sampling technique. Use any good book on statistical methods.

13.3.2. Analyze a technique for counting bacteria or any other microscopic objects present in great number. Point out exactly what the assumptions, the data and the inferences are.

13.3.3. The measurement of time can be reduced to counting periodic events such as day-night cycles. Can this procedure exhaust the continuum of time intervals assumed by physical theory? And is the choice of sequence of regular (periodic) events arbitrary?

13.3.4. Expound a technique for the determination of the Avogadro number or any other universal constant, and point out (i) the underlying hypotheses and (ii) some checking procedures. Use any good book on experimental atomic physics.

13.3.5. Comment on the assertion of H. Dingle, in *Scientific American*, **202**, No. 6 (1960), that the Avogadro number "actually stands for a combination of measurements with thermometers, pressure gauges, balances and so on", rather than for the number of molecules in a certain unit amount of gas in a certain state. Take into account (*a*) the meaning of the Avogrado number in atomic theory, and (*b*) the existence of several mutually independent methods for measuring that constant.

13.3.6. Counting, like measuring, is subject to error, the more so if random events rather than regularly spaced events are counted. The expected range of error in the count of N random events is $\sigma = \sqrt{N}$, the spread of counts around the mean. The counts themselves are distributed according to the Poisson law (see Fig. 13.6). Is counting a purely empirical operation? *Alternate Problem:* Is denumerability identical with actual countability?

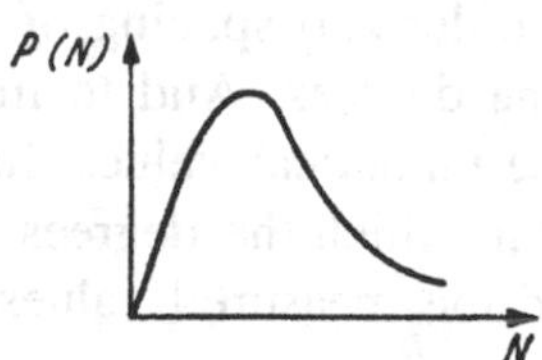

Fig. 13.6. The Poisson distribution: a control of the counts of random facts and a criterion of randomness.

13.3.7. Imagine a case in which counting a set of facts could change the cardinality of the set. Invent a way of correcting for this error.

13.3.8. Discuss the measurement of probabilities by observing relative frequencies and relying on the law of large numbers.

13.3.9. Discuss the view that (i) counting is always an empirical operation, even when its objects are the roots of an equation; (ii) it obeys universal rules independent of subject matter, and (iii) it is prior to every mathematical theory. *Alternate Problem:* Study the gradual independization of pure mathematics from measurement. See H. S. Carslaw, *Introduction to the Theory of Fourier Series and Integrals*, p. 31: "To make the theory of number independent of any geometrical axiom and to place it upon a basis entirely independent of measurable magnitude was the object of the arithmetical theories associated with the names of Weierstrass, Dedekind and Cantor".

13.3.10. Different theories are obtained for collections of particles of the same kind if the particles are assumed to be different in some respect or to be identical. (Boltzmann's, Bose-Einstein's, and Fermi-Dirac' statistics, corresponding to the three cases in Fig. 13.4.) Most physicists say that such differences—which can be tested empirically result from our inability to distinguish (e.g., label) the particles. Examine this interpretation.

13.4. Scale and Unit

To quantify a property of a concrete system is to map the set of degrees of the property (the $\dot{r}$-amounts) into a set of numbers (the r's) in such a way that the order and spacing of the numbers reflect the order and spacing of the degrees. And to measure is to effectively determine some of those numerical values. In other words, a magnitude is the precise way in which the degrees of a property are represented by numbers, and the measured values constitute a sample of this number set.

Consider the length degrees attained by a metal rod subject to heating. The rod expansion is one aspect of a complex process and the length increase one trait of this process. The only exact way of describing such a length increase is to represent it numerically, i.e. to coordinate the set of length amounts ($\dot{r}$-values) with a set of numbers (r-values). This coordination is done conceptually by introducing the quantitative concept of length, i.e. the function "Length of the body x in the state y on the scale $s=r$", where r is a nonnegative real number. Measuring the length of a given rod in a given thermal state and on a given scale consists in assigning the numerical variable r a definite value upon performing an empirical operation such as comparing the mensurandum with the length of a standard body such as a ruler. Upon performing this operation the observer perceives a mark m^* on the

Concrete System	*Physical length* (objective degree)	*Conceptual length* (real number)	*Instrument reading*	*Measured length* (rational number)
■■	$\dot{c}_1$	c_1	m_1^*	$m(\dot{c}_1)=[c_1]$
■■■■	$\dot{c}_2$	c_2	m_2^*	$m(\dot{c}_2)=[c_2]$
■■■■■■	$\dot{c}_3$	c_3	m_3^*	$m(\dot{c}_3)=[c_3]$

Fig. 13.7. The mappings of length degrees, and of their instrumental counterparts, on numbers.

measuring instrument, which he interprets as the image of a number $m(\dot{c})$, itself an estimate of the actual value c of the measured length. The value c itself is unknown in most cases; the assumption is that the measured value $m(\dot{c})$ is a fraction $[c]$ near c (see Fig. 13.7). In short, measurement is the empirical counterpart of quantification or measure determination, and it consists in interpreting certain conventional marks (e.g., numerals) as numbers providing in turn a more or less faithful image of the amounts or degrees of a property. Measurement, then, is anything except a direct and complete mapping of such degrees on numbers.

*The complexity of measurement can best be appraised by disclosing its structure. One starts by assuming the existence of a *factual* relational system $\dot{\mathcal{R}}=\langle \dot{R}, \dot{\leq}\rangle$ consisting of the set $\dot{R}=\{\dot{r}\}$ of degrees of a physical property, such as weight, and the concrete ordering relation $\dot{\leq}$, e.g., "less heavy than or as heavy as". Then one introduces the *conceptual* relational system $\mathcal{R}=\langle R, \leq\rangle$ consisting of a subset R of the continuum of real numbers and the arithmetical relation "less than or equal to". The corresponding magnitude—e.g., weight—is a *mapping* of $\dot{\mathcal{R}}$ into $\mathcal{R}$, such that it reflects the physical order of $\dot{R}$. Thus far quantification.

*Now call $\mathcal{M}^*=\langle M^*, \dot{\leq}\rangle$ the factual relational system consisting of a set $M^*=\{m^*\}$ of marks such as the streaks on the dial of a meter, ordered by the physical relation $\dot{\leq}$ (e.g., "to the left of or coincident with"). And call $\mathcal{M}=\langle M, \leq\rangle$ the corresponding set of partially ordered rational numbers, itself a subset of the reals. The net result of a run of measurements of a given property consists in establishing a *partial mapping* of the empirical relational system $\mathcal{M}^*$ into the conceptual relational system $\mathcal{M}$, which is in turn a partial image of $\mathcal{R}$, itself an image of $\dot{\mathcal{R}}$ (see Fig. 13.8). Eliminate any of the four systems, $\dot{\mathcal{R}}$, $\mathcal{R}$, $\mathcal{M}^*$, or $\mathcal{M}$, or any of the three mappings, and no measurement remains. In particular, if any of the factual components is eliminated,

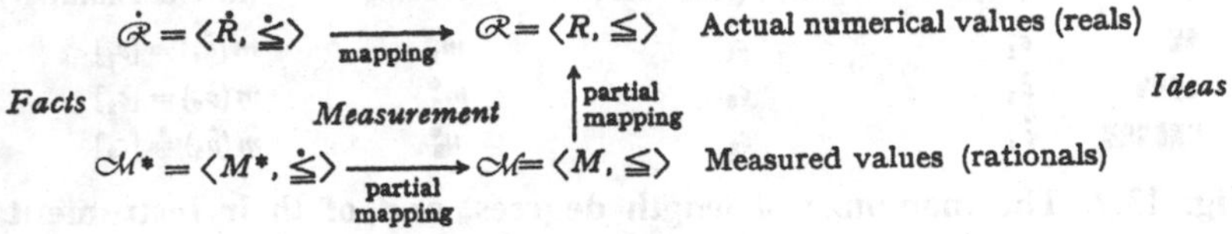

$\dot{\mathcal{R}} = \langle \dot{R}, \dot{\leq} \rangle \xrightarrow{\text{mapping}} \mathcal{R} = \langle R, \leq \rangle$ Actual numerical values (reals)

Facts *Measurement* $\uparrow$ partial mapping *Ideas*

$\mathcal{M}^* = \langle M^*, \dot{\leq} \rangle \xrightarrow[\text{partial mapping}]{} \mathcal{M} = \langle M, \leq \rangle$ Measured values (rationals)

Fig. 13.8. Measurement as a correspondence between the degrees of a property and the instrument readings via numbers.

the conceptual ones are robbed of their referents; if any of the conceptual components is left out, no link remains between the object of measurement and the instrument.*

The interval in which the degrees of a property are represented, together with the ordering and spacing of the representatives, constitute a measurement *scale*. Thus the Celsius and the Fahrenheit are two different temperature scales: every value of the temperature of a body is differently represented on either scale. Actually, for every pair constituted by a magnitude (e.g., time) and a device for measuring it (e.g., a watch) there are two scales, one conceptual and the other material. For example, the time of day can be represented (ambiguously) on the [0,12] number interval, the conceptual scale employed in designing and building most watches; the corresponding material scale is, of course, the watch dial. In general, a *conceptual scale is* the ordered interval on which the numerical values of a magnitude are represented, whereas the corresponding *material scale is* the ordered set of marks, such as the numerals on an instrument, the reading of which enables us to assign numerical values to the magnitude concerned. (In the diagram of Fig. 13.8 $\mathcal{R}$ is the conceptual scale and $\mathcal{M}^*$ the corresponding material scale.)

In principle, a number of different scales can be associated with any given magnitude; only the magnitude-instrument pair has a definite conceptual and a definite material scale built-in. The passage from one scale to another is called a *change of scale* or rescaling. One such *change of scale* is the breaking down of the time continuum, which stretches from minus infinity to plus infinity, into 12 hour periods, another familiar change of scale is the reduction employed in the drawing of maps and graphs. *A change of scale is a function f from one set R of numbers to another set R', i.e. $r'=f(r)$, such that the correspondence is one-to-one. If f is continuous, the biuniqueness of the mapping results in the continuity of the image R' of R. The usual

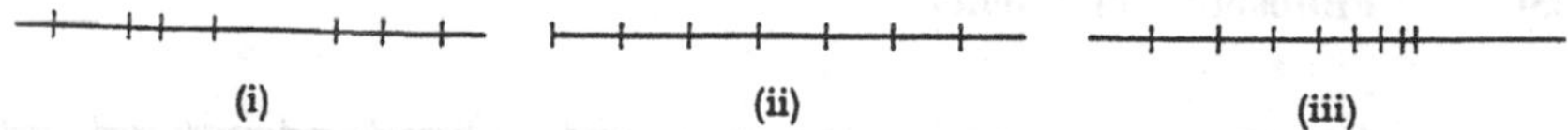

Fig. 13.9. A linearly ordered array of marks, however chaotic, affords an ordinal material scale (i). A set of orderly spaced marks constitutes a material metrical scale. Two metrical scales: uniform (ii) and logarithmic (iii). The third maps the second via the log function; neither maps the first.

changes of scale are linear, i.e. of the form $r'=ar+b$, where a and b are definite real numbers. But some instruments are gauged in a logarithmic scale, i.e. they do not show the values of the corresponding magnitude but their logarithms i.e. $r'=\log r$. And for low temperature research the $1/T$ temperature scale (i.e. the change of scale $T'=1/T$) is more convenient although it is neither linear nor continuous.*

As long as the spacing of the marks on a material scale satisfies no known law, the scale is *ordinal* or topological. Ordinal scales are, of course, those involved in the quantification and measurement of ordinal properties. As soon as a lawful (not necessarily constant) spacing among the marks is stipulated, a *metrical scale is* at hand (see Fig. 13.9). For most purposes the equal spacing or *uniform* scale is the most convenient. But any given metrical scale can be transformed into any other scale of the same kind, and the choice of scale is a matter of practicality, not of truth.

With the building of finer and finer material scales, smaller and smaller numerical intervals can be represented and consequently more accurate measurements can be performed. Thus, a partition of a yardstick into 1,000 parts will supply a finer scale than a division of the same interval into 100 parts. In general, a material scale associated with a given magnitude will be *finer* than another material scale associated with the same magnitude if and only if the set of marks of the former includes that of the latter. While the accuracy of measurements can be improved by building finer and finer scales and, at the same time, by devising more sensitive and stable instruments for discriminating among the marks, there will always remain a gap between an interval Δr of a magnitude and the physical mark m^* that represents it on the material scale: whereas the numerical interval is a set (an infinite set if the magnitude is continuous), the mark is an individual, and whereas the former is an idea the latter is a fact.

A metrical scale is incomplete unless its *zero* or origin is specified:

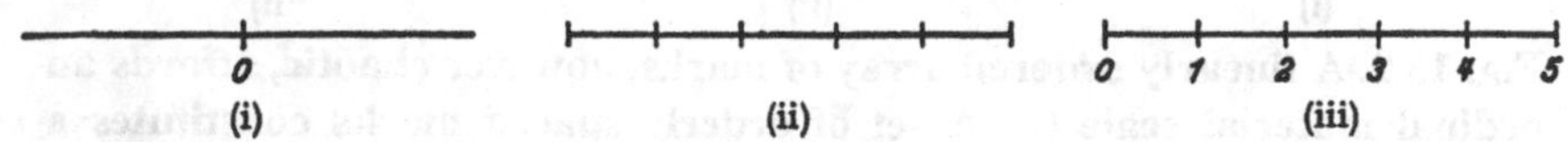

Fig. 13.10. A scale with a zero but with no unit (i). A scale with a unit but no zero (ii). A completely specified metrical uniform scale (iii).

otherwise we do not know where to count from, or at least there is an ambiguity and consequently two observers might be unable to compare, hence to control their results. The choice of the zero of a scale depends on the magnitude and on the body of knowledge resorted to. In the case of altitude, time, and empirical temperature, the origin is conventional. We may refer altitudes to sea level or to the center of the earth or to any other base line. Similarly, time may be counted from any event of interest, e.g. from the instant the process or the observation concerned begins. And the zero of the scales of empirical temperature may be chosen to be the temperature at which any clearly perceptible and easily reproducible change occurs, such as the freezing (or the boiling) of water. In all these cases the scales have *relative zeros,* i.e. origins that can be shifted arbitrarily and the choice of which is based on practical convenience rather than on theory. On the other hand length, age and thermodynamic temperature are measured on scales with *absolute zeros,* i.e. origins that cannot be shifted arbitrarily. Lengths are counted on the $[0, \infty)$ metrical scale because, by definition, length is a nonnegative number. Age scales must have an absolute zero, sometimes called the fundamental epoch, because by definition the age of x is the time elapsed since x came into being, and no time was taken up by x before it emerged as such. Finally, the choice of the zero of thermodynamic temperature is based on (classical) physical theory. Absolute zeros, in short, are not at our choice but are dependent on definitions and laws belonging to some theory or other; in most cases they correspond to the vanishing of the corresponding property (as understood in some body of knowledge).

A zero is necessary but not sufficient for having a metrical scale: the scale will remain incomplete unless a unit or basic interval is adopted (see Fig. 13.10). The reason is, of course, that the execution of a measurement boils down to finding out empirically how many *units* (or subunits) go into a given quantity. Accordingly, the adoption of a metrical scale for a magnitude involves a choice of unit: of the *concept of unit* for the conceptual scale, and of a standard physical object, i.e. a *material unit* for the material scale. Thus the hour is a time unit represented, on most watches, by a 30° arc: the former is the conceptual unit and the latter the material unit. In the material scale the unit can be an interval, spatial or not, between two given marks, such as streaks or sounds. The unit is not necessarily the smallest scale interval: the basic interval on a scale, e.g. the interval between two successive streaks, may represent a unit, a subunit, or a multiple of the basic unit. In any case, without knowing what the unit on a scale is, the operator might assign any number to the magnitude he measures. And once a scale-cum-unit system has been agreed on, the numeric r can be thought of as the number of basic intervals or units of the given magnitude. This gives rise to the usual way of expressing magnitudes, namely

$$P=ru \qquad [13.6]$$

where u is the name of the unit in which the property P is reckoned. This way of writing is harmless as long as it is kept in mind (i) that P is not a number but a function, (ii) that the arguments of P are the thing variable x and the scale variable s, and (iii) that the product 'ru' is symbolic since u is not a number.

We agree then that measurement proper—by contrast to quantification or measure—requires a choice of units. (An amazing progress in experimental psychology can be expected as soon as every practitioner of this science realizes this much.) But before we can propose a unit for a given magnitude we must answer two questions: (i) To what family does our magnitude belong, i.e. what is its dimension?, and (ii) is it a fundamental or a derived magnitude?

The first problem is raised by the variety of magnitudes that are basically of the same kind. Distance, height, depth and wavelength are just different applications of the length concept: they belong, as we shall say, to the same *magnitude family*. What characterizes this family is the common dimension L. This was probably not realized when the fathom was introduced along with the yard, and the fluid ounce along with the cubic foot. In order to emphasize that one and the same concept underlies a given magnitude family, and so eliminate a source of confusion and time waste, a single *basic unit* is adopted in science for all the members of a given magnitude family. In other words, we have the following *Convention:* Magnitudes with the same dimension shall be reckoned and measured with the same basic unit. The methodological consequence of this convention is obvious: an analysis of the dimension of the magnitude concerned should precede the determination of its units. No such analysis proper is required if the magnitude is fundamental or simple, as in the case of length or time. But most magnitudes are derived or complex: they are introduced by formulas (empirical or theoretical) that link them to fundamental magnitudes, and such formulas provide the basis for a dimensional analysis. Thus, the elementary definition of "velocity" is "$v=_{df}d/t$", where t is the numerical value of the duration taken by something to cover the distance d. Since the dimension of d is L, i.e. $[d]=L$, and that of t is $[t]=T$, the dimension of v is $[v]=[d]/[t]=L/T$. Similarly, from the definition of the acceleration, "$a=_{df}d/t^2$", we get $[a]=[d]/[t^2]=L/T^2$. The dimension of force is found by analyzing Newton's law "$f=ma$"; the outcome is: $[f]=[m][a]=ML/T^2$. All of these dimensional formulas, which result from a dimensional analysis, presuppose a body of theory. In other words, dimensional analysis is an analysis *of* formulas that either belong to or presuppose some theory.

Our second task prior to the choice of units is the distinction between fundamental and derived magnitudes, which distinction is relative to the theoretical context. In mechanics length, mass and duration are taken as independent primitive concepts and as *fundamental magnitudes* in terms of which every other magnitude is reckoned and measured. Additional fundamental magnitudes, and accordingly additional basic units, are added in other sciences whenever they are not reducible to mechanical magnitudes: this was the case of temperature in thermology and of electric charge in electrodynamics. (It is instructive to imagine what the fundamental magnitudes and consequently what the basic units would be for a rational being deprived of mass. The mere idea of an intelligent field, however wild, should teach us not to regard the status of fundamentalness presently enjoyed by certain magnitudes as anchored in nature and therefore final.)

The choice of units is conventional in the case of the fundamental magnitudes, but it always consists in adopting a certain materialization of the given concept. Thus, the length of the King's foot may be chosen as unit length, the King's inertia as mass unit, and the minimum lapse between two royal sneezes as the duration unit. In all three cases, the procedure is one of *referition* or sign-physical object relation (recall Sec. 3.6). More precisely, a concept is coordinated to a physical object (thing or fact) conventionally chosen as a *standard* or material unit. This kind of referition was called *coordinative referition* in Sec. 3.6 and is often, and misleadingly, characterized as a definition.

It has sometimes been supposed that, since the basic units of the fundamental magnitudes are adopted by convention, the measured values of physical magnitudes, and perhaps their referents as well, are arbitrary and subjective. Such a conclusion must indeed be drawn unless the distinction between a property and its measured values is drawn. If this distinction—made in Sec. 13.2—is accepted, it is easy to understand that measured values, by contrast to the properties themselves, should have a conventional component: after all magnitudes are not traits of concrete systems but conceptualizations of them. Length is not conventional but the *choice of a standard* of length (or duration, or mass) is conventional and therefore every *measured value* is conventional to the same extent. Similarly, the choice of a way of representing human faces is conventional but the faces themselves are not conventional—save in a colloquial sense of 'conventional'. *The objectivity of the concrete property as distinct from its numerical values in the various possible unit systems is expressed by the following law of *invariance under unit changes:*

$$P(x)=ru=r'u' \qquad [13.7]$$

which can be read: The magnitude P representing a certain property of an object x is the same whether reckoned in units u or in units u'. (This postulate of *objectivity* is subsumed under the more general axiom of *covariance,* according to which properties and their relations are independent of human conventions.) An immediate consequence of the preceding postulate is the law of unit conversion:

$$r'=r\,(u/u'). \qquad [13.8]$$

That is, if the numeric of a property, reckoned in units u, is r, then the numeric of the same property reckoned in units u' equals the former numeric times the *conversion factor* u/u'. The conversion factor involved in the equality "100 cm=1 m" is u/u'=1/100, and the conversion factor enabling us to switch from seconds to minutes is 1/60. Unit conversions would be lawless if the objectivity postulate [13.7] were not assumed.*

Standards, i.e. material units of fundamental magnitudes, are chosen conventionally but not arbitrarily: they must be fairly precise, constant and easily reproducible, in order to make objective comparisons through time and space possible. This is why a certain metal bar (the standard metre) came to replace the various medieval feet and, later on, a certain light wavelength took the place of the metal bar. Similarly standards of time, mass, voltage and other magnitudes were developed. Every advanced country has a bureau of standards engaged in the preservation, production, and refinement of material units or standards. In some cases nature itself, as it were, supplies such constant and repeatable units. Thus, the mass and charge of the electron and the velocity of light in vacuum function as "natural" units in atomic physics. And if physical space had a constant curvature radius this would be the natural length unit. Needless to say, such natural units satisfy the condition of constancy better than man-made standards, but they usually supplement rather than replace the latter, because they are not always to the measure of man. Thus it would be no more practical to measure the charge in a condenser by comparing it with the electron charge than to measure geographical distances in fermis (1 *fermi* = 10^{-13} cm) or plot land areas in barns (1 *barn* = 10^{-24} cm^2).

Once a measurement unit has been determined by the choice of a material standard, the measurement of any value of the corresponding magnitude—for example the measurement of the duration of a process—may be made to consist in comparing the given quantity with the unit as represented by the standard. Measurements of this kind are often called absolute, all other comparative. (At other times, a measurement is called absolute if it boils down to measuring geometrical and mechanical magnitudes alone.) But these are misnomers, both because every measurement involves a comparison and because basic units are conventional; *standard* and *nonstandard* measurements would be less misleading names.

After a basic unit has been chosen and a materialization of it has been constructed, multiples and submultiples of either can be generated at will by successive multiplication and division by a given number. In this way a *system of units,* not just a set of units, is generated. If the conversion factor is 10, a decimal metrical system is developed. In mechanics the centimeter, gram and second are usually adopted as basic units out of which all other units of the same fundamental magnitudes are built: one therefore speaks of the *CGS* system. In electrical engineering the *MKSA* (metre, kilogram, second, ampere) system of basic units is usually employed.

The adoption of a system of basic units for the fundamental magnitudes of a discipline uniquely determines the units of the derived magnitudes. This is not done in a theoretical vacuum but via definitions and law statements belonging to some theory. Thus, the unity of velocity is introduced on the basis of the definition "$v=d/t$" and depends on the choice of distance and duration units. And the unit of force is usually introduced on the basis of Newton's law "$f=ma$", namely thus: the force unit, called the dyne, is the force that gives a unit acceleration (1 cm/sec^2) to a body of unit mass (1 g); that is, 1 *dyne* = 1 g-cm/sec^2. (In the foot-pound-second system the force unit will be the *poundal*, a monster equal to 1.3825×10^4 *dyne*.) Notice that 'dyne' is the *name* of the force unit rather than the force unit itself, which is the force the numerical value of which in the *CGS* system is unity. Notice further that unit names are treated as if they were numbers, which they are not. For example the pressure unit or microbar, is defined thus: *microbar*=dyne/cm^2=g-cm/sec^2-cm^2=g-cm^{-1}-sec^{-2}. In general, derived units are obtained by multiplication or division of basic units of fundamental magnitudes without introducing numerical factors, and using theoretical formulas (definitions or law statements). A set of basic units and derived units obtained in this way is called a *coherent system of units*. The *CGS* and the *MKSA* are coherent systems, whereas the set {cm, lb, sec, poundal} is incoherent.

Material standards of derived units can, of course, be built for practical purposes, and so referitions of derived units can be added to their specification in theoretical contexts. But from a fundamental point of view such referitions are dispensable. In all rigor, basic units of *fundamental* magnitudes alone must be introduced through referitions of the coordinative kind, because they are based on no prior concepts. In other words, standards (material units) are needed only for the measurement of properties denoted by *undefined* concepts. In turn, only foundational research can determine which concepts in a theory are primitive or fundamental and which not.

Units, even units of fundamental magnitudes, are not proposed in a vacuum: the conceiving of a new unit may require considerable substantive knowledge, and proposing it may elicit hot controversy. Controversies about the advantages of the various unit systems do not, of course, concern their truth, which none of them could possibly possess since they are not sets of statements although they do presuppose such sets (namely, theories). But such controversies are not always senseless: a system of units should be (i) *coherent,* in the sense that every item in it should be either fundamental or derived from fundamental units by multiplication or division; (ii) *theoretically convenient,* i.e. adequate for theoretical discussion, and (iii) *practically convenient,* i.e. adequate for experimental and/or technological purposes. While coherence must be required from every system of units, no system can possibly meet the conflicting desiderata (ii) and (iii) and thus satisfy both the pure and the applied scientist. The theoretical physicist will prefer the *CGS* system to the *MKS* system in mechanics whereas the engineer will show the converse preference, each with his own good ground. But neither will choose the inch-stone-wink system: conventions need not be absurd.

Two examples will make clear the amount of knowledge required to introduce a unit. First example: suppose we wish to measure bluffing, that curious pattern of animal behavior. If we are operationalists we may start by choosing a standard of bluffing. We may, for instance, pick up a remarkable bluffer call him Farcia—and stipulate that he, or rather his peculiar behavior, will serve as the bluffing unit—the *farcia.* But having the farcia—or some more practical subunit, say the milifarcia—we still do not have a complete bluffing scale because we have not stipulated what the zero of the bluffing scale is. And we shall be unable to choose a zero unless we form a clear concept of bluff, which requires in turn some theory of bluff. Hence we had better reverse our path, starting by building a quantitative theory of bluff. In such a theory, bluff would conceivably be treated as a derived magnitude—unless it happened to be a phenomenological theory—and so the problems of the zero and the unit of the bluff scale would be shifted to the zero and unit of the scales of the fundamental magnitudes, say ambition, incompetence and dishonesty. Suppose further that imaginary theory were based on a set of postulates relating the mutually independent concepts of ambition A, incompetence I, and dishonesty D, all of them metrical. Assume further that the scales of A, I and D were all uniform metrical scales ranging between 0 and 1. We would then introduce the units of ambition, incompetence and dishonesty via referitions, since those are primitives in the theory. Suppose we decided on Caesar, Nero and Cicero as material units or standards of ambition, incompetence and dishonesty respectively, and called the corresponding concepts *caesar, nero* and *cicero.* We might then introduce the concept of bluff into the theory by means of the definition:

$B =_{df} A \cdot I \cdot D$; in vernacular: bluff equals ambition times incompetence times dishonesty. Minimal bluff would then correspond to either zero ambition or to maximal competence or maximal honesty (i.e., if $A=0$ or $I=0$ or $D=0$, then $B=0$). And maximal bluff to $A=I=D=1$. Thereupon the farcia would become a derived unit, namely: farcia=caesar-nero-cicero. Ours was, of course, a bluff of a theory, but in any case the point is that the *concept* of a unit must be introduced before an adequate materialization or standard is built, particularly so in the case of derived magnitudes.

Second example: before information theory was invented, the search for a unit to measure the information content of messages seemed as hopeless as the introduction of a bluff unit. One might have thought of measuring information content by the number of signs, or the time required to understand it, or the total length of codework necessary to decipher the message, and so on. The eventual adoption of the *bit* was based on a definite theoretical formula ($I= \log_2 n$, where n is the number of simple signs employed in communication), which in turn was the quantitative expression of the idea that a transmission of information takes place only if the uncertainty or ignorance of the receiver is thereby decreased (recall Sec. 1.3). It was only natural to adopt as the unit information—the bit—the least amount of information necessary to remove uncertainty in the simplest situation, namely that in which prior to the message reception there had been two equally likely possibilities, such as the sex of a newborn child.

Let us pause and collect the items we have so far picked up. These are: (i) the *mensurandum* or property of the concrete system we wish to measure; the mensurandum is supplied by nature or by society but its existence is often suggested by theories; (ii) a clear quantitative (metrical) *concept* of the mensurandum, i.e. the magnitude that is supposed to represent the objective property; this concept should, so far as possible, be immersed in some scientific theory and logically analyzed in terms of object variable(s) and numerical variable(s), in order not to miss some relevant aspect; (iii) a *conceptual scale* and a *material scale* on which the magnitude can be reckoned or measured; the scale, if metrical, will have a privileged mark—the origin or zero—and (iv) a *unit* of measurement, belonging to some coherent unit system.

These four items are necessary but insufficient to undertake a measurement operation: we must first devise a measurement technique and design and build a measurement equipment. To this task we now turn.

Problems

13.4.1. Study the problem of choosing a conceptual and a material scale for some magnitude not dealt with in the text. *Alternate Problem:* Examine the problems of scale that arise in connection with the introduction of negative absolute temperatures. Take into account that when this scale extension is performed the mapping of thermal states on numbers ceases to be order-preserving: in fact, negative temperatures are made to correspond to the hottest bodies rather than to the coldest; an "infinitely hot" body is assigned a negative temperature and, upon cooling down, its temperature will jump to a high positive value, which may in turn decrease to nearly zero.

13.4.2. A set of degrees of a property is mapped into a set of numbers in such a way that every member of the first set has only one numerical image. On the other hand the measured value of a magnitude is itself a set (an interval) if the error (plus or minus σ) is added to the central value $\overline{m}$; hence a set of measured values is a set of sets. What are the correspondences between this latter set and (i) the set of actual values and (ii) the set of marks? *Alternate Problem:* According to thermodynamics no system can be cooled down to absolute zero. Yet the values of several magnitudes at that temperature are evaluated and, moreover, the origin of the thermodynamic temperature scale is precisely 0 °K. That is, the assumption of an objective temperature zero is made in thermodynamics and underlies temperature measurement, yet at the same time it is shown that the absolute zero is empirically unattainable. Solve this conundrum.

13.4.3. Psychologists usually speak of *scales of measurement* meaning what natural scientists call 'magnitudes', 'variables', or 'measurable quantities'. For example, the former call ordinal magnitudes, such as hardness, ordinal *scales. Is* this advantageous? On the other hand, psychological measurements are notoriously slack in the matter of units. Why? *Alternate Problem:* Do the meaning and the truth value of scientific constructs depend on the choice of scale?

13.4.4. Does geometry make any provision to fix a distance unit? Does theoretical mechanics fix the unit mass? And does atomic physics suggest natural units? And would it be possible and desirable to introduce a unit system based on atomic physics? See W. W. Havens, Jr., "Modern Physics has its Units Problems", in C. F. Kayan, Ed., *Systems of Units* (Washington: American Association for the Advancement of Science, 1959).

13.4.5. Two independent basic units of time are in use: the mean solar second and the ephemeris second. Usual clocks "keep" mean solar time, and position measurements in field, by surveyors and navigators, rely on mean solar time; on the other hand, astronomers reckon time in ephemeris seconds because the mean solar second is not exactly the same for all years (i.e., the motion of the earth is not completely regular). Comment on this duality. *Alternate Problem:* Examine the procedures for establishing time standards and their peculiar difficulties due to the immateriality and elusiveness of time. See, in particular, H. Spencer Jones, "The Measurement of Time", *Endeavour*, **4**, 123 (1945) and G. M. Clemence, "Astronomical Time", *Reviews of Modern Physics*, **29**, 2 (1957).

13.4.6. From the fact that the velocity unit equals the unit distance over the unit duration many textbooks infer that the *concept* of velocity can be defined as distance per unit duration or even as "space per unit time". Examine this inference. *Alternate Problem:* From the fact that law statements are *invariant* under unit conversions it follows that no unit need be specified for theoretical purposes, this being why the discussion of units has been postponed to this moment. Does it follow that the whole language of science is unit-free?

13.4.7. Examine the proposal to adopt the *darwin* as the unit of rate of evolution. See J. B. S. Haldane, *Evolution*, **3**, 51 (1949).

13.4.8. Are fundamental magnitudes connected with the possibility of "defining" them in an operational way? In other words: does the identification of a magnitude as fundamental or derived depend on the way of measuring it or on its place in some scientific theory?

13.4.9. N. R. Campbell, in his influential *Foundations of Science* (1920; New York: Dover, 1957), called *fundamental magnitudes* those "the measurement of which does not depend on any other magnitude". Can fundamentalness be diagnosed outside some theoretical context?

And are the most accurate measurements of length, mass and duration (fundamental magnitudes) in fact fundamental in Campbell's sense (i.e. simple, direct, not further reducible)?

13.4.10. A clock in motion relative to a given frame is slower than a clock at rest relative to the same frame. This result of special relativity is sometimes interpreted as a dilatation of the time *unit* attached to the moving clock. If this interpretation were correct, would it be possible to measure the difference between the times "kept" by the two clocks? Hint: Recall that only the numerics of magnitudes occur in law statements that relate magnitudes with the same dimension.

13.5. Techniques

Measurement techniques are among the distinctive traits of modern science. They depend on the nature of the mensurandum, on the knowledge at hand, on the accuracy requirements, and on the operator's ingenuity and dexterity. That measurement techniques should depend primarily on the nature of the mensurandum in its relation to human observers, is clear: it is one thing to measure accessible distances, quite another to measure interatomic or interplanetary distances. Although the same *concept* of distance is involved in all three cases, the procedures for measuring its numerical values are very different: in the case of bodies that can be handled directly the measured value can be read on a material scale such as a graded ruler, whereas in the case of practically inaccessible bodies related magnitudes are actually measured and the mensurandum is found with the help of formulas. Here we have a first dichotomy with an epistemological import: measurements can be direct or indirect.

A *direct measurement is* made by comparing the mensurandum with a standard or with the units on a material scale and counting the number of times the unit is contained in the mensurandum. An epistemologically interesting problem in this case is the one raised by the coordination or congruence between the mensurandum and the marks on the material scale. Suppose the length of a body is measured by placing a number of yardsticks end to end. From a semantical point of view this procedure amounts to establishing material correlates or interpretations of the arithmetical signs '+' and '=': the former is made to correspond to physical juxtaposition ($\dot{+}$) and the latter to physical congruence or coincidence ($\doteq$). Furthermore, it is assumed that the material model obeys the laws of arithmetic as interpreted in that way. For instance, that the juxtaposition of yardsticks is commutative ($x \dot{+} y \doteq y \dot{+} x$) and that lengths are not altered in the process of measurement ($L(x \dot{+} y)=L(x)+L(y)$). It is also assumed that the results of these operations are independent of space and time; in particular, that they do not depend on the orientation of the bodies. But these are all very specific *physical hypotheses* that might be false. Thus it could well be that space, or at least the region of space where a measurement is performed, were not isotropic but that lengths depended on the orientation—as is in fact the case of the dimensions of a zinc rod placed in a field of ultrasonic waves. It might also occur that the yardsticks were compressed or expanded by the juxtaposition process. These various assumptions can in principle be checked and they are found to be very closely realized by rigid bodies—not, however, by marshmallows. But quite aside from the thorny problem of the existence of perfectly rigid bodies, the point is that, upon performing any length measurement, we tacitly make a number of assumptions; some of these are about the nature of the bodies involved (in particular the measuring instruments) and others regard matter and spacetime in general. In short empirical operations, and particularly measurements, cannot be made an absolute starting point for the construction of science, because they presuppose a number of hypotheses.

That measurement presupposes some amount of theorizing—at least if one wishes to know what he is measuring and how accurately he is doing it—was argued in the preceding sections and was dramatically illustrated by duration measurement. The particular kind of congruence involved in duration measurement is simultaneity. While the simultaneity of close perceptible events is easily established, ascertaining the simultaneity of distant events requires at the very least the hypothesis that the signals involved in establishing a time order propagate with a finite velocity that depends neither on the region of spacetime nor on the direction of propagation. The special theory of relativity, which has occasionally been hailed as an exercise in operationalism, actually started by presupposing Maxwell's theory of light signals and by constructing a definition of the simultaneity concept in agreement with this theory though out of tune with the measurement practices prevailing at that time—which practices presupposed classical physical theory.

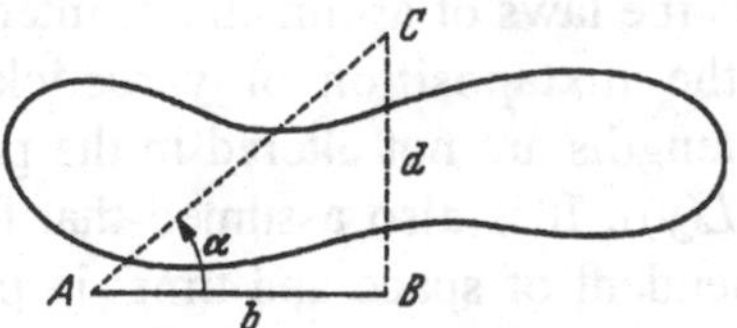

Fig. 13.11. Measurement of distance to inaccessible point.

A further point of interest in connection with direct measurement is the question whether angles, lengths, times and weights are directly measurable—in the case of empirically accessible systems—because they are directly observable. If this were so, direct measurement would be just exact observation. But it is not so: what we observe is not the length of a body, but the given body—or rather its outside—and a standard body or some material scale. Upon measuring we observe the relation of coincidence or congruence of the ends of the side that is being measured with certain marks on the scale of the measuring rod. Whole concrete systems—things, events and processes—rather than their separate properties are observable; on the other hand no thing is measured: only some properties of things are measurable.

Let us now turn to *indirect measurement*, i.e. measurement involving both direct measurement (of something else) and computation. Consider the problem of measuring the distance to a point lying across a lake: see Fig. 13.11. We first build a right-angled triangle consisting of an arbitrary base *b* traced on the ground and the light rays going from the inaccessible point *C* to the ends *A* and *B* of the base—which rays inform us, of course, about the existence of something at *C*. Our triangle is physical rather than mathematical: the "points" *A*, *B*, *C* are extended bodies and the "sides" are material systems as well. Next we measure the length *b* of the base and the angle a between the base and the "line" *AC*; the former we do with a tape and the latter with a goniometer. With the data *b* and α, and elementary geometry, we can infer the value of the unknown *d*. To this end we make the assumption that our *physical* triangle is a concrete model of a *mathematical* Euclidean triangle—or rather that it is a close enough realization of it. We make the additional assumption that our measured values are close enough to the actual values—i.e. that the *empirical* triangle set up in the process of measurement coincides reasonably well with the physical triange. The two assumptions are approximately rather than exactly true: a more refined measurement procedure (even with the same technique) is conceivable, whereby a closer agreement between the three triangles is attained; also, a non-Euclidean geometry might have to be used for triangles of a cosmic size. In any case we make certain assumptions leading to the adoption of a piece of physical geometry underlying our measurement technique. We now feed the measurement data into this piece of geometry, namely thus: we compute the required distance d from the formula "$d=b$ tg α", making use of a table to find the value of tg α for the particular value of α found by measurement. In every case we replace in the formulas the best values obtained from measurement, i.e. the mean values, rather than the full pair ⟨average, error⟩.

Angles, lengths, durations and weights are directly measurable in many cases, that is, they can be measured by comparing them with units on material scales or with standards. Thus durations can be measured by counting the swings of an oscillating system adopted as a time standard. But direct measurements are the exception rather than the rule in pure science. Moreover, for purposes of recording it is convenient to reduce every measurement to angle and/or length measurements. This reduction is done via definite formulas that link the various nongeometrical magnitudes, on the one hand, with lengths and angles on the other. In particular, such statements may establish relations among nonobservable magnitudes and observable angles and lengths. In other words, law statements and definitions are often used in measurements as objectifiers of nonobservable properties and relations; it is by means of such formulas that the material scales of electrical instruments are built.

Ex. 1. The duration of a process may be made to correspond to the change in position of a pointer on the scale of a chronometer. In this case the linear relation "$t=k\ \alpha$:" between times and angular displacements is employed. The indications cease to be right when this relation fails, as is the case of worn-out clocks or of the revolutions of our planet around the sun, which fit a more complicated law. *Ex. 2.* The most common temperature measurements are based on the temperature dependence of the size of bodies particularly sensitive to temperature changes, such as mercury and alcohol. In this case too a linear relation is assumed to hold between the length and the temperature of the thermometric substance, a relation that objectifies the temperature. Again, this relation is far from being conventionally posited: it is a partially true law statement that fails outside limited intervals. *Ex. 3.*

The strength of an urge such as hunger can be measured by the quantity of gastric juice secreted by the subject at the sight of food. In turn the quantity of gastric juice may be reckoned as volume and, in turn, volume measurements can be reduced to measuring lengths along tubes of a known constant section.

Why are angles and lengths usually chosen as independent variables for measurement rather than durations and masses? There seems to be no reason for this other than the accident that man's most evolved sense is sight. Nonhuman rational beings are conceivable with a more acute perception of duration or even of electric charge than of size and shape. The centrality of angle and length in measurement does not seem to be due to an ontological priority or physical fundamentalness of space but rather to a trait of our perceptual apparatus—but this conjecture may prove to be quite wrong. A number of magnitudes are not related to geometrical magnitudes via law statements occurring in some fundamental theory, whence those formulas are insufficient for measurement purposes. For example, electric circuit theory provides formulas linking the voltage and the current in a circuit but none of these is a geometrical magnitude and consequently that theory is "operationally meaningless"—whatever this expression may mean. If we wish to measure those magnitudes we must search for law statements containing both the mensurandum and some geometrical magnitude. This is a task of *instrument design*.

Suppose we wish to measure the intensity of electric current, a typical unobservable. By just looking at the circuit in which there is a current we shall be unable to attain our goal: we shall have to set up an observable effect of the current on some system other than the circuit itself. In other words, we must set up an *interaction* between the system subjected to measurement and some measurement device, whereby the former can modify the state of the latter in an observable way. In the case of the electric current, a system that may serve as an indicator of what goes on in the circuit is a magnetic needle, because electric currents are surrounded by magnetic fields which interact with magnets. More precisely, the angular displacement of a magnetic needle will be an objective indicator of the current intensity. The precise relationship between the two variables is established by the theory of electrical meters.

Certain conditions will have to be met for the needle displacement to constitute a *reliable* objective indicator of the electric current. First,

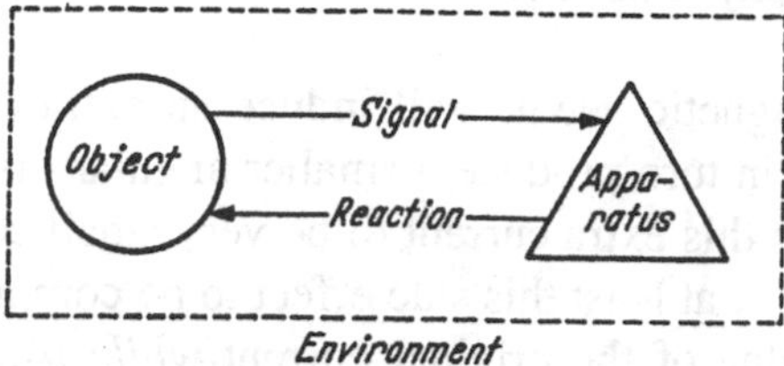

Fig. 13.12. The partial isolation of the object-measuring instrument complex from the rest of the world. The enclosure (dotted line) must reflect or absorb most incoming signals.

the needle and, in general, any measuring instrument, should record the signal under consideration, with no appreciable interference from the rest of the world and, in particular, with no intervention of the operator: otherwise we would not be measuring just what we wish to measure but something much more complex. In order to focus on the mensurandum and minimize external disturbances we must begin by insulating the object-instrument complex. For example, we shall enclose our electrical instruments in a Faraday cage, or we shall place our dog in a Pavlov observation cage. In any case, our instrument will record the outcome of its interaction with both the object under measurement and the rest of the world, and we shall design the set-up so that the second interaction becomes negligible with respect to the first one (see Fig. 13.12). If our apparatus has no automatic operating and recording device we shall have to take the readings while the interaction is on, thereby running the risk of disturbing the object in an unpredictable way and consequently adulterating the result. This is one reason for preferring throughout automatic instrumentation. In any case, exact measurements are not a special case of the subject-object interaction: the cognitive subject is not supposed to disturb the objective object-apparatus interaction but to set it up and to interpret its outcome. Ideally, the observer steps in once the interaction is over, in order to collect and interpret the signs left on the instrument by its interaction with the object. Hence the object of epistemology is not the object of measurement but rather the whole complex object-measuring instrument.

A second condition to be fulfilled by an ideal instrument design is that the *reaction* of the apparatus on the object be either negligible or computable. In the case of electric current measurements the movements of the magnetic needle will induce an extra current in the circuit, which will in turn produce a smaller angular displacement of the needle. We want this extra current to be very small compared with the original current, or at least this side effect to be computable, so that we can infer the value of the original current *while it is not being measured.* This actual undisturbed value is not attainable through measurement alone but with the help of theory. Every experimentalist takes such a disturbance into account and attempts to assign it a definite numerical value. In the case of electrical measurements with instruments more refined than a modest magnetic needle, the internal electric resistance of the apparatus is separately measured and taken into account to correct the measured value. In general, we may say that the original effect is distributed between the measured effect and the measurement disturbance, and that a good instrument design both minimizes the disturbance and supplies means for estimating it. Calling $m(E)$ the raw measurement result and ΔE the measurement disturbance—e.g., the fraction of current dissipated in an ammeter—we may say that the actual value of the effect is $E=m(E)+\Delta E$. For a strict empiricist only $m(E)$ is meaningful, E and ΔE being metaphysical misfits. The fact that scientists are most keen on corrections of this kind and that they adopt the corrected values E rather than the $m(E)$ values suggests that they do not practice an empiricist philosophy of empirical procedures.

*Atomic measurements raise difficulties of their own into which we cannot enter here because they require a quantum theory of measurement—an almost nonexistent subject. We shall only remark, first, that atomic measurements are often part of atomic *experiments,* as they consist m measuring some interaction between a microobject and a macroscopic apparatus that can introduce important modifications in the value of some property, so much so that ΔE can become comparable with E itself and therefore mask it completely. Second, that a suitable measurement of an atomic variable is such that this interaction does not alter the value of the mensurandum, however much it may alter all the others. In other words, the measurement design must be such that the value obtained by measurement is the same as the value the property is supposed to have had *before* measurement—which clearly presupposes that the object had that same property before it was brought to light by measurement. And this ruins the Copenhagen interpretation, according to which properties are not possessed by objects but are conjured up by the operator upon measurement. Be that as it may, the quantum interactions affect certain other properties associated with the mensurandum, and the theory is not yet in a position to compute the exact value of such disturbances although it can estimate their lower bound. The current theoretical limitations on the prediction of every aspect of the instrument-object interaction has been exalted to an impossibility in principle, thus barring the very attempt at analyzing the measurement interaction. Thirdly, spectroscopic and other measurements are not experiments proper and therefore do not involve an appreciable disturbance of the apparatus on the object, so that the quantum theory of measurement does not apply to them—which does not inhibit the orthodox from making the sweeping generalization that every quantum-mechanical formula refers to some experiment or other. The whole subject is in its infant stage and therefore immature for engendering long-lived philosophical children. The philosopher must limit himself to pointing out this circumstance, i.e. that the quantum theory of measurement is an unsettled problem, and to uncovering presuppositions, such as that the very care taken to avoid changing the mensurandum during the measurement operation pressupposes the objective existence of the mensurandum before its measured value is obtained—which in turn contradicts the subjectivistic (operationalistic) philosophy espoused by the Copenhagen school.*

A third requirement to be met by a good instrument design is that it be based on *definite assumptions* concerning the relation between the mensurandum and the observed effect—and moreover on reasonably well corroborated assumptions. Suppose an instrument designer claims that a certain apparatus of his "defines" humidity. He would be asked 'How can we know your instrument in fact measures what you claim?'. If the designer were unable to offer an *independently checked theory* explaining how his device does in fact measure humidity rather than, say, timidity, he would be dubbed as a psychological tester rather than as a physical scientist. Anyhow it is at this point where geometrical magnitudes are usually introduced in instrument design. In the case of an ammeter, its material scale and consequently our readings of it are based on a definite law that maps currents onto pointer angles. In other words, an ammeter design is based on a law statement that relates the current intensity to the angle deflection of a small magnet (or rather electromagnet) checked by a spring; that is, the instrument design is based on a theoretical formula "$i=f(\alpha)$" that objectifies the unobservable current intensity i as an observable angle α. This relation is not found in either the theory of electricity or in mechanics separately but is set up in the theory of electrical instruments, which uses fragments of electrodynamics and mechanics. There are alternative ways of measuring electric currents but all of them rest on definite hypotheses concerning the mensurandum-observable index relation.

Good instrument design, in short, involves a number of fragments of theories: those regarding the relation between the mensurandum and observable physical magnitudes. In this way a number of logically independent theories become related. For instance, although classical mechanics is independent of optics, the *tests* of the former involve in some way or other a number of hypotheses of the latter—and conversely. The bulk of the body of theory entering in instrument design will be the larger the more indirect and the more accurate the measurement. Suffice it to recall the theories involved in measuring atomic masses: their function in measurement is to provide the links of a long chain going from the transobservational concept "atomic mass" all the way down to descriptive concepts such as "oscillation amplitude of a luminous spot on the screen of an oscilloscope".

The design of instruments culminates the construction of a measurement technique. (The human observer may be regarded in this respect as a special kind of instrument, and his training for performing a scientific observation as the analogue of instrument building.) The end result of a piece of instrument design is a set of blueprints or diagrams representing the ideas conceived by the instrument designer. There is a gap between the blueprint conceived by the scientist (eventually helped by a technologist) and the actual apparatus built by the technician or by the scientist himself: only very coarse designs are accurately realizable. The gap will become apparent on determining the performance of the instrument (calibration). The apparatus may not even be realizable at the moment for lack of either technological knowledge or practical know-how. Even if realizable, a number of unforeseeable problems are likely to come up which cannot have a perfect solution: vacuum leaks will occur, some materials will wear out too rapidly, the division of the material scale will be inaccurate, and so on. Many such shortcomings may be remedied by adding additional pieces of apparatus such as screws for level regulation, voltage stabilizers, and so forth. But all these material correction devices will, in turn, have some fault or other. Perfect measurement instruments are therefore unattainable. All we can obtain, and do strive for, is *improved* instruments. According to what was said above, improved instruments will be based on the specifications of improved instrumentation, which in turn requires improved instrument theory, which ultimately depends on improved basic theory. Theory, then, is the basis of measurement and not the other way around.

It is only after the required measuring instruments have been designed, built, and calibrated, that measurement proper can begin. We shall skip the many technical details of instrument manipulation—which have no philosophical significance save for pragmatism—and shall leap to their outcomes.

Problems

13.5.1. Analyze a direct measurement technique.

13.5.2. Analyze an indirect measurement technique.

13.5.3. Give a short description of any of the meters you are familiar with, such as the car speedometer or the domestic gas and electricity meters, and point out the elements of physical theory their design and reading involve.

13.5.4. Discuss the significance of dating techniques, such as the radiocarbon technique, for palaeontology and anthropology. See, e.g., D. Brothwell and E. Higgs, Eds., *Science in Archaeology* (New York: Basic Books, 1963). For the tests conducted to confirm the suspicion that the Piltdown Man had been a forgery, see the entertaining book by I. Adler, *Stories of Hoaxes in the Name of Science* (1957; New York: Collier Books, 1962).

13.5.5. Are laws found by measurement or are they rather tested by measurement involving other laws? *Alternate Problem:* Examine the technique for measuring (actually, computing) planet-sun distances and evaluate the function of theories in it. Remember that at least three theories enter any astronomical computation of that kind: mechanics (e.g., laws of motion), gravitation theory (chiefly Newton's law of gravitation), and optics (e.g., light aberration).

13.5.6. Are measured properties produced by measurement or are they objectified and estimated by it? *Alternate Problem:* Analyze the null method of electrical measurements, in which no pointer readings other than that of the zero value are taken. In this technique, voltages in different parts of a circuit are adjusted until they balance, so that no current flows. The application of this technique depends on theoretical formulas that are independently checked by meter readings. A similar technique is employed in psychophysics.

13.5.7. Why are different techniques for measuring the same quantity devised? Could we not settle on the most accurate technique and forget about the others?

13.5.8. From the fact that it is convenient to reduce all measurements to angle and/or length measurements, it has been inferred that all magnitudes are *au fond* angles or lengths. Is this conclusion correct? *Alternate Problem:* Speculate on the possibility of there being nonhuman rational beings sensitive to infrared waves (emitted by all hot bodies) but insensitive to the kind of light we react to. What kind of measurement instruments would they probably use?

13.5.9. Make a detailed methodological analysis of any standard manual of psychological measurement.

13.5.10. Study the problem of measurement in the light of quantum mechanics. See D. Bohm, *Quantum Theory* (Englewood Cliffs, N.J.: Prentice-Hall, 1951), Ch. 22; J. A. Wheeler and W. H. Zurek, Eds., *Quantum Theory and Measurement* (Princeton: Princeton University Press, 1983); M. Cini and J.-M. Lévy-Leblond, Eds., *Quantum Theory Without Reduction* (New York: Adam Hilger, 1990). In particular, investigate whether it is correct to make reference to measurement in the axioms of the quantum theory, since the theory must be at hand before it can be applied to a discussion of measurements in the light of that theory—whence the quantum theory of measurement comes, if at all, last rather than first in any treatise on quantum mechanics.

13.6. Upshot

Once a measurement equipment has been designed, built and tested (see the preceding section), measurement proper can commence. The

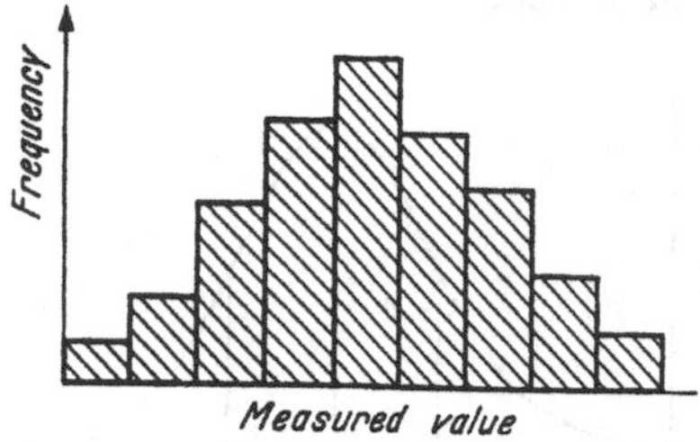

Fig. 13.13. An (imaginary) histogram or frequency distribution displaying the results of a run of measurements. The ordinate shows the number of roughly identical measured values.

raw results of a run of measurements will be a set of numbers: thus a precise mass measurement may yield 100 slightly different numbers i.e. 100 answers to the original question (see Fig. 13.13). These raw data will have to be processed and condensed in a search for a single answer as only a *refined* datum is subjected to interpretation and computation and only such single refined information can be fed into a theory in order to test it or to derive further data. In other words, the final report on a run of measurements will not be a set of raw measurements but a single statement representing the whole run. This representation is done via two statistical concepts: the average $\overline{m(r)}$ of the set of measured values and the scatter or standard deviation σ of the set around the mean. Thus, the Mariner II exploration of Venus (1962) yielded, among others, the following measurement outcome: "The mass of Venus is 0.81485 times that of the Earth with a 0.015 per cent standard error". In this case the average was 0.81485 u, where 'u' designates the mass unit, and the standard error was σ= (0.015/100)×0,81485≅0,00012. In general, a value estimated on the basis of a run of measurements will be stated in the form (see Sec. 13.2)

$$m(\dot{r}, s)=\overline{m}(\dot{r})\pm\sigma. \qquad [13.6]$$

Let us now deal with the process of data processing and its end result. The main stages of data processing are, as we saw in Sec. 12.4, standardization, systematization, the study of errors, and the computation of the "exact" (average) value. The standardization of the readings is their reduction to a uniform ideal environment (standard pressure and temperature conditions), so that the data can be compared with values obtained by other workers. Data systematization is the

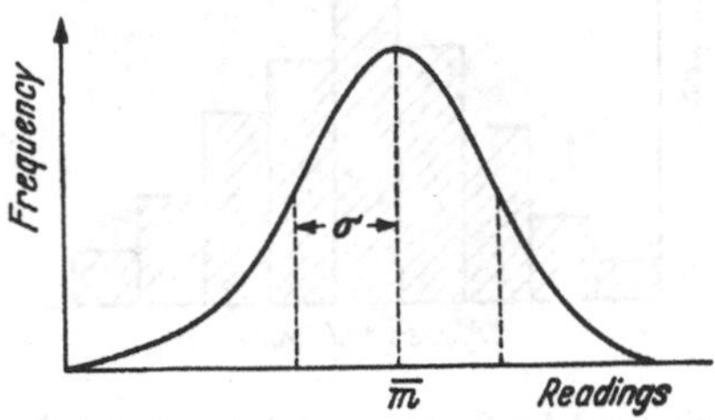

Fig. 13.14. The Gaussian distribution: a theoretical curve approximated by most quantitative data. Compare with Fig. 13.13.

ordered display of data in the form of tables, frequency distributions (histograms), and so forth; it may go as far as summing them up and generalizing them into empirical curves or empirical functions. We have dealt with data reduction and systematization in Sec. 12.4; let us now take a look at the last two stages of data processing.

Once systematized, measurement data are examined in search for experimental errors. Observation and measurement errors are of two kinds: systematic and random. *Systematic errors* are due to deficiencies in the design or execution of the measurement. For example, the lighting system may overheat the object under measurement producing a systematic bias towards larger lengths, or the fluctuations of the atmospheric pressure may produce changes in a gas volume. In the first case the error source can be eliminated or greatly weakened; in the second case the correction can be computed by means of a theoretical formula. Systematic errors, then, are those which result from overlooking variations in the environment (deficient insulation of the object-apparatus complex) or from irregularities in the equipment: they show up as one-sided deformations of the actual value and can in principle be removed or corrected for by means of theory. Systematic errors can often be discovered by checking the set-up or by controlling the results by independent techniques, or with the help of theory. If systematic errors are found out the whole set-up had better be reexamined, readjusted and retested, and the run of measurements subsequently repeated. No thoughtless increase in the number of measurements can make up for systematic errors; this applies, of course, to any kind of mistake. With nonsystematic or random errors it is the other way around.

Systematic errors can most easily be discovered and tamed by examining the distribution of a large number of measured values; these usually spread symmetrically only if random or nonsystematic influences are at play. If, on the other hand, a certain constant influence has been acting during the measurements, then a one-sidedness or skewness will show up in the distribution of measured values. More precisely, a numerous set of measurement values free from systematic error will most often lie near a symmetrical bell-shaped curve like Fig. 13.14. (Actually this curve is theoretical: it is obtained on the assumption of strict randomness for an infinite population of raw data and with the help of probability theory.)

*The most frequently found values are close to the mean or "exact" value $\overline{m}$ around which all others rally: $\overline{m}$ is the most probable value. The extent of the spread of the readings around the average $\overline{m}$ is given by the half-width σ, called the *standard error* or mean-square error. The probability that an individual reading lies within $\overline{m}-\sigma$ and $\overline{m}+\sigma$ is roughly 0.68 and the probability that it lies within $\overline{m}-2\ \sigma$ and $\overline{m}+2\ \sigma$ is 0.95. The *probable error* ε is roughly 0.68σ. The *exact error* is, of course, as unknown as the exact value. But we can compute the *probability* that the actual but unknown error will exceed the computed probable error. Thus for a bell-shaped distribution the odds that the exact error exceeds σ are even, and there is one chance in 4.5 that it exceeds 2σ. The probable error of the probable error decreases with the probable error itself and with the increasing number of observations: it is given by $0.48\ \varepsilon / n^{\frac{1}{2}}$.*

If there were no random influences acting all the time in and on the object-apparatus complex, there would be no random errors, i.e. we would have σ=0. A good measurement technique is not one which ignores such errors but one which minimizes the error width 2σ. Can this desideratum be pushed to the limit of error-free measurements? Before taking up this important question we must introduce a distinction between accuracy and precision. The *accuracy* of a measurement depends on the technique and on the systematic error: an accurate or exact measurement is one which (i) measures just what it purports to measure and (ii) is practically free from systematic error. The two conditions are mutually independent: the first depends on design, the second on execution. Thus an intelligence test may be applied with practically no systematic error yet it may happen that it does not measure natural intelligence but information fund or memory, in which case it will not yield exact results. The *precision* of a measurement free from systematic error depends both on the technique and on the random error: the smaller the σ the larger the precision. (More precisely, the precision h of a set of readings is given by $h=1/\sigma 2^{\frac{1}{2}}$.) Thus, the precision of time measurements by means of chronometers is one-tenth of a second, whereas that of masers and lasers is of one part in 10^{11}.

Suppose now we have been able to trace the systematic errors of a measurement to removable or computable disturbances, or to inaccuracies in design or execution, and that a theoretical analysis of our technique shows that it actually measures what it purports to measure. Suppose, in short, that we have performed *accurate* measurements: could we eliminate the random errors and so attain *complete precision* (σ=0)? In order to answer this question we need not wait for future improvements in measurement techniques: our present knowledge is sufficient for this. The answer is a complex one. If the mensurandum can take as sole values integral multiples of an elementary unit (such as the electron charge or the unit of angular momentum), complete precision is attainable in principle although it may have to wait for technical improvements. If, on the other hand, the mensurandum is a continuous magnitude such as position or linear momentum, there are two natural limitations to precision: (i) the finest scale division (least count) sets a lower limit to the error range, and there are technical limitations to increasing the fineness of the scale division; (ii) the random error cannot be decreased below a certain threshold which is the background "noise" or random fluctuation (e.g., the Brownian motion) inherent in every macroscopic system. The "noise" can be reduced by improving the insulation of the system—in particular by remote control—and by operating it at a low temperature, but no such measure will eliminate the random fluctuations altogether. In short, there are limits to the decrease of the noise-to-signal ratio.

The following morals seem to flow from the foregoing. First: the experimental error should not mask the measured value. That is, the systematic and the random errors together should not be as large as to hide the signal to be detected. (In short, $Min\ \{\overline{m}\} \gg \sigma$.) In the beginnings of a research line, though, hundred per cent and even larger errors may occur and accordingly only orders of magnitude can be estimated. This was the case with many early atomic physics measurements and is still the case with the astronomical measurements by means of which cosmological theories are tested. Second: experimental errors, if purely random, are not *errors* proper, i.e. mistakes: unlike systematic errors, which stem from overlooked distorting factors, random errors are in the nature of things rather than an illustration of human frailty. *This is not just a verbal question: different interpretations of quantum mechanics are obtained if the standard deviations Δq and Δp occurring in Heisenberg's relations are read as (i) subjective uncertainties, or (ii) objective random fluctuations of the system itself, or (iii) effects of the coupling of the microsystem with the measurement apparatus. Third: contrary to current belief, the discreteness or quantization of certain magnitudes, such as the angular momentum, far from being the source of irreducible measurement errors, renders their precise (error-free) measurement possible at least in principle. In this case the random disturbances can be kept lower than the threshold needed for inducing a transition from one value of the magnitude to another, which is not the case when the magnitude can vary by arbitrarily small amounts. The impreciseness inherent in the measurements of continuous magnitudes is somewhat compensated by the preciseness attainable in principle in the measurement of magnitudes with a discrete spectrum of values. Spectroscopy was the earliest illustration of this.*

Let us return to the processing of raw data displayed in the histogram or error graph. The form of the graph, and particularly the peak and the width of the spread, will depend on the number of measurements. It is desirable to take a large number of readings in order to make room for the largest possible number of mutually counteracting random deviations: we should welcome a large variety of random errors. As the number of measurements performed with one and the same equipment increases, the center of the distribution of values, i.e. $\overline{m}$, should approach a definite value—apart from occasional fluctuations for short streaks. Otherwise the equipment would be manifesting a slant in a definite direction. This convergence of the average or "exact" value to a definite number is called *internal convergence.* Any good measuring device should yield internally convergent results, i.e. the averages should be fairly stable.

A necessary condition for obtaining internal convergence is that the successive measurement operations made with the equipment be practically independent from one another. In other words, a necessary condition for stabilizing the center of the distribution of measured values is that no individual measurement exerts an influence on any of the successive measurements—i.e. that the equipment has a bad memory. If, on the other hand, the state of the measuring instrument is critically dependent on some of the previous results, then the various outcomes will not be statistically independent but will be correlated. In other words, the time of recovery (relaxation time) of the measuring device must be shorter than the time interval between measurements. Consequently, the task of following up the successive states of an object—e.g. the successive positions of a particle—at very close intervals may be very difficult if not impossible. Anyway, certain instruments will recover quicker and more completely than others, but there can be no guarantee that all scars (particularly the thermal and magnetic ones) left by one measurement act will be effaced before the next reading is taken.

Improved instrument design will improve the statistical independence of the measurement acts and consequently the precision. In short, advances in technique are accompanied by a process of *external convergence* whereby random errors are progressively being diminished: the new error curves are slender than the older. Classical physics was confident that the process of external convergence would steadily decrease the random errors to zero, or at least that no obstacle to an indefinite shrinking would be met. The discovery of thermal fluctuations in the first place, and thereafter of quantum fluctuations, has forced to the adoption of a more modest goal. All we wish is to be able to determine the bounds of such fluctuations and, at a later stage, perhaps to analyze them into more fundamental processes.

This completes our account of measurement. The standard account, by focusing on the construction of measurement scales and on the handling of instruments, is apt to miss the conceptual richness of the measurement procedures. In the preceding account, on the other hand, measurement proper was found to be one link of a rather long chain of conceptual and empirical operations, namely: Clear *conception* of both the mensurandum and the goal of measurement with the help of theory—*Objectification* of the mensurandum via law statements—Conception of an ideal measurement *scale* (including zero and unit if metrical)—Realization of a *material scale* mirroring the conceptual scale—Equipment *design* based on relevant theories—Equipment *building* (often with the help of fragments of technology based on pieces of pure science)—Equipment *testing* (idem)—Equipment *correction (idem)*—Measurement—*Standardization* of readings with the help of theory—*Systematization* of data—Search for *systematic errors*—Computation of the "exact" (average) value and the standard error (with the help of mathematical statistics).

The analysis of measurement has taken us far away from the simplistic look-at-the-pointer-and-conclude doctrine. An examination of experiment will take us even farther.

Problems

13.6.1. Make as complete a statistical analysis of a run of measurements as you can. *Alternate Problem:* A precision chronometer is fast by one hour a day. What, short of repairing it, can be done to obtain exact results?

13.6.2. How would you compare the precisions of different instruments designed for the same purpose? *Alternate Problem:* What can be done with a measurement equipment that shows nonconvergent behavior?

13.6.3. Why are measurements performed? Are measurement outcomes inherently valuable? And is the mere execution of measurements scientific research? *Alternate Problem.* Would you organize a physics institute on the basis of the measurements that there are in physics—e.g., chronometry, thermometry, and so forth?

13.6.4. Is every science concerned with measurement? And does measurement take up all or most of scientific research? See N. R. Campbell, *Foundations of Science* (1920; New York: Dover, 1957), p. 267: physics "might almost be described as the science of measurement". *Alternate Problem:* Examine the opinion, popular among historians of science, that modern science originated with measurement and developed as a result of it.

13.6.5. Pick a measurement report from some scientific journal and see (i) whether it makes sense outside every theoretical context and (ii) whether its truth value and significance can be estimated apart from every theory. *Alternate Problem:* The most precisely known physical constant does not refer to observable things or events: it is the Rydberg constant occurring in atomic theory. What does this suggest regarding the doctrine that man's most precise knowledge is the knowledge of himself?

13.6.6. Are measurements repeated (i) because repetition effaces systematic errors or (ii) because averaging over large sets of numbers improves the quality (accuracy) of the measurement, or (iii) because the variation in the outcomes is an index of the reproducibility, or (iv) in order to increase the number of favorable instances and so the degree of confirmation, or (v) because random errors tend to compensate, or (vi) because a large number of readings provides a reliable estimate of the precision? Caution: these possibilities are not mutually exclusive. *Alternate Problem:* Discuss the following paradox. An ideal measurement design will yield exactly reproducible results. But the greater the sensitivity of the apparatus, the more faithfully will it record individual differences and changes in time—consequently the lower will the reproducibility be.

13.6.7. Optical microscopes cannot discriminate among two objects unless these are at least 10^{-5} cm apart. This natural limitation of the resolving power is imposed by the wavelength of light. It was first overcome by the use of *X* rays, which have a much shorter wavelength, and later on by electron microscopy, which, using even shorter waves (matter waves), can distinguish objects 10^{-7} cm apart. None of these developments could have been suspected before our century; in particular, the electron microscope could be invented only after de Broglie conjectured the reciprocal relation between wavelength and momentum. Draw some moral concerning the caution with which pronouncements about unsurpassable limits on precision must be made.

13.6.8. Make a thorough examination of the usual derivation of the Gaussian or normal curve. Take into account that the formula refers to an ideal model of a set of measurements, in fact to an infinity of measurements subject to random fluctuation. Show why such an ideal model has to be built instead of just making a straight inductive generalization from histograms.

13.6.9. A precise length measurement requires a precise temperature measurement, which in turn requires a precise length measurement. Is this circular? Hint: consider a sequence of alternate length and temperature measurements made with increasing precision. See R. Carnap, *Physikalische Begriffsbildung* (Karlsruhe: G. Braun, 1926), p. 36. *Alternate Problem:* Every set of data, whether statistical or derived from measurement, is affected by irregularities that tend to mask whatever regularity may hide in them. Such accidental irregularities are screened out by various *smoothing* (or adjusting) techniques. The earliest among these procedures were groundless: they consisted in forcing the data to crowd along simple curves, such as parabolas. Examine how data smoothing has passed from the empirical stage, dominated by the simplicity creed, to its present theoretical stage, dominated by the probability theory. See the pioneer work of E. Whittaker and G. Robinson, *The Calculus of Observation,* 4th ed. (London and Glasgow: Blackie & Son, 1944), Ch. xi, or L. Jánossy, *Theory and Practice of the Evaluation of Measurements* (Oxford: Clarendon Press, 1965).

13.6.10. Measurement *acts* are neither true nor false: they can be correct or incorrect, or right or wrong, according as they abide by or violate the relevant rules stipulated in advance. From this it has been concluded that measurement outcomes, as expressed by measurement sentences of the type of 'The measured value of x with the equipment so and so equals y with a standard error z', are neither true nor false. Examine this view. *Alternate Problem:* Measurements can only yield finite numbers (moreover fractions) in a finite amount. Does this justify the programme of eliminating infinity and continuity from factual theories—as suggested by philosophers diffident of infinity and continuity because they are transempirical concepts? And does the lack of empirical counterpart of these concepts render them factually meaningless? (For the various kinds of meaning see Sec. 3.5.)

Bibliography

Baird, D. C.: Experimentation: An introduction to measurement theory and experiment design, Chs. 2 and 3. Englewood Cliffs: Prentice-Hall 1962.

Blalock, H. M., ed.: Measurement in the social sciences. New York: Macmillan, 1974.

Bohm, D.: Quantum theory, Ch. 22. Englewood Cliffs, N.J.: Prentice-Hall 1951.

Bunge, M.: Foundations of physics, Ch. 2, Sec. 5. Berlin-Heidelberg-New York: Springer-Verlag 1967.

——— A mathematical theory of the dimensions and units of physical quantities. In: M. Bunge, Ed.: Problems in foundations of physics. New York: Springer, 1971.

——— On confusing 'measure' with 'measurement' in the methodology of the behavioral sciences. In: M. Bunge, Ed.: The methodological unity of science. Dordrecht-Boston: Reidel, 1973.

Campbell, N. R.: Physics: The elements. London: Cambridge University Press 1920. Reprinted as Foundations of science, Part II. New York: Dover 1957.

Churchman, C. W.: A materialist theory of measurement. In: R. W. Sellars, V. J.

McGill and M. Farber (Eds.): Philosophy for the future. New York: Macmillan 1949.
Dumond, J. W. M., and E. R. Cohen: Fundamental constants of atomic physics. In: E. U. Condon and H. Odishaw (Eds.), Handbook of physics. New York: McGraw-Hill Book Co. 1958.
Ellis, B.: Basic concepts of measurement. Cambridge: Cambridge University Press, 1968.
Feather, N.: Mass, length, and time. Edinburgh: Edinburgh University Press 1959.
Forest Palmer, A. de: The theory of measurements. New York: McGraw-Hill 1912.
Isis, **52**, Pt. 2 (1961): issue devoted to the history of quantification.
Jánossy, L.: Theory and practice of the evaluation of measurement. Oxford: Clarendon Press 1965.
Jeffreys, H.: Scientific inference, 2nd ed., Chs. iv to vi. Cambridge: Cambridge University Press 1957.
Kayan, C. F. (Ed.): Systems of units. Washington: American Association for the Advancement of Science 1959.
Lenzen, V. F.: Procedures of empirical science, vol. 1, No 5 of the International encyclopedia of unified science. Chicago: University of Chicago Press 1938.
Margenau, H.: The nature of physical reality, Ch. 18. New York: McGraw-Hill Book Co. 1950.
———, and G. M. Murphy: The mathematics of physics and chemistry, Secs. 13.29–13.37. New York: Van Nostrand 1943.
National Bureau of Standards: Experimental statistics. Washington: Government Printing Office, 1966.
Neumann, J. V., and O. Morgenstern: Theory of games and economic behavior, 3rd ed, Sec. 3.4. Princeton: Princeton University Press 1953.
Schultzer, B.: Observation and protocol statement. Copenhagen: Munksgaard 1938.
Stein, P. K.: Measurement engineering, vol. 1, 3rd ed. Phoenix: Sietn Engin. Service, 1965.
Whittaker, E., and G. Robinson: The calculus of observation, 4th ed. London and Glasgow: Blackie & Son 1944.
Wilson Jr., E. B.: An introduction to scientific research. New York: McGraw-Hill Book Co. 1952.

14

Experiment

Of all kinds of human experience, scientific experiment is the richest: to observation it adds the control of some factors on the strength of theoretical assumptions and, if precise, it involves measurement. Scientific experiment, when performed methodically, designed on the basis of theory, interpreted with its help and aimed at testing ideas, is said to be conducted according to the experimental method. And the experimental method is in turn often regarded as distinctive of modern factual science. A study of scientific experiment is therefore of interest to the scientist, the philosopher, and the historian of ideas.

14.1. Planned Change

If we hear a bird singing without having watched for it we are having a spontaneous experience. If we listen intently to it, even though we may not actually hear the song, we are having a directed experience. And if in addition we record the song and take note of the circumstances accompanying it, our experience becomes an observation that may be put to a scientific purpose. In none of these cases, however, is an experiment proper being performed: by definition, *experiment is* the kind of scientific experience in which some *change is deliberately provoked,* and its outcome observed, recorded and interpreted with a *cognitive aim.* An experiment in bird song would be, for example, to rear singing birds in solitude since birth in order to study the influence of learning on singing. The mere caging of a bird, without such a purpose in mind, is not an experiment but just an experience—particularly for the bird.

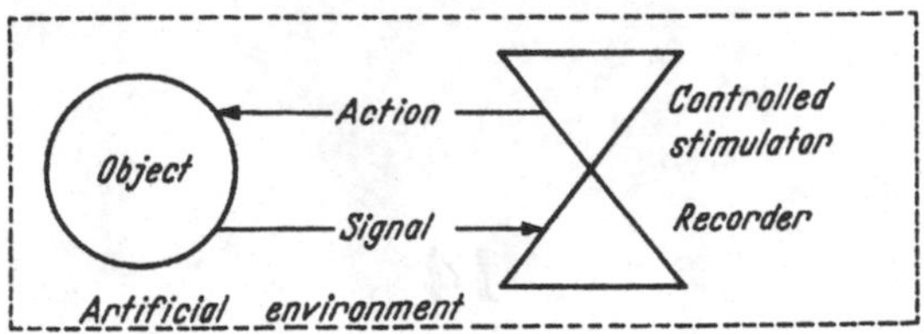

Fig. 14.1. Experiment: study of an object's reaction to controlled stimuli. Often, recorder=subject.

We saw in Sec. 12.2 that observation is analyzable into three components at the very least: the object of observation (embedded in its environment), the observer, and a communication channel transmitting signals between the two. Measurement introduces a fourth factor: the measurement device. And in experiment the object is surrounded by a more or less *artificial environment,* i.e. one which is in some respects under the operator's control—preferably a remote control (see Fig. 14.1). The experimenter's control is exerted both on the stimuli aimed at eliciting the object's reaction and on the latter. Stimuli may be exerted directly, as when an organism is subjected to an electric discharge, or indirectly, as when the environmental moisture is varied. And the control of stimuli may consist just in recording or measuring them, or in a deliberate variation of kind or intensity, as when the flow of liquids is controlled by handling a faucet or by an automatic control device. In the first case we may speak of *passive control,* in the second of *active control.*

If the control of inputs and outputs is not made with quantitative precision, the operation is called a *qualitative* experiment. In a *quantitative* experiment the values of certain magnitudes—those regarded as relevant or suspected to be so—are measured; if the measurement precision is poor one may talk of *semiquantitative* experiment. In other words, when the presence or absence of the various variables or factors are taken into account without measuring them, the experiment is qualitative even if—as in the case of the so-called factorial design—it is planned with the help of statistical theory. Most exploratory experiments, concerned with discovering new facts or with giving new theories a preliminary test, are qualitative or semiquantitative: only if they are either definitely favorable or altogether inconclusive is it worth while to attempt the attainment of quantitative accuracy. For example, Oersted's experiments on the interactions of currents and magnets were qualitative: he had only a qualitative hypothesis to test, namely that all the "forces" of nature were interlinked. Ampère, who had conceived a quantitative hypothesis to account for Oersted's experiment, designed and performed quantitative experiments concerning the same facts. The design of quantitative experiments is clearly more complex than that of qualitative experiments but not necessarily more ingenious: the use of measurement instruments presupposes that the variables concerned have been objectified and that measurement techniques have been evolved, whereas a qualitative experiment may "demonstrate" (i.e. objectify or render manifest) certain variables and relations for the first time. Here again originality may be more valuable than accuracy. Pedagogic moral: Qualitative experiments made with home-made equipment are more instructive than accurate experiments performed with ready-made black boxes.

Whether quantitative or not, experiments involve scientific constructs—concepts, hypotheses, and theories. Most often they involve a number of fragments of theories. In an experiment aimed at gathering new information, theories will intervene both before and after the execution of the experiment, i.e. in the experimental design and in the interpretation of the experimental results. In either stage both substantive theories (e.g., optics) and methodological theories (e.g., the theory of errors of measurement) intervene. Obviously, the intervention of hypotheses and theories is even greater when the experiment is aimed at testing a hypothesis or a theory, since in such a case the experimental set-up will have to materialize a long chain beginning in unobservable facts, such as the scattering of electrons by atoms or the distraction of a thought flux by an external stimulus, and ending in observable facts such as the motion of a pointer or a gesture. In no case is an experiment performed in a conceptual vacuum, if only because experiments are made to help solve problems born in a body of ideas. This is what makes the popularization of scientific experiments so hard: the observable facts can be presented to everybody, but the motive ideas to very few. In short, what is at stake in every experiment is a set of *ideas* concerning facts that the experiment produces or keeps under control: this emerges clearly from an examination of any scientific experiment. We shall choose the epoch-making experiments conceived and performed by Lavoisier to test his suspicion that mass is neither lost nor gained in the course of chemical reactions despite appearances to the contrary.

As early as 1630 J. Rey had found that the weight of tin increases when roasted (see Sec. 5.5). Rey was able to make this discovery because he formulated the new problem of following up (controlling) the weight variations accompanying such processes. And this he did, in turn, because he conceived the programmatic hypothesis that weight might be a variable relevant to chemistry. Moreover, Rey did not rest content with stating the phenomenological hypothesis "Tin increases its weight upon roasting" but went on to advance a hypothesis concerning the mechanism of this process: he assumed that the increase in weight of the lime or "calx" (oxide) relatively to the corresponding metal was due to the intake of air. Several other cases became eventually known in addition to the one investigated by Rey but they were all brushed aside as unimportant anomalies because they contradicted the accepted views, in particular the phlogiston theory proposed almost one century after Rey's experiments. The phlogiston theory was related to the Aristotelian hypothesis concerning fire, the igneous element, and had accordingly the backing of orthodox science and metaphysics. The theory stated that the limes, such as metal ores (our oxides), were all simpler than the corresponding metals because they were found in nature (notwithstanding the ugly exceptions of mercury and gold)—which was in turn just an instance of the general ontological principle according to which nature is simple, man being the source of all complexity. Now, if the limes are simpler than the corresponding metals, then something had to be added to the former in order to convert them into metals in the foundries: this something was called phlogiston, the fire-producing. Whenever a metal and fire were involved, the process had to fit either of the following patterns (laws):

$$\text{Metal heated in air} \rightarrow \text{Lime} + \text{Phlogiston} \qquad [14.1]$$

$$\text{Lime} + \text{Phlogiston} \rightarrow \text{Metal}. \qquad [14.2]$$

The phlogiston theory coordinated in a simple way a large mass of data, i.e. it had both simplicity and coverage. Consequently few were bothered by the "anomaly" that a number of limes were known to be heavier than the corresponding metals or by the fact that the theory did not disclose the mechanism of roasting. The concern of chemists at that time was not so much to test and evolve (complicate) their hypotheses as to confirm over and over again a simple theory that enjoyed the backing of tradition; in the worst of cases phlogiston could be assigned a negative weight—in case anyone was worried about

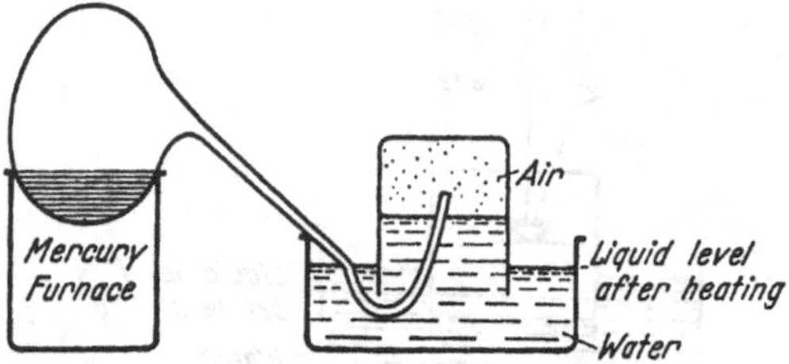

Fig. 14.2. The oxydation of mercury is accompanied by consumption of air.

weights at all. What more than a simple coverage could people ask for? And why should we reproach the chemists of two centuries ago for being dadaists and dataists if some of the more eminent theoretical physicists of our own time behave in exactly the same way with respect to the theories of the day? In any case quantitative experiments, in particular experiments involving the control of weight, played little or no role in that context because the weight *concept* did not occur in the prevailing *theory;* and it could play no role in it because it was a phenomenological theory rather than a mechanistic one. Chemistry was to remain semiquantitative and phenomenological as long as it made no contact with the science of the immediate lower level—physics. It was left to Lavoisier to begin this process of digging into the chemical reaction until reaching its physical root. Lavoisier showed that the neglect of weight was wrong and he built and tested an altogether new chemical theory (1772–1778) of combustion processes. The key concept of this theory was that of oxidation or combination with oxygen: this is a specific concept of chemistry, but one which was to be clarified with the help of physical theories. Let us briefly recall and examine the two classical experiments that supported Lavoisier's ideas.

Experiment 1: Oxidation of mercury (Fig. 14.2). If a sample of mercury is heated in a retort connected with an air chamber, the following facts can be observed to occur. First, on the free surface of mercury a layer of reddish powder (*mercurius calcinatus per se*) is formed. Second, the volume of air enclosed in the chamber decreases down to a minimum which does not vary even if the mercury is further heated. Third, the reddish powder formed on the surface of mercury is heavier than the metal out of which it was formed—as found by weighing before and after the experiment.

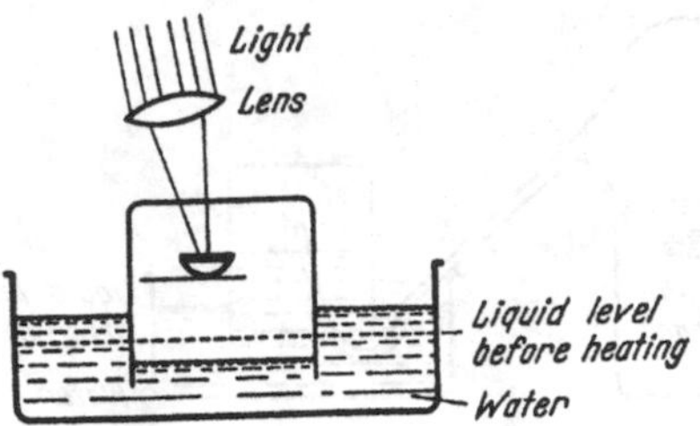

Fig. 14.3. The decomposition of red powder is accompanied by the release of gas.

Experiment 2: Decomposition of the reddish powder (Fig. 14.3). The red powder obtained in the foregoing experiment is heated in a hermetically closed glass container by means of an intense light beam. The following facts can be observed. First, pure mercury is gradually formed out of the red powder. Second, part of the liquid is dislodged—which is interpreted as the formation of some gas. Third, if this gas is added to the gaseous residue of the previous experiment, a mixture identical with common air is formed.

Thus far the descriptions of the experiments; if Lavoisier had stopped here he would not have fathered modern chemistry. Notice the following traits of these experiments. First, one is the inverse or complement of the other. Second, fairly closed systems were built in both cases. (In experiment 2 light must be reckoned as part of the system.) That is, most variables are kept constant, particularly so the total weight, since no transport of materials is allowed. Third, three physical variables are controlled because thought to be relevant: temperature, metal weight, and gas volume. Fourth, in no case is it merely a question of observing the spontaneous workings of nature: the temperature is deliberately changed. Fifth, the gas volume, which is imperceptible, is objectified and controlled by means of the liquid level.

Lavoisier explained the outcomes of his twin experiments by introducing the following assumptions—and these ideas rather than what he observed were revolutionary.

Hypothesis 1 (*on combustion*): The roasting of a metal consists in its oxidation or combination with oxygen, which is one of the components of atmospheric air. In the converse process, of formation of metal out of limes (oxides), oxygen is liberated. The patterns of the reactions in which heated metals take part are, then, the following:

$$\text{Metal} + \text{Oxygen} \xrightarrow{\text{heat}} \text{Lime (oxide)} \quad [14.3]$$

$$\text{Lime (oxide)} \xrightarrow{\text{heat}} \text{Metal} + \text{Oxygen} \quad [14.4]$$

which are to be contrasted with the patterns [14.1] and]14.2] indicated by the official phlogiston theory. (An instance of the first general law is the reaction schema "2 Hg $+O_2 \rightarrow$ 2 HgO", which is exemplified in turn by experiment 1. An example of the second pattern is the converse reaction, which is exemplified by experiment 2.) It should be noted that the oxygen concept was as much of a construct as the phlogiston: in fact, the assumptions that oxygen exists and that it takes part in certain chemical processes were not "derived from experience" but were posited and they gave rise to heated argument. Epistemologically there is no difference between the phlogiston theoreticians and Lavoisier: both accounted for experience by means of constructs. The difference is a methodological one: Lavoisier's hypotheses were testable and, moreover, they happened to pass the tests. It is therefore mistaken to hold that what distinguishes modern science is that it shuns hypotheses concerning unobservables.

Hypothesis 2 (*on mass conservation*): The total mass of an isolated system is conserved in the course of its chemical transformations. Let 'A_i' denote the mass of the *i*-th reacting substance and 'B_j' that of the *j*-th reaction product. Then h_2 can be symbolized thus: $A_1+A_2+ \ldots +A_m=B_1+B_2+ \ldots +B_n$. An instance of this general law is the following: "$O_2 + 2\ H_2 \rightarrow 2\ H_2O$", the corresponding ponderal equation being: $(16+16)g+2\cdot(1+1)\ g=2\cdot(2\cdot1+16)\ g=36g$.

It will be noticed, first, that h_1 is qualitative and typically chemical, whereas h_2 is a quantitative hypothesis referring to a physical aspect (the mass) of chemical reactions. Second, whereas h_1 refers only to combustion processes, h_2 is universal in the field of chemistry. Third, h_1 is representational to some extent: it accounts for the mechanism of roasting; h_2, on the other hand, is a phenomenological hypothesis as long as it is not made part of the atomic theory of chemical reactions.

The above experiments do not establish the universality claimed for h_2 since they concern the transactions of the particular pair mercury-oxygen: after all, the conservation of mass might be an idiosincrasy of this couple. Lavoisier therefore tested h_2 in several other cases by methodically using the balance. And once he regarded h_2 as sufficiently corroborated—by the standards of his time—he jumped without hesitation to the general conclusion that the conservation of mass holds universally, i.e. for every chemical reaction taking place in a closed system. This bold jump has enabled chemists to *use* the principle of mass conservation in a heuristic way. Thus, if the total mass of a system in which a chemical reaction takes place changes, the variation in mass is attributed to the violation of the isolation condition, i.e. the assumption is made that a certain quantity of matter has leaked through it—a leak that has passed unnoticed to observation, not to theory. This has often resulted in the discovery of substances which are not directly accessible to the senses; oxygen itself was discovered in this way and so was nitrogen.

A philosopher might remark that, even after testing the mass conservation principle in a couple of million reactions, one particular reaction might be found which altogether falsified this principle: after all Lavoisier's is just a hypothesis with an inductive support, i.e. an empirical hypothesis, and there is no reason why mass could not be gained or lost in the course of chemical reactions. This objection would have been valid while Lavoisier's principle remained a stray hypothesis, but it has ceased to hold ever since it was theorified. In fact, Lavoisier's law h_2 can be deduced from the basic assumptions of the chemical atomic theory concerning the mechanism of the combination and dissociation of atoms. This theorification of h_2 has given it a much stronger support than the one it could gain from millions of new empirical confirmations, because it has incorporated it into the net of law statements that embraces the whole of physical reality. (For law explanation see Secs. 9.3 and 9.5.)

Yet the very same theory that explains Lavoisier's law effects a small quantitative correction to it. In fact, the total mass of a molecule is somewhat less than the sum of the masses of its component atoms. This difference—the mass defect—goes into the binding energy of the atoms in the molecule and is therefore of a paramount theoretical importance although it is not detectable in the chemical laboratory, the scales of which are insufficiently sensitive for this purpose. If atomic masses were strictly additive there would be no stable compounds: every combination would result from a chance encounter of indifferent atoms and would therefore be unstable; as it is, there are binding forces to which binding energies correspond.

A first metascientific moral of the above story is that mass is objectively nonadditive although the *measured value* of masses are additive in the context of the present measurement technology. In other words mass is physically (objectively) nonadditive though empirically additive—much as all Eskimos look the same to unexperienced Europeans. (With length it is the other way around: we attribute every empirical nonadditivity of length measurement to defects of the measurement procedures.) Second: scientific laws do not cease being corrigible even after abundant confirmation—this being one reason for classing them among hypotheses. Third: confirmation is never final, if only because it may use new techniques or ideas, as well as new, more exacting standards of rigor. Fourth: law statements are not being perpetually subjected to deliberate tests but, once they seem reasonably well corroborated, they are employed in explanation, prediction, and action—including experiment. Corrections, small or large, often come accidentally either in the course of their application or upon a completely unexpected linking with some new theory—as was the case of the mass defect that spoiled the accuracy of Lavoisier's law, and which was suggested by relativity theory, which is relevant to chemistry only via the physical theories that underlie chemical theories.

Let us take a closer look at the key to experiment: control.

Problems

14.1.1. Elucidate the concepts of experience and experiment. *Alternate Problem:* Comment on C. Bernard's *An Introduction to the Study of Experimental Medicine* (1865; New York: Schuman, 1949), Ch. 1, where he discusses the differences between observation and experiment.

14.1.2. Image trying to explain to Aristotle any modern scientific experiment. Do not forget to state the problem, the hypotheses, and the interpretation of the outcome.

14.1.3. Write in a complete form the equation of mass balance (Lavoisier's h_2), i.e. add the clauses that the system is to be isolated composed of two or more reacting substances, and so on. And see whether the complete formula refers to any of the possible experiments that could test the equation, or rather to an objective pattern.

14.1.4. Would Lavoisier have had conducted the experiments described in the text had he no axe to grind? Discuss in general the defensive phrase "I have no axe to grind". Does it reflect scientific objectivity, scientific impartiality, poverty of ideas, or dread of them? *Alternate Problem:* Study the guiding function of hypotheses in Lavoisier's and Priestley's experimental researches on "dephlogisticated air" (oxygen) and compare it with their avowed philosophy of science.

14.1.5. The popular belief—shared by Aristotle—that plants eat soil was put to the test by J. B. van Helmont in the 17th century. By weighing the amount of soil consumed by a plant, he found that the latter's increase in weight was many times the former. To this end he grew plants in pots, watered and sunned them, and used a commercial scale. Point out the conflicting hypotheses involved and the variables that were controlled and those that were not. And discuss whether it would be worth while to add a pot without a plant as a control or whether an individual can be its own control. If possible, speculate on the reasons that may have delayed this test for so long. *Alternate Problem:* Study the contribution of philosophy to the establishment of conservation laws.

14.1.6. Sociologists often claim that sociological research based on records or on questionnaires about past facts are not just empirical but experimental: they call such observations *ex post facto experiments.* In this sense the genesis of mountains, the emergence and extinction of biological species, the emergence of the modern state, and other processes that have left records would be the object of experimental science. See E. Greenwood, *Experimental Sociology* (New York: King's Crown Press, 1948). Subject the expression '*ex post facto* experiment' to critical scrutiny. *Alternate Problem:* Why can observations be better rigged to suit a hypothesis than experiments?

14.1.7. Make a thorough analysis of a scientific experiment. Suggested sources: experimental science textbooks and J. B. Conant, Ed., *Harvard Case Histories in Experimental Science* (Cambridge, Mass.: Harvard University Press, 1957), 2 vols., and M. L. Gabriel and S. Fogel, *Great Experiments in Biology* (Englewood Cliffs, N.J.: Prentice-Hall, 1955)

14.1.8. Is it possible to perform experiments in geology, evolutionary biology, archaeology and sociology? Hint: Before answering in the negative become acquainted with some recent work in either of these fields.

14.1.9. Analyze any of the following experiments: (i) the *Kon Tiki* expedition led by T. Heyerdahl in 1947, which was planned to test the hypothesis that a migration from Perú to Polynesia (particularly the Easter Island) was physically possible; (ii) the crossing of the Alps with elephants under J. M. Boyte (1959) to test the story about Hannibal's crossing of the Alps with elephants. What, if anything, did these experiments prove? And can they be regarded as experiments in history? In general: is experimental history possible?

14.1.10. Examine the possibility of building an experimental asthetics concerned with (i) aesthetic valuation (perhaps in the form of estimating subjective utilities) and (ii) the psychology of aesthetic experience. See D. E. Berlyne, Ed., *Studies in the New Experimental Aesthetics*. New York: Wiley, 1974.

14.2. Control

It is convenient to distinguish the experimental *approach* from both the experimental *method* and the experimental *techniques*—the special procedures for the implementation of the experimental method. Craftsmen and scientists adopt an experimental approach to certain problems in the sense that they try to solve them by deliberately modifying certain objective conditions and checking their assumptions and procedures, correcting them if necessary in the light of fresh experience and fresh ideas. Whether in the craftsman or in the scientist, the experimental approach is characterized by an effective and self-correcting control of things and ideas referring to things. But the problems, hence the aims and specific means of the scientist differ from those of the artisan: the latter's problems are chiefly problems of doing—e.g., repairing a machine—whereas the scientist deals mainly with problems of knowing—e.g., searching for the "principles" of a machine, i.e. the laws on which it is designed. Whereas the craftsman and the technician pick up knowledge in order to act, the scientist acts in order to know: there where the former applies rules that are more or less well-grounded, the latter applies or tests hypotheses (see Ch. 11).

In the narrow sense of the word, the experimental *method* consists in subjecting a material system to certain stimuli and in observing its reactions to them in order to solve some problem concerning the stimulus-response relation; the problem may be to test a definite hypothesis about that relation or to conceive it on the basis of the input-output data. In principle, the experimental method can be applied in any field of factual science dealing with the present. But the effective application of the experimental method will depend on the peculiarities of the problem: it is one thing to study the effects of gamma rays on plant growth and quite a different thing to study the mechanism responsible for the said effects: different problem systems call for different special experimental methods or experimental *techniques.* Accordingly, the employment of the experimental method requires the invention of a variety of experimental techniques—as many as kinds of experimental problems.

We are not saying that the special experimental techniques are peculiar to *disciplines* but rather to experimental *problems.* Thus the historian may use physical techniques in order to gather historical data or to test historical hypotheses: he may resort to the radiocarbon dating technique in order to find the age of an object, or to *X*-ray analysis in order to uncover the early history of a palimsest or a painting. An evidence being an empirical information about a physical system, physical techniques are apt to be employed in every factual discipline regardless of the nature of the *referents* of the hypotheses involved. (For the reference-evidence distinction see Sec. 8.4.) To say that the sciences of man have no use for physical techniques because they do not deal with purely physical systems is to confuse reference with evidence. Whether mental states are just states of the brain or not, any scientifically utilizable piece of evidence in favor or against any hypothesis concerning mental occurrences will have to be of a physical nature—a bodily motion or some other objectifier of internal states—hence one which can be controlled by physical means. Otherwise the experimental method will not be applicable to the investigation of mental states. In short, however mentalistic the psychology of the higher functions may be as regards its referents, it must be methodologically behavioristic if it is to be scientific.

However diverse in detail, experimental techniques consist in procedures for the manipulation and observation—in short, the *control*—of those variables which, for some reason, are regarded relevant to a

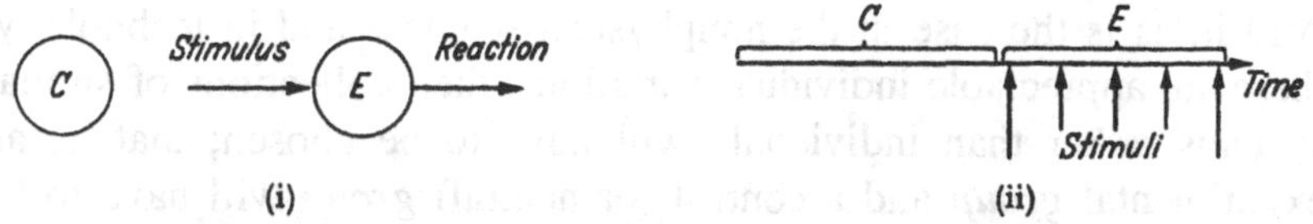

Fig. 14.4. In (i) two similar systems are observed simultaneously: the control C and the experimental system E. In (ii) the system before the stimulus is applied serves as control of its own future states.

given study. Both the conditions of *occurrence* of the facts under study and the conditions of *observation* must be controlled. The latter control is particularly necessary if the observation process disturbs the object, but it is always necessary since what we observe are phenomena, not completely objective facts independent of our mode of observation. Accordingly, the invention of experimental techniques presupposes hypotheses concerning the factors or variables that may be operant in a given situation: both the variables characterizing the facts themselves and those that characterize the operator's mode of observation (such as the coordinates and stains he uses). Unless such a list of suspects is made, no provision for their control can be made. Among all suspect variables, those which are deliberately varied by the experimenter are called the *independent* variables, whereas the variables the values of which change as a consequence of the former's changes are called the *dependent* variables. (The pragmatic notion of dependence is involved here: see Sec. 6.1.)

Now, in order to assess the differences introduced by changes in the values of the independent variables, some witness or control system is needed, some system in which no such deliberate changes are introduced. In particular, the witness or *control system*—not the control technique, which is a procedure, not a thing—may be the system while it is not under the influence that is assumed to make some difference to the *experimental system* (see Fig. 14.4). For example, in inquiring into the effect of the irradiation of a piece of matter (a polystirene bottle or a living tissue) by neutrons, certain variables are measured before and after the irradiation so that the past states of the system play the role of control. This technique can be used when the system does not change appreciably before the experiment begins. Otherwise a different technique must be used: two separate systems must be set up simultaneously, the experimental and the control (or neutral) ones.

And if, as is the case in the nonphysical sciences and in technology, there are appreciable individual variations, then collections of similar systems rather than individuals will have to be chosen; that is, an experimental *group* and a control (or neutral) *group* will have to be formed, in such a way that the only appreciable difference between them lie in the stimuli that are under the experimenter's control and act on the experimental group but not on the control group. (Recall Fig. 14.4.)

Suppose we wish to find out whether, and if so how much, a given painting prevents the rusting of iron. We choose two equal pieces or iron or, better, two sets of perceptibly identical pieces of iron one of which (*E*) is given a coating of painting whereas the other (*C*) is left unpainted. The two groups are then subjected to the same environmental conditions and, moreover, to such that will normally produce the rusting of uncoated iron—unless the aim is to cheat someone, in which case the most benign conditions are chosen. We next observe the differences, if any, among the control and the experimental group. Finally, from this observation we infer some "conclusion" concerning the effectiveness of the painting. If the experiment has been qualitative, the conclusion will be qualitative as well—e.g., of the form "Coating with our painting reduces rusting substantially". Experiments in which one or more factors (variables) are varied at a time and differences are recorded qualitatively—i.e. in which the presence or absence but not the amount of an effect is recorded—are called factorial experiments, or experiments with *factorial design.* Most experiments involving the measurement of a number of values also involve varying one factor (the independent variable) at a time; their plan is called *functional design.* Quantitative "conclusions" can be gathered from functionally designed experiments alone: there can be no quantitative answers to nonquantitative questions.

The control and the experimental groups must be *homogeneous* to begin with, i.e. they must be nearly equal in all relevant factors. Such a homogeneity is the more difficult to obtain the more complex the individual systems and the more exacting our approach, but in any case near homogeneity in the relevant aspects must be obtained if the results are to be significant, i.e. due to the experiment rather than to the choice of experimental subjects. Different techniques are used to control the initial values of the relevant variables so as to attain approximate group homogeneity. There are two kinds of techniques for

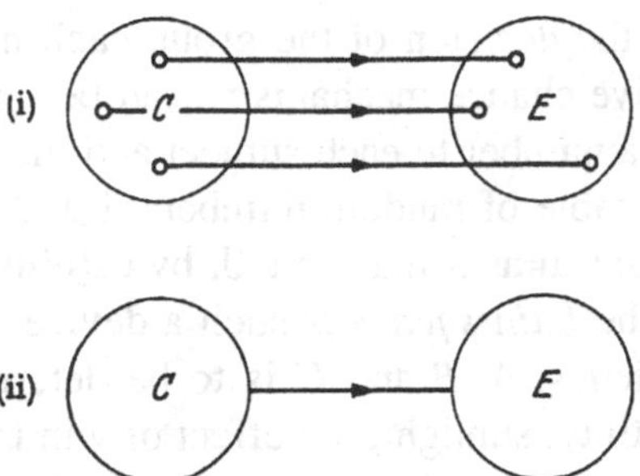

Fig. 14.5. Individual control: every member in *C* is matched to one member in *E* (i). Statistical control: *C* and *E* are matched collectively, as wholes (ii).

the formation of groups: individual and collective. *Individual control is* achieved by the simultaneous pairing of individuals in both groups: every member of the control group is matched to one nearly equivalent member in the experimental group. For example, for every twenty years old college girl in the control group another twenty years old college girl is assigned to the experimental group. Simultaneous pairing is difficult and expensive when the group size is large, in which case statistical control is called for (see Fig. 14.5). There are two main kinds of *statistical control:* control of distributions, and randomization. The *control of distributions is* performed when certain statistical parameters are equated, namely the averages, spreads (standard deviations), and perhaps a few other collective properties (e.g., medians and modes). In other words, the individual compositions of C and E are disregarded and only some of their statistical properties are kept under control. For example, in order to study the influence of television watching on scholastical performance, two samples of children with the same age average and the same average grades, as well as with the same average deviations from these averages, will be taken.

Simultaneous pairing and the control of distributions have their peculiar advantages and disadvantages, but they share one shortcoming in either case the formation of the *C* group and the *E* group may be inadvertently biased. For example, we may unwittingly allot the strongest, or the ugliest subjects to an *E* group that is to receive a tough treatment, fearing that the weaker or the more beautiful will suffer. This ever-present possibility is ruled out if the two groups are formed *at random.* We may, for example, flip a coin for each pair of subjects leaving to chance the decision of the group each member of the pair will join. Alternative chance mechanisms can be imagined, as well as the assignment of a number to each subject and the subsequent choice with the help of a table of random numbers. Randomization can also be attained, paradoxical as it may sound, by careful arrangement. The design known as the *latin square is* such a device. Suppose the yield of three plant varieties, *A*, *B* and *C* is to be determined. In order to randomize soil fertility, sunlight, the effect of winds, humidity, and so on, this field layout may be adopted:

$$\begin{matrix} A & B & C \\ B & C & A \\ C & A & B \end{matrix}$$

In this way all three varieties are given the same opportunities—which is the characteristic of an unbiased experiment design. By means of randomization all the variables (mostly unknown) that are not under control become randomly distributed, so that the differences in performance between the *C* group and the *E* group can quite confidently be attributed to the variations in the independent variables alone. (Moreover, the probability theory can then be applied to the control of inferences from the results of the experiment.) Otherwise we might suspect that the differences in performance are due to a chance accumulation of privileged individuals in one of the groups. This, called the *null hypothesis,* can be refuted only if the groups have been built at random. The *randomization technique* (R. A. Fisher), by minimizing the possibility of biased choice, provides us with a homogenization procedure that contributes largely to the objectivity and toughness of tests particularly necessary in the sciences of man. Randomization, though, is not an alternative to simultaneous pairing but may be used together with it: in this case individual matching will bear on the observable variables that are explicitly taken into account whereas randomization will bear on the variables that are not explicitly taken into account but which may make a difference. The most complete group control is thereby achieved: one which is individual in some respects and statistical in all others.

Once the *C* group and the *E* group have been formed, the experiment can start: the stimuli can be applied to *E* and the performance of *E* in relation to the behavior of *C* will be evaluated, whether quantitatively or not. The differences in performance can be significant or not:

this is for statistics to decide. Finally some "conclusion" will be "drawn" from the outcome of the experiment—some conjecture tentatively advanced on the basis of the significant differences between *C* and *E;* this inference is controlled by statistics, the science of the control of jumping from data to low-level "conclusions".

An experiment involves consequently the following operations:

1. *Choice of problem.*
2. *Identification of variables* suspected to be relevant, hence in need of control.
3. *Design* or planning of the experiment.
4. *Formation of homogeneous C and E groups.*
5. *Application of stimulus* to *E.*
6. *Observation and/or measurement* of performance of *C* and *E.*
7. *Judgment of significance* of differences in behavior between *C* and *E.*
8. *Inference* concerning relation(s) among the independent and the dependent variables.
9. *Control of previous inference* by relevant theory when available.

Stages 3 and 6 to 9 will concern us later on. Let us now say a few words on stage 5—the application of the stimulus. This may be "positive", i.e. an external agent acting on the experimental group, or "negative", i.e. the deprivation of some item normally occurring and accordingly present in the control group. The latter is very often the case in the nonphysical sciences, particularly when the function of a substance, an organ or an institution is studied. Thus, if the function of the cerebral cortex is under study, the experimenter may, e.g., train a group of rats to some task such as discrimination of visual form and will then remove all or part of their brain cortex and will compare their performance to that of a normal or control group, which can be the same rats before the surgical operation. If the learning ability is not substantially altered upon the removal of the cortex, it is "concluded" that the latter is not essential for learning—at least not for learning of the kind envisaged.

In some cases the stimuli cannot be applied to actual systems either because of technical limitations or because of moral restraints. In such cases experiments on concrete *models* can be performed. Thus, e.g., engineers will build small-scale models of dams, railway nets, planes, and so on, and will subject such models to stimuli of the desired kind and will then extrapolate their findings to the original system; likewise, the pharmacologist will use rats or pigs as models of men. When not even such concrete models can be obtained, a conceptual model is imagined which mimics certain traits of the real system, and certain changes in them are effected (in thought) and the corresponding reactions are estimated with the help of some theory. The design of a conceptual model is called *modelling* and the subjection of the model to imaginary changes is called *simulation.* A modelling-simulation pair is called a *thought experiment.*

Thought experiments are more and more extensively performed in technology. For example, the acoustic qualities of an auditorium (or the fidelity of a wave transmission system) can be investigated by first modelling it on paper and then simulating the system and its behavior on a computer—in either case with the help of theory. Also, the probable behavior of a human community in the face of disaster can be modelled and simulated on a computer on a probability basis suggested by a previous statistical study of the corresponding real community and certain empirical generalizations concerning human behavior in such circumstances.

Thought experiment, and particularly the modelling and simulation on computers, is taking the place of much actual experiment in applied science as long as verisimilar law statements are available. It could not be otherwise, as the computer is given both the input data and the input-output relation, its sole task being to find the output: the computer makes simulation feasible, rapid and inexpensive. But no simulate replaces reality. Consequently, thought experiment can replace actual experiment when it does not aim at *testing* a theory but at *applying* it: it is ideally suited to the test of concrete systems (things or humans) on the strength of ideas that have already passed the test of actual experiment. It has therefore a brilliant future in applied science (technology), none in pure science, since it neither creates theory nor tests it.

Now that we are clear on the centrality of actual control to experiment, we may take a look at the planning of experiment.

Problems

14.2.1. Draw a parallel between the procedure pattern of a plumber called to fix a fault and that of a scientist who has conceived an experiment.

14.2.2. Analyze the experiment by which it is "demonstrated" (exhibited) that plants exhale water vapour. Two pots, one with a plant and the other without, are left overnight in a closed glass container each. Next morning moisture is observed on the walls of the container with the plant. *Alternate Problem:* If mental states were irreducibly private, i.e. unobjectifiable, the study of the mind could not be approached from the outside but only from the inside (introspection) and no experiment proper could ever be performed in psychology. Is this the case? If not, what justification is there for the philosophy of mind?

14.2.3. Comment on some of the following cases of incomplete control. (i) R. Boyle neglected controlling the temperature in his experiments on gas expansion, simply because it had not occurred to him that temperature might be a relevant variable. (ii) L. Thomas discovered by chance that rabbits injected with papain had their ears collapsed. He made numerous observations in order to find the cause of this but found nothing until, several years later, he compared sections of floppy and normal ears: this showed a change in the matrix of the cartilage, which had previously been regarded as an inactive tissue. See B. Barber and R. C. Fox, "The Case of the Floppy-Eared Rabbits: An Instance of Serendipity Gained and Serendipity Lost", in B. Barber and W. Hirsch, Eds., *The Sociology of Science* (New York: Free Press, 1962).

14.2.4. Before the Salk and the Sabin polio vaccines were adopted in the U.S.A. they were tried on a nation-wide scale. Two large groups of children, of the order of one million each, were chosen: to one of the groups the vaccine was given, to the other distilled water was injected. Neither the children nor their parents were told what they had actually received. Why was this precaution taken?

14.2.5. In experimental medicine placebos (dummy pills or distilled water) are often given to the control group in order to equalize the two groups with respect to psychological factors: in fact anything, even plain water, can improve the patient's condition by suggestion. Consequently the two groups of patients, the one under treatment (E) and the neutral one (C) are being acted on: in a sense both are experimental groups. Discuss the placebo effect and try a redefinition of "control group" which copes with it. Hint: think in terms of relative (or differential) stimuli or experimental agents rather than in terms of absolute changes in the manipulable variables.

14.2.6. Clinical tests yield some "false positives" and "false negatives". How does one know this? And what is the consequence of this for epistemology?

14.2.7. Why is randomization less used in psychology than in agronomy? Is there any reason aside from the fact that men are rated lower than land?

14.2.8. It is often asserted that there would be no science if experiments were not indefinitely repeatable and if they gave every time a different set of results. If this were true physics would presuppose that the universe is spatially and temporally uniform, i.e. unchanging both in respect of place and time, as claimed by the steady-state theory of cosmology: see H. Bondi, *Cosmology,* 2nd ed. (Cambridge: University Press, 1960), pp. 11–12. Is the initial assumption true, i.e. does science presuppose the indefinite repeatability of particular conditions?

14.2.9. Study the nature and function of thought experiment—such as Heisenberg's gamma ray microscope—in pure research. See K. R. Popper, *The Logic of Scientific Research* (1935; London: Hutchinson, 1959), Appendix *xi, where the following functions are mentioned: (i) critical—for the refutation of hypotheses; (ii) heuristic—to help conceive hypotheses, and (iii) apologetic—to defend theories from criticism.

14.2.10. Study the nature and function of modelling and simulation in applied research. In particular, evaluate the place of computers in it and the claim that thought experiment will eventually displace actual experiment.

14.3. Design

The planning or design of an experiment depends on the experimental problem it is hoped to help solve. In other words, experiment design depends on the question that is being asked and on the context of the question, i.e. on the available theoretical background, on the techniques at hand, on the kind of data expected, and on the required accuracy. It is one thing to answer, in the absence of a theory, the question whether two psychological variables are in fact correlated—to test a programmatic hypothesis of the form "x covaries with y"—and quite another to test to five per cent accuracy a precise numerical law occurring in mathematical psychology. Experiment designs must be tailored after experimental questions, and these are asked within a definite framework of knowledge.

Now, since experimenting boils down to controlling variables, experimental problems can be classed—from a methodological point of view—according to the kind of control involved. The kind of control will in turn reflect not only the rigor aimed at but also certain objective properties of the experimentally manipulable variables. Variables suffering large statistical fluctuations—such as soil fertility and intelligence—will require a statistical control, while if such fluctuations are bounded within narrow limits—as is the case with distances and durations—a precise control of variables is possible.

In all experimental work statistical techniques will be used in the course of *evaluating* the experimental results, in particular when the time comes to estimate the mean value and the mean scatter around it. When all the relevant variables are individually and precisely controlled statistics enters in the final stage alone, hardly in experiment design. But when the control of variables is itself statistical, as is so often the case in behavioral science and in applied research—e.g., engineering and agronomy—statistical techniques are used both in the evaluation of the experiment and in its planning. It is therefore reasonable to distinguish two broad classes of experiment design and accordingly of experimental test: *nonstatistical* and *statistical.* Their distinctive features will best be understood through some distinguished cases. We shall present six cases, the first three of nonstatistical design and the last of statistical design.

Experiment 1. Production of aminoacids under possible primitive earth conditions. S. Miller, *Science* **117**, 528 *(1953). Problem:* Could life have emerged spontaneously from nonliving matter? *Hypothesis* to be tested: The organic compounds which constitute the building blocks of organisms were formed spontaneously at a time when the terrestrial atmosphere was composed of methane, ammonium, water vapour, and hydrogen; and the synthesis of organic compounds was activated by electric discharges and ultraviolet radiation. *Test* proce-

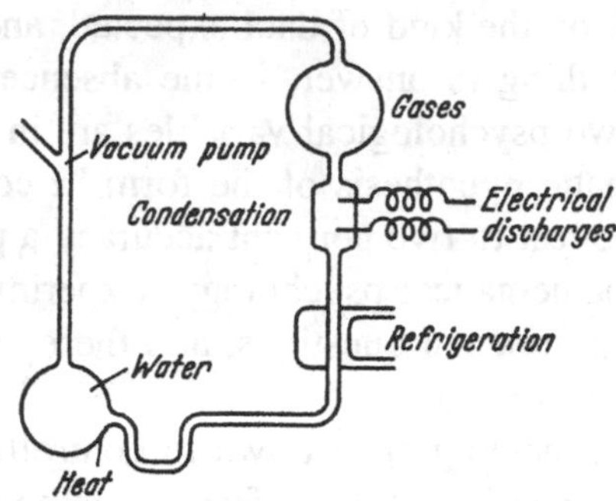

Fig. 14.6. Model of primaeval atmosphere. The gas chamber contained CH_4, NH_3, and H_2.

dure: A small-scale model of the hypothetical primaeval atmosphere was built and operated (see Fig. 14.6). The contents was treated with a germicide in order to exclude the possibility that microorganisms produced the organic compounds. After one week the water had become turbid and red. Identification tests showed that three different amino acids had been formed, among them alpha-alanine: CH_3–CH–NH_2–COOH.

Experiment 2. Cure of neurosis by reciprocal inhibition. J. Wolpe *British Journal of Psychology* **43**, 243 (1952). *Problem:* Is neurosis conditioned? *First hypothesis:* Neurosis can be induced experimentally. *Test* procedure: Cats are conditioned to search for food pellets at the sound of a buzzer. Once they have learned this behavior they receive electric shocks whenever they move to seize the food. As a consequence they exhibit increasingly neurotic behavior, mainly anxiety and refusal to eat. *Second hypothesis:* Feeding might inhibit such neurotic reactions just as these had inhibited feeding (reciprocal inhibition). *Test* procedure: The same cats are now fed in the absence of anxiety-evoking stimuli, then in the presence of small increasing doses of such stimuli. Result: After a number of sessions no traces of neurosis remain. *Checking:* In order to make sure that the neurotic reactions had been eliminated and did not lay dormant ("repressed"), the food-seeking response to the buzzer was experimentally extinguished. Once the cats had unlearned this association they were offered food in conjunction with the auditory stimulus. Result: No inhibition on eating was noted—i.e., the cats were cured.

Experiment 3. Group influences on the formation of norms and attitudes: M. Sheriff, *Sociometry* **1**, 90 (1937). *Problem:* Does prevalent opinion influence perception? *Hypothesis:* If opinion has any influence on perception, it will be the stronger, hence the easier to detect, the less clear-cut and spontaneous the judgments of perception. *Test* procedure: The subject is placed in an unstable situation, first isolated and then in a group; the order is then inverted and the difference in behavior is noted. An adequate unstable situation is found to be one in which the subject is placed in a dark room with only a small point-like light source. The subject has no fixed frame of reference and suffers the illusion that the light spot is moving when it is actually at rest. The description he gives of the movement is strongly influenced by the opinions uttered by other subjects in the same group, some of whom will be confederates of the experimenter. The convergence of opinion will be the greater when the subject begins the experiment in a group situation without having had the chance to form an individual norm.

Let us now describe three experiments in the design of which statistics plays a central role, to the point that they are called *statistical tests.*

Experiment 4. Genetic basis of learning ability: R. C. Tryon, in F. A. Moss, Ed., *Comparative Psychology* (New York: Prentice-Hall, 1942). *Problem: Is* learning ability inborn or acquired? *Hypothesis:* If learning ability is inborn it can be improved by artificial selection. *Test* procedure: A biologically uniform group of rats is selected and trained to maze learning. On the basis of their performance the rats are divided into two groups: the quick and the slow learners. Individuals within the same group are then mated and their offspring subjected to the same treatment: the quick are mated with the quick and the slow with the slow. After eight generations two strains are obtained which exhibit a significant difference in maze-learning ability. Inference: learning ability is hereditary. (However, see below.)

Experiment 5. Dependence of performance on motivation: H. J. Eysenck, *Scientific American* **208**, No. 5 (1963). *Problem:* Is motivation relevant to performance, and if so how? *Hypothesis:* The larger the reward expectation the better the performance. *Test* procedure: Two randomly constructed groups of applicants to a highly prized job are assigned a task, the performance of which can be measured quantitatively. The subject must try to keep a stylus on a small target in a rapidly revolving turntable. An electric meter measures the time-on-table. The control group is told that the test outcome will have no bearing whatever on their standing, whereas the experimental group is persuaded that the outcome will influence the final decision. In this case the stimulus is an idea. Result: The high-drive group's performance is, after a while, significantly higher than the performance of the low-drive group. Since all the other relevant variables (innate ability, experience, etc.) have been randomly distributed, the sole independent variable is drive and the sole dependent variable is performance. Moreover, the relation between the two becomes a means for measuring drive strength.

Experiment 6. Clairvoyance: G. R. Price, *Science* **122**, 359 (1955). *Problem:* Is the hypothesis of clairvoyance true? *Hypothesis:* If clairvoyance is real, it will show up despite precautions taken to avoid delusion and fraud. *Test* procedure: Packs of cards are shuffled automatically a very large number of times (knowing how hard it is to imitate perfect randomness starting from order). The cards are placed in metallic containers which are closed by welding; the weldings are microphotographed. The containers are then presented to the clairvoyant, who is to say what he sees inside them—without, of course, opening the containers. The reports are recorded on a tape. The containers are then sent to the experimenter, who checks their weldings and opens them. The cards are extracted mechanically and their actual order is compared with the clairvoyant's reports. The whole operation is filmed. This experiment was never performed: the best known parapsychologists assaulted the proposal as being insulting but did not meet the challenge.

All of the previous experiments can be criticized through and through: from the design to the interpretation through to the actual execution. For example, the inference from Experiment 4, that learning ability is hereditary, is inconclusive in the light of the experiments showing that the learning ability of animals reared in isolation is far below normal (R. Melzack and W. T. Thompson, 1956). The two experiments together suggest that learning ability (or intelligence) is due to both genetic and environmental factors. No single part of a scientific experiment is above criticism and this is precisely one reason for calling it *scientific* rather than a mystic experience—another reason being that every scientific experiment is designed on the strength of some scientific theory or other and with the help of rules of the scientific method.

Let us now focus on the inference concerning the significance of the *C-E* differences.

Problems

14.4.1. Discuss the precautions taken in the experiments summarized in the text. Are they designed to facilitate the confirmation of the hypotheses involved or to avoid their cheap corroboration?

14.4.2. Think of possible applications of the outcome of Experiment 3 to education (rearing of yes-men). *Alternate Problem:* Design an experiment to test a psychological hypothesis—e.g., that man seeks to minimize external stimuli. See D. E. Berlyne, *Conflict, Arousal and Curiosity* (New York: McGraw-Hill, 1960).

14.4.3. Examine the widespread idea that experiment is to be equated with the set of actual operations (e.g., meter readings) demanded by experimenting.

14.4.4. Quantitative experiments involve measurements. Is the converse true of all kind of measurement? What about measurements in atomic and nuclear physics involving atom smashers?

14.4.5. In factorial design all of the suspect variables or factors are included, in order to allow for possible interactions among them. Study this technique in connection with the desideratum of the independent variation of factors. See R A. Fisher, *The Design of Experiments,* 6th ed. (Edinburgh and London: Oliver and Boyd, 1951), Ch. vi.

14.4.6. Design an experiment to test the popular idea that rabbit legs, or any other amulets, give good luck. Recall that you are not to endanger the lives of your subjects.

14.4.7. That noise interferes with learning is a hypothesis belonging to ordinary knowledge. In order to test and refine it, ascertaining how much does noise affect learning, some experiments have to be conceived and performed. Plan one or two.

14.4.8. Design an experiment to find out the percentage of students willing to cheat in exams if they are offered low risk opportunities. Discuss the value of the questionnaire technique in this case. *Alternate Problem:* Design an experiment to test an ESP hypothesis. See C.E.M. Hansel, *ESP: A Scientific Evaluation* (New York: Scribner, 1966).

14.4.9. H. Hertz could hardly have designed his experiment for the production and detection of electromagnetic waves had it not been for Maxwell's electromagnetic theory, which gave him something to test and suggested how to test it. Similarly, the design of possible gravitational wave generators and detectors would not occupy the fantasy of some contemporary physicists were it not for Einstein's theory of gravitation. Study this latter situation and decide whether the theory is advanced enough to suggest ways for its own test. For an optimistic view see J. Weber, *General Relativity and Gravitational Waves* (New York: Interscience, 1961), Ch. 8.

14.4.10. Most testing in engineering, military science, agronomy and personnel and educational psychology is not testing of ideas but testing of things or persons. Study the differences between the two kinds of testing and determine whether the intervention of ideas in technological testing constitutes guarantees that the ideas themselves are being tested. Recall Ch. 11. *Alternate Problem:* Discuss the claim that laboratory work consists in making and reading "inscriptions". See B. Latour and S. Woolgar, *Laboratory Life: The Social Construction of Scientific Facts* (London-Beverly Hills: Sage, 1979), and M. Bunge, "A critical examination of the new sociology of science", *Philosophy of the Social Sciences* 21:524–560 (1991), 22: 46–76 (1992).

14.4. Significance

Once an experiment has been performed its outcome must be interpreted and evaluated. In the first place we must find out whether the stimuli did indeed make any difference, i.e. whether the differences between the experimental system and the control one—there will always be some differences—are *significant* or are the effect of random (uncontrolled) variations. Only in case the differences are real or significant can we ask the further question about their relation to the applied input. And once such an input-output relation is found we should try to explain it either with some available theory or by build-

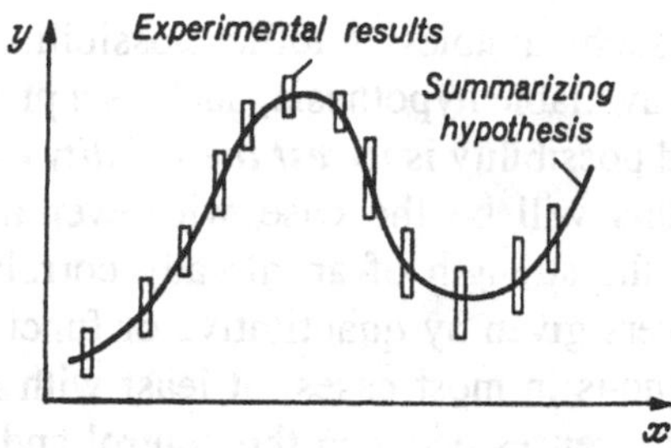

Fig. 14.7. Summary of experimental results concerning two variables: the independent or control variable x and the dependent variable y.

ing a new theory. Let us handle the problem of significance in the first place.

Suppose a quantitative *nonstatistical experiment* has been performed in which every measurement has been carefully repeated a number of times. The first thing to do with the mass of raw data is to subject them to a statistical analysis in order to uncover possible systematic errors (see Sec. 13.6). If none are discovered or if some have been detected and corrected, and a whole new run of measurements has been made, then the first task facing the experimenter is to condense the heap of raw data into a much smaller set of propositions, which can in turn be displayed in a table, summarized (and generalized) in an equation, or visualized in a graph. In Fig. 14.7 an imaginary graph has been drawn in which the error ranges (the standard deviations around the mean values) have been indicated by rectangles; the bases of the rectangles represent the standard errors in x whereas their heights visualize the standard errors in y. A few isolated "points" may lie far off the general trend. These deviants may be genuine or they may be due to systematic errors—caused, say, by the operator unawarely moving a piece of apparatus. If an isolated point lies beyond three standard error units in the corresponding frequency graph—see Fig. 13.14 in Sec. 13.6—it can be rejected as an experimental error. Perhaps the deviant is not such: ulterior research may vindicate it; but a decision must be reached at every stage, and the mentioned decision rule is quite tolerant.

Once the experimental data have been sifted, any of the following things can be done with them. One is to suggest a *new quantitative hypothesis* concerning the variables under control—that is, a new law statement relating such variables. Another possibility is to use the data to *test or retest* an available hypothesis, such as a prediction computed in a theory. A third possibility is to *test the validity of the experimental technique* itself: this will be the case whenever a new experiment design is tried on the strength of an already corroborated theory. In any case, the answers given by quantitative or functional experiments are quite unambiguous in most cases, at least with regard to the significance of the differences between the control and the experimental systems. The more thought is invested in the ideas to be tested the less work is required to assess the significance of the outcome of the test: this is one of the advantages of quantitative over qualitative (e.g.; factorial) experiment.

Suppose now we are given the outcomes of a *statistical experiment,* whether quantitative or qualitative. Since both the control and the experimental systems or groups are subject to statistical fluctuations, some differences between the two are bound to appear even if no stimuli act on the "experimental" group. For instance, in testing a hypothesis concerning the efficiency of a fertilizer, we may find that a plot treated with fertilizer gave a larger crop than the control plot, but the difference could be a chance coincidence produced by the accumulation of uncontrolled facts, such as differential soil fertility. In order to conclude to a significant (systematic, nonrandom) difference we need not only a good experiment design making allowance for the uncontrolled variables—e.g., dividing the land in accordance with the Latin Squares technique (see Sec. 14.2); we need, in addition, a statistical *test of significance.* It is only when this test has been passed that we can go over to compute the value of the correlation, if any, between any two variables (e.g., crop yield and quantity of fertilizer).

There are several significance tests (chi-square, t, F, etc.). In all of them the differences between the values of the dependent variable (e.g., the crop yield) in the control and in the experimental groups occur. Let us call $C_i–E_i$ these differences. We are interested in their absolute values or, equivalently, in their squares rather than in the signs of these differences if our aim is just to make sure whether they are random or not. Moreover, the ratios $(C_i–E_i)^2/C_i$ rather than the absolute values are of interest—and indeed all of them taken together, i.e. added. A good measure of the differences between the results regarding C and those concerning E is therefore the following sum, which is called χ^2 (chi-square):

$$\chi^2 = {}_{df}\Sigma_i (C_i - E_i)^2 / C_i. \qquad [14.5]$$

A small value of χ^2 suggests that the observed difference is not significant, while a large value is interpreted as a real association between the dependent variable (e.g., crop yield) and the independent variable (e.g., fertilizer quantity). 'Small' and 'large' being vague words, we need more precise rules if they are to be used as decision procedures for asserting or denying the significance of the differences found between C and *E. A* definite decision procedure is supplied by statistics but we cannot enter into further technicalities: suffice it to say that the question '*How* significant is the given difference?' can be given a definite answer.

In any case, if χ^2 is below a certain minimum there is no systematic difference between *C* and *E:* the differences may be real yet random rather than the effect of the stimulus. In this case one says that the *null hypothesis* or hypothesis of randomness has been established. If so, infinitely many possible relations between the two variables are disposed of with one stroke and the experimenter is free to turn to some other pair of variables. But if χ^2 surpasses that minimum, the null hypothesis is refuted and a *programmatic hypothesis is* established—namely, that the dependent and the independent variables concerned stand in some systematic relationship or other. The precise form of such an association is left to ulterior research—in case the experimenter is aware that he has just left an exploratory stage behind and that his aim, the finding of the pattern, is still ahead.

The research strategy in the case of statistical experiments can be summed up in the following way:

1. *Formulation of rival working hypotheses:*
   h_0 (*null hypothesis*): no systematic *C–E* difference: no law.
   h (*programmatic hypothesis*): systematic *C–E* difference: law.
2. *Execution of experiment.*
3. *Significance test.*
4. *Inference*: if h_0 is confirmed look for different variable pairs. If h_0 is refuted, conclude to the systematic (but still unknown) relationship between the variables and go over to next step.
5. *New problem:* compute the degree of systematic association between the two variables.

Should the *C–E* difference be significant according to any of the significance tests, the next task will be to compute the amount of association or covariation between the two variables, i.e. the extent to which x and y go together, so that an increase in x tends to be associated with an increase of y (positive correlation) or with a decrease of y (negative correlation). There are various measures of association or strength of relation, one of the commonest being the coefficient of correlation. *The *covariance* or mutual *relevance of* two random variables x and y is defined as $Cov(x,y) =_{df} \overline{(x-\bar{x})(y-\bar{y})} = \overline{(x \cdot y)} - \bar{x} \cdot \bar{y}$. For one variable, i.e. $x=y$, this formula reduces to the square of the standard deviation. Usually the *correlation coefficient is* taken as a measure of the extent of mutual relevance and is defined as $r(x, y) =_{df} Cov(x, y)/\sigma_x \cdot \sigma_y$, where $\sigma_x^2 =_{df} \overline{(x-\bar{x})^2}$ and similarly for σ_y.* The coefficient of correlation $r(x, y)$ between two variables x and y is a number in the $[-1, +1]$ interval. If $r=-1$, all the points $\langle x, y \rangle$ lie on a descending straight line; if $r=+1$ they lie all on an ascending straight line; if $r=0$ the two variables are entirely disassociated and the point set is unpatterned (no law); in intermediate cases the points cluster more or less definitely within ellipses (see Fig. 6.3 in Sec. 6.2). In some cases the distribution of dots may suggest a functional relation between x and y, but this will occur only if the variables do not fluctuate wildly, which in turn requires their precise control.

Significance tests and correlation analysis should not be the terminus of a run of experiments, because they can end up only in *coarse hypotheses* of the form "x is relevant to y" or "x is strongly (slightly), positively (negatively) correlated with y". This should be a hint and a stimulus for further research. Unhappily a disproportionately large number of experiments in the sciences of life and culture stop at this point, i.e. precisely when the most interesting problems can be stated. One of them is whether the obtained correlation is real or fortuitous. A correlation, even if high, may be accidental: that is, two variables may show a strong association during a certain period, but this association may be a chance conjunction. If no further experiments are possible in this case, theory may help by indicating whether a systematic association is at all possible. For example, if two minerals have been observed to go together in a given deposit, geochemistry will be called to task in order to tell whether the rocks "belong together" and the same association is to be expected in all rocks of the same kind. But if an experiment is possible we may eliminate one of the two variables—or keep it constant—and watch whether its former partner disappears or stops changing.

But even if the correlation between two variables is real and sustained, it may be *direct* or *indirect,* i.e. it may reflect a direct association or it may be brought about by a third underlying variable. Thus the social status-scholastic achievement correlation, which is high and positive, is the effect of additional opportunities rather than an instance of the direct correlation "The higher the better". In such cases the correlation is, for some spurious reason, called spurious. Whether it is so or not can be told by further research alone. In short, then, we have the following kinds of correlation:

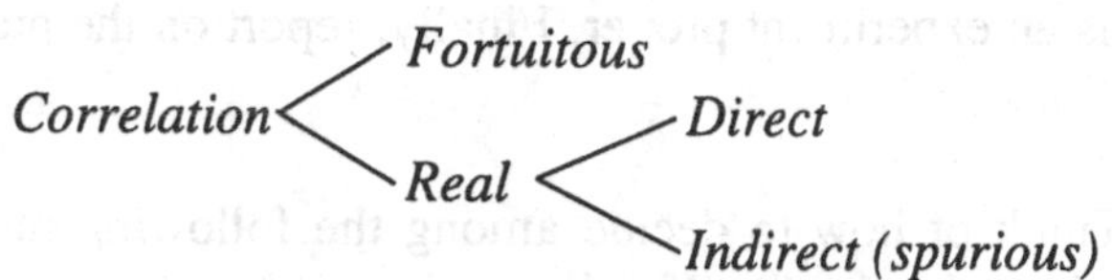

Once a correlation has been found it must, then, be checked by further experiment and/or by theory. But, again, a checked correlation should not be the goal of research but the basis for posing a further problem, namely: What is the *mechanism* underlying the correlation, i.e. what brings it about and in accordance with what laws? For instance, how does the effective fertilizer act on the plant's metabolism to increase the plant size—or how does noise act on the nervous system inhibiting the subject's learning process? Correlations, in short, are just stepping stones: the desideratum is a set of functional relations representing some patterns. These patterns may exist only at a level deeper than the level of the correlated variables, or it may be that the latter are compound variables, i.e. lumps that a theoretical analysis alone can disentangle (as has been the case with "conductivity" and "intelligence" among others). In short, scientific experiment is a torch in the search for pattern. But it is not the sole torch and it is not infallible. That experiment is not conceived and interpreted in a conceptual vacuum but in a body of ideas both scientific and philosophical has been argued at length before; let us now tackle the problem of the fallibility of experimental tests.

Problems

14.4.1. The early partisans of the hypothesis that intelligence is inherited submitted as evidence a number of distinguished family trees such as the Bernoullis, the Bachs and the Darwins. The environmentalists rejoined that a favorable and fairly constant family environment could account for the same intellectual inclinations and successes. In short, one and the same body of evidence (family trees) supported two mutually contradictory hypotheses. The procedure chosen to decide between them was to abandon the ambiguous family tree and to make follow-up studies of twins. The correlation between the performances of identical twins (siblings with the same genetic equipment) was found to be significantly higher than for fraternal twins. Analyze the so-called twin method: point out the variables involved and determine whether it is an experiment proper. Finally, report on the present state of the question.

14.4.2. Think of how to decide among the following mutually incompatible hypotheses. (i) "If religion is a deterrent to crime, then religiosity must be negatively correlated with crime". (ii) "If criminals need and seek the assistance of religion more than other people, then there must be a positive correlation between religiosity and criminality" (iii) "If certain religions make the individual egoistic and allow him to expiate his faults easily, there must be a positive correlation between criminality and religiosity of certain kinds". For actual statistical studies of this problem see M. J. Moroney, *Facts from Figures* (London: Penguin, 1956), pp. 303ff.

14.4.3. A manager, wishing to improve the work output, increases the lightning, paints the workrooms with bright colors, improves the sanitary conditions, and offers economic incentives. As a result the output does go up. Is he justified in concluding straight away that the increase in output was caused by the changes he introduced? Take into account that the same increase might have occurred in control groups—as in fact it did in the early experiments on productivity conducted in the 1920's.

14.4.4. The following theorem is proved in mathematical statistics: "If two random variables are independent then their correlation coefficient is zero". Suppose an experiment "tells" us that a certain correlation coefficient is zero: can we conclude that the corresponding variables are mutually irrelevant?

14.4.5. Survey the various significance tests and their limitations.

14.4.6. Survey the various association measures and point our their peculiar advantages.

14.4.7. What determines the choice among significance tests and among association coefficients? Or is this choice arbitrary?

14.4.8. Study the ambiguous inferences of systematic relationships from correlations. See H. A. Simon, "Spurious Correlation: A Causal Interpretation", *Journal of the American Statistical Association*, **49**, 467 (1954), and H. M. Blalock, *Causal Inference in Nonexperimental Research* (Chapel Hill: University of North Carolina Press, 1964).

14.4.9. Among the misconceptions concerning the applicability of the statistical methodology one finds the belief that it can only be applied when the object itself behaves randomly. Discuss this contention and show that a sufficient condition for the applicability of statistics is that *our information* be a random sample of a large population of possible data. Finally, see whether this condition is in fact satisfied in scientific experiment, and recall Rule 1 in Sec. 10.4.

14.4.10. Discuss the meaning of correlation coefficients and standard deviations in quantum mechanics. See M. Bunge, "A survey of the interpretations of quantum mechanics", *American Journal of Physics* 24: 272–86 (1956). *Alternate Problem:* The task of the methodologist is usually described as that of describing the current scientific procedures. May he go as far as criticizing them and proposing alternative ones? If so under what conditions?

14.5. Testing the Test

When we fail to cut the steak with the knife we do not jump to the conclusion that the steak is hard: we first test the knife for sharpness by applying it to something of known hardness, such as a crumb of bread, or we use an independent test of hardness, such as direct biting. In either case we *check* our test procedure with an *independent* technique rather than blindly believing its results. This precaution we take in ordinary life we may neglect in other matters: the medieval tests for witchcraft were not independently tested, and the ink blot test for personality traits is applied in the same uncritical fashion. In science, on the other hand, empirical procedures, both observational and experimental, are carefully scrutinized.

Test procedures or experimental techniques should be tested or validated both *theoretically and experimentally:* experimentally, because they may not work at all, and theoretically because, even if they work, this may be due to some reason other than the one advanced to advertise the test. The theoretical test of an experimental technique consists in explaining, in terms of independently corroborated theories, how the procedure works. Even a successful technique must remain suspect until its successes—and limitations—are accounted for in terms of scientific laws. In pure science, at least, when a new experimental technique is proposed it is analyzed in order to determine, on the basis of the available knowledge, whether it can work; this is why scientists rarely waste their time with testing quack techniques.

Thus, when radiocarbon dating was first proposed, it was well grounded on current knowledge: it had been known that organisms take up ordinary carbon C^{12} and radioactive carbon C^{14} in a constant ratio; furthermore, the law and rate of decay of C^{14} was known. This knowledge explained both the success of the technique and its limitation to dating events not much further back than 5000 years, the half-life of radiocarbon. In spite of this theoretical foundation, the radiocarbon test was tested by confronting its results with the results of alternative techniques. For example, the age of the Stonehenge megaliths was estimated by analyzing a sample of charcoal from a ritual pit in the same place, and the result (1800 B.P.E.) was found consistent with the result found independently on astronomical evidence. Still another test was possible in the case of objects taken from ancient Egyptian tombs, the age of which was accurately known from historical documents. And in the case of trees, the radiocarbon dates were compared with the tree-ring chronology. In this way it was possible to find that the average error of the radiocarbon technique was about five per cent, a margin that was subsequently reduced.

A more recent case of *cross checking is* provided by the analysis of xenon tetrafluoride—or, for that matter, of any interesting new compound. The chemical composition formula "XeF_4" was first inferred (1962) from the ratio of fluorine to xenon consumed in the reaction. This hypothesis was then tested by recovering the xenon—specifically, by analyzing the products of the reaction "$XeF_4 + 2\ H_2 \rightarrow Xe + 4HF$". The same hypothesis was independently tested by a number of nonchemical methods, such as determining the molecular weight of XeF_4 by means of a mass spectrometer and by *X*-ray analysis. The various tests provided a mutual control and, in addition, new kinds of evidence—i.e. they disclosed new properties of the compound. In general, the more valuable sets of experimental techniques are those which are the more independent, both because they can control one another without circularity, and because they yield the larger variety of data.

A number of empirical procedures employed in medicine and clinical psychology are so far theoretically unfounded and have not been checked with alternative techniques; even worse, some of them have been found inconsistent with theory and with competing techniques. This is particularly the case of the intuitive procedures for assessing personality—such as the interview, the psychoanalytic interrogatory, and the projective tests. Their propounders have never tested these supposed tests, either theoretically or empirically: they have oracularly proposed them without bothering to make follow-up studies of the subjects, which alone might tell whether the subjects did in fact behave as prophesied on the basis of such alleged tests. When these intuitive procedures have been so tested they have exhibited either a low correlation—smaller than 0.5—or a systematically negative correlation between clinical prediction and actual performance. In other words, clinical guesses made with intuitive—groundless—procedures are usually no better than those that any of us can make by flipping a coin; very often they are below chance—which suggests that they are based, in the best of cases, on false hypotheses.

A good test should satisfy the following conditions to a satisfactory degree:

1. *Validity* or adequacy: the test should measure or check what it purports to achieve. E.g., a hearing test should not be interpretable as an understanding test, and a test of computational skill should not be mistaken for a test of mathematical ability.

2. *Foundation:* the test should be consistent with what is known and, if possible, current scientific theory should explain its successes as well as its limitations. Otherwise how are we to know that it is valid?

3. *Objectivity:* the performance, recording and interpretation of the test should be essentially independent of the operator. Test for this condition: the test results should not vary significantly from operator to operator.

4. *Stability or reliability:* the test should test always the same thing. Test for this condition: the results obtained after a short interval should show a high degree of agreement (self-correlation) with the prior ones.

5. *Agreement with other tests:* test procedures, just as scientific laws and scientific norms, must be backed up or disclaimed by other members of the body of scientific knowledge rather than stand in isolation. If a new test is proposed, the older and established ones will, as it were, watch and evaluate its performance. Test for this condition: the correlation with the results of accepted concurrent tests should be high.

In short, empirical procedures should be *scrutable*, analyzable, comparable, criticizable, and in principle improvable. The same holds, of course, for every one of the results of empirical tests: even if the technique itself is valid, it may be wrongly performed or interpreted. Only naive philosophers have blind faith in experimental results: the good experimenter will mistrust, hence keep under control and check once and again the design, the apparatus, the performance, the hypotheses involved, and the interpretation of the results.

Reports on scientific experience—observation, measurement, or experiment—are, then, as fallible as techniques and hypotheses. Their fallibility, however, is not such as would satisfy the radical skeptic, because they can all be improved. Sometimes the failures themselves point out the way to effective correction—on condition that such failures be explained by theory. We are, once again, driven to the conclusion that empirical procedures are worthless if unaccompanied by scientific theory.

Problems

14.5.1. List the kinds of mistakes that can be made in experimenting. See W. B. Cannon, *The Way of an Investigator* (New York: W. W. Norton, 1945), Ch. xi, and E. B. Wilson, *An Introduction to Scientific Research* (New York: McGraw-Hill, 1952), pp. 119ff. What becomes of the popular idea that experiments can never go wrong?

14.5.2. Since a reliable test must be objective, would it improve reliability to build machines for giving and interpreting tests?

14.5.3. What would you do if presented with a set of white gaseous filaments moving in space and emitting a frightening voice? Would you believe it is an apparition of the underworld, or would you reject this interpretation on the ground that it does not fit into your scientific knowledge? In other words, would you accept talking ectoplasms as a reliable technique for the disclosure of ghosts?

14.5.4. Discuss the claim that certain experiences—mystical, metaphysical, or scientific—are final.

14.5.5. Should we have blind faith in the experimental techniques employed by the most outstanding scientists of our time?

14.5.6. What is more valuable: intelligent experiment design or skill in experiment execution? And how are the two related?

14.5.7. How are chemical tests validated—e.g. on what basis is litmus paper adopted for the identification of acidity? *Alternate Problem:* Report on some of the twelve hundred critical examinations of current psychological tests contained in O. K. Buros, Ed., *The Sixth Mental Measurements Yearbook* (Highland Park, N. J.: Gryphon Press, 1965).

14.5.8. Discuss R. A. Fisher's remark that experiment design and experiment interpretation are interdependent because an ill-designed experiment is susceptible to a number of interpretations or yields results that are inconclusive in some respects.

14.5.9. Parapsychologists hold that their "experiments" must be performed in a friendly atmosphere not in the presence of hostile or even skeptical critics, because the "sensitives" may then lose their extraordinary abilities. Examine this claim. In particular, decide whether this condition does not render parapsychology empirically irrefutable.

14.5.10. In order to test the hypothesis that higher animals—e.g., rats—choose their nourishment not only on account of its metabolic value (set of physiological variables) but also on account of its taste (psychological variables) the following experiment is performed. Sugar water of different sugar concentration, hence of varying sweetness, is offered to normal rats and to rats whose sense of taste has been destroyed by removing the taste areas in their brains. Result: whereas the normal rats show a marked preference for a medium concentration, the injured rats show no preference at all. Problem: How can we independently ascertain that, in fact, the *taste* areas have been removed?

14.6. Functions

Why are experiments made: what is gained by experimenting? In the first place every experiment is, metaphorically speaking, a direct and precise questioning of reality in so far as every set of inputs ("questions") elicits a set of outputs ("answers") and therefore poses to the theoretician the problem of accounting for (explaining) a set of "question"–"answer" pairs. Secondly, many such "questions" are not spontaneously posed in a natural environment: induced radioactivity, induced genic mutation and purposive maze training hardly occur in a state of nature. Experiment, by enriching the set of naturally occurring facts, is apt to disclose deep and unsuspected aspects of things—as when a friend is asked to risk something. Thirdly, whereas spontaneous facts are usually extremely complex, experimental facts are simpler and therefore more tractable. The control of variables, peculiar to experiment, brings the object near its theoretical model, by focusing on a few variables, by reducing their ranges, and minimizing the disturbances and irrelevances—in particular the "noise" characteristic of purely observational data—present in every natural setting.

For several good reasons, then, experiment has become essential to modern factual science. But this is not to say that experiment is an *alternative* to theory, or the *basis* of it, or that it *leads the way* in factual science. Every experiment is a search for an answer to a question born in a body of ideas, it must be planned and it must be interpreted—in contrast to the blind haphazard trial and error—and both the design and the interpretation of experiments require more or less elaborate systems of hypotheses. The execution of an experiment is the realization of a plan conceived on the strength of certain assumptions, and what it yields is a ciphered message that cannot be decoded outside a body of knowledge. Whether in generating new ideas or in checking them, experiment is a *means* rather than an end; it can be a goal in itself for the individual scientist, not however for science as a collective enterprise. In short, no experiment is possible in the absence of ideas: it is a materialization of a set of ideas and it is performed in the service of ideas—only, not to fix them but to test and enrich them.

Psychoanalysts have made a number of observations—but not a single experiment—and have raised a number of interesting—but fuzzy—questions. Their empirical material is useless both because they have made no use of control techniques and because they lack a consistent, thoroughly testable and corrigible theory by means of which their observations can be interpreted (explained). The ancient astrologists have rendered a greater service to science by making a large number of correct observations and by inventing a number of devices for astronomical observation and computation. But, again, their theory was wrong to the extent to which it was testable, and it was utterly incorrigible (dogmatic). And the alchemists surpassed both the psychoanalysts and the astrologists by giving a number of correct descriptions of substances and processes, as well as by designing useful techniques and apparatus that are still in use. But their theory was too simple to be true, their operations much too blind to be useful, and they did not care to test or improve their ideas: like the psychoanalysts and the astrologists, their chief concern was with illustrating their fixed views and applying them to practical ends. All three were *empirical* disciplines in the sense that they used experience; but none of them were *experimental* in the modern sense of the word because they had no consistent, testable, corrigible and reasonable (externally consistent) theory to design and interpret their operations.

The moral of the story of pseudoscience is clear: no mass of observations or even experiments is valuable unless conducted and interpreted in the light of a theory consistent with the bulk of scientific and metascientific knowledge, and capable of learning from experience. Accordingly experience, and in particular experiment, is insufficient: it is a means for posing problems and testing proposed solutions to them. An empirical operation, whether a stray observation or a controlled experiment, may give rise to an interesting problem or even to an interesting conjecture on condition that it takes place in a body of knowledge. And scientific experience has a test value only if it is relevant to a definitely stated hypothesis that cannot be twisted to agree with any outcome whatsoever. In short, scientific experience is valuable to the extent to which it is soaked with ideas—ideas which the experience is capable of controlling and enriching.

The absence of interaction between theory and experiment was responsible for the failure of the Greeks to go much beyond Archimedes and for the barrenness of the few observations and experiments made during the Middle Ages: they were either misguided by wrong theories or they were self-contained, in the sense that they aimed just at finding out how certain things were and, in the best of cases, what happened if certain changes were provoked—the type of experiment recommended by classical behaviorism. Mere observation and casual experimentation aiming at gathering data rather than evidence relevant to some view were, in fact, the best that the schoolmen could produce: this is what Albertus Magnus, Petrus Peregrinus, and Rogerius Bacon did. That situation is sadly paralleled by much contemporary psychological and social research, in which theories are speculative and empirical operations are not systematically guided by theories but rather by the desire to see what happens.

As to the experiments conducted between the 15th and the 17th centuries by craftsmen, gunnery experts, surgeons, alchemists, and instrument makers, they had not a chiefly cognitive aim but rather a practical goal: those men wanted to know whether certain man-made things and procedures worked, not whether certain theories were true. Neither the scattered observations of the schoolmen nor the habitual technical experimenting of the superior craftsmen have given rise to modern experimental science: neither of them employed the *experimental method.* This is not a procedure to see what happens if one empirically changes the values of certain variables—for example if one reduces by one the number of legs of a fly. The experimental method is the way factual *hypotheses* are empirically *tested*—namely, by rigorously *controlling* both the relevant variables and the inferences "drawn" from (suggested by) the outcomes of the operation. He who applies the experimental method has some idea to test and some further ideas to design the test.

Accordingly the radio amateur who plays around with circuitry trying intuitively to improve the performance of a set without explicitly using and testing any relevant physical theory is not an experimentalist: like Leonardo da Vinci, like the gunners with whom Galilei collaborated at the Venice arsenal, or like the military surgeon Vesalius who made anatomical drawings in the battlefield, the radio amateur is a skilled craftsman. Doing things has, in itself, no cognitive value: production is useless for the advancement of knowledge unless there is some idea to be tested by production—for example some hypothesis in operations research. Mere production, with ignorance of the mechanism or without raising new problems of knowledge, is cognitively useless however practically valuable it may be. In short, for science both pure and applied production is valuable as long as it is a test of ideas, not a substitute for them. A genuine experimentalist is not a craftsman or a gadgeteer but a scholar capable of conceiving (designing) experiments and experimental techniques for testing hypotheses.

Such differences between experiment and the experimental method, between the skilled craftsman and the experimentalist, between production and research, and so on, are at stake whenever the origins of modern science are debated and whenever the roles of technicians and technologists in the modern laboratory are evaluated. These questions are, therefore, of interest to the historian of ideas and to the science manager no less than to the pure scientist.

The pure scientist is interested in ascertaining what makes an experiment valuable: under what conditions it can solve or pose a significant research problem. In particular, he wants to find out whether a given experiment is fairly conclusive: for, if it is not, he will have to repeat or redesign it. An experiment may be said to be *inconclusive* with regard to a given set of hypotheses either if it is irrelevant to them or if its outcome is consistent with both the given set and some rival set of hypotheses. The case in which the experimental error masks the expected effect falls of course into the latter class. That there should be inconclusive experiments is easily understandable in view of (i) the number of error sources, (ii) that so much depends on the interpretation of the results, and (iii) that any set of experimental outcomes can in principle be derived from a number of alternative theories.

But are there *conclusive* experiments, i.e. such that they unequivocally and conclusively decide in favor or against some hypothesis? In particular, are there *crucial* experiments, i.e. experiments enabling us to unambiguously and definitively decide between two rival hypotheses such as, e.g., the null (or randomness) hypothesis and its negate? The answer to this question depends not only on the experimental design and the experimental outcome but also on the nature of the hypothesis or theory questioned by the experiment. It is one thing to test an existential hypothesis and quite a different task to check a universal one. Again, it is one thing to probe a qualitative hypothesis and another to test a quantitative one. Likewise, it is not the same to subject an isolated hypothesis to the test and to test a theory with an unlimited number of testable consequences. In short, the answer concerning the value of an experiment must be sought in an analysis of (i) the experimental design, (ii) the experimental results or data, and (iii) the logical relation between the data and the constructs (hypotheses and theories) to which the data are supposed to be relevant. Having dealth with (i) and (ii) we must now attack (iii), that is, the problem of the logic of empirical test or, as is also called, the question of scientific inference. This will be the subject of the next chapter.

Problems

14.6.1. Dig up the ontological and epistemological presuppositions of the experimental method. In particular, find out whether it assumes the objective existence of the objects of experiment.

14.6.2. Does the existence of idealess experimenters refute the thesis that scientific experiments are concerned with ideas about things rather than with things? *Alternate Problem:* Do social programs count as scientific experiments?

14.6.3. Myriads of tests were performed by physicians on prisoners in Nazi concentration camps: subjection to extremes of hunger, cold and fatigue, as well as vivisection. The alleged aim was "to see what happened" rather than to test definite hypotheses. Were those scientific experiments? *Alternate Problem:* Study the ethical problem of medical experimentation on humans: is it justified, and if so under what circumstances? See the Nuremberg Code in *Science*, **143**, 553 (1964).

14.6.4. Pedagogical techniques have seldom been subjected to the test of observation, much less to experiment. Why are they used, whereas on the other hand fertilizers and germicides are not thrown to the market untested? *Alternate Problem:* Is experience sufficient to confirm (refute) ideas or does confirmation (refutation) require some theoretical elaboration?

14.6.5. Discuss the radical empiricist thesis that experiment provides direct answers to questions, with no need for hypotheses and theories, whence a sufficiently advanced science is purely experimental. *Alternate Problem:* Why do so many behavioral scientists identify research with observation and experiment?

14.6.6. Discuss the deductivist thesis that the sole role of experiment is to refute hypotheses, i.e. to trim or eliminate ideas. *Alternate Problem:* For Kant and the neo-Kantians no ordinary experience was possible outside a categorial frame (e.g., the intuitions of space and time) given once and for all. Could this thesis be modified to apply to scientific experience?

14.6.7. The theory of evolution was not fully credited until the 1940's, after two revolutionary changes: (i) the synthesis of Darwin's theory with genetics, and (ii) the deliberate production and observation of adaptive changes in bacteria and fruit flies, whereby evolution was brought within the reach of the experimental method. Comment on this development. *Alternate Problem:* Describe a psychoanalytic laboratory—should you find one.

14.6.8. Elaborate on the dichotomy

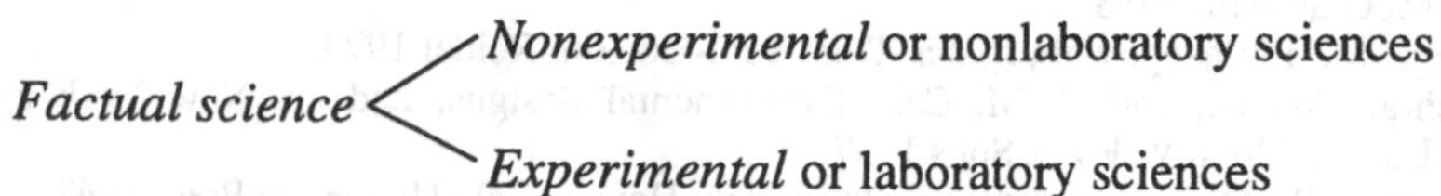

Take into account that until recently geologists performed no experiments on small scale models—e.g., to simulate foldings and erosion—that the study of evolution was empirical but nonexperimental, and that sociology and economics had not begun to experiment on small groups. *Alternate Problem:* What is wrong with philosophical psychology: that it deals with subjective phenomena or that it eschews experimental control? See J. Piaget, *Sagesse et illusions de la philosophie* (Paris: Presses Universitaires de France, 1965), Ch. IV.

14.6.9. Study the function of ordinary experience and experiment proper in mathematics, where it can occur not only by suggesting ideas but also by helping to find approximate solutions to problems for which physical models can be built. In particular, examine (i) the role experience may have played in Archimedes' mathematical work, (ii) the employment of the Monte Carlo or random sampling techniques for computation, and (iii) the use of computers in pure mathematics.

14.6.10. A number of distinguished historians of science, notably P. Duhem, J. H. Randall, Jr. and A. C. Crombie, have maintained that modern science was born in the Middle Ages; Duhem gave even the birth date: 1277. The evidence they have offered in favor of this hypothesis is triple: (i) the experiments performed in Paris in the 13th and 14th centuries and in Padova in the 16th century; (ii) the nonAristotelian physical hypotheses advanced from the 13th century on, and (iii) the methodology elaborated by the Padova school. Discuss that thesis. *Alternate Problem:* Several historians of science, among them A. Mieli, have regarded Leonardo da Vinci as a scientist because he performed experiments and even—like Roger Bacon long before—praised experiment as a source of knowledge. Discuss this view.

Bibliography

Atkinson, R. C. et al.: Stevens' handbook of experimental psychology, 2nd ed. Reading, Mass.: Addison-Wesley 1988.

Baird, D. C.: Experimentation: An introduction to measurement theory and experiment design, chs. 4 to 6. Englewood Cliffs: Prentice-Hall 1962.

Blalock, H. M. and A. B. Blalock, Eds., Methodology in social research. New York: McGraw-Hill 1968.

Bunge, M.: Philosophy of physics. Dordrecht-Boston: Reidel 1973.

Cochran, W. G., and G. M. Cox: Experimental designs, 2nd ed. New York and London: John Wiley & Sons 1957.

Conant, J. B.: On understanding science. New Haven: Yale University Press 1947.

— (Ed.) Harvard case histories in experimental science, 2 vols. Cambridge, Mass.: Harvard University Press 1957.

Cox, D. R.: Planning of exponents. New York and London: John Wiley & Sons 1958.

Fisher, R. A.: The design of experiments, 6th ed. Edinburgh and London: Oliver and Boyd 1951.

— Statistical methods and scientific inference. Edinburgh and London: Oliver and Boyd 1956.

Gabriel, M. L., and S. Fogel (Eds.): Great experiments in biology. Englewood Cliffs, N.J.: Prentice-Hall 1955.

Galison, P. L. : How experiments end. Chicago: University of Chicago Press 1987.

Greenwood, E.: Experimental sociology: A study in method. New York: King's Crown Press 1945.

Hacking, I.: Representing and Intervening. Cambridge: Cambridge University Press 1983.

Moroney, M. J.: Facts from figures. London: Penguin 1956.

Yule, U., and G. Kendall: An introduction to the theory of statistics, 14th ed London: Griffin; New York: Hafner 1950.

15

Concluding

Man is a conjecture-making animal: he does not stop inventing and trying out hypotheses and jumping to bold "conclusions" about their worth. The scientific attitude is not to forbid such jumps but to keep them under control. In this, the last chapter, we shall investigate how such a control is exerted and we shall see that it is severely limited: there are no fool-proof criteria for jumping to the right "conclusions" unless these are conclusions proper, i.e. deductive consequences of some set of premises. The problem of nondeductive scientific inference is a hard one if only because of the gap between ideas and the facts they are supposed to represent: indeed, how shall we judge whether an idea "fits" its referent? Furthermore, if the idea happens to be a hypothesis, how can we compare it with empirical evidence if the two do not share the same concepts and are consequently couched in different languages? Even the most elementary procedures whereby we test our ideas concerning facts involve grave and hardly solved philosophical problems, such as the problem of truth and the problem of scientific inference—i.e. the problem of ascribing truth values to scientific ideas. The purpose of this chapter is to examine those procedures and to uncover the underlying philosophy.

15.1. Inferring

Inference is the passage from one set of propositions to another: the former may be called the premise class, the latter the conclusion class.

Like other human activities, inferring can be successful or not. Unlike other human activities, inferring can be valid but barren, as in the case of "*p*, therefore *p*", or invalid but fruitful, as in the case of many a hasty yet plausible generalization and analogy of science. Just as arriving at a class of true conclusions is not a validity criterion, so validity or formal correctness does not guarantee the fruitfulness of an argument: validity and fruitfulness are mutually independent properties. The ideal is, of course, to marry fertility to rigor—but this is to be strived for rather than required from the start, as no train of thought could probably begin to form if constrained by severe standards of rigor. We must face the truth that in all departments of culture both valid and invalid inferences are performed and needed: that the more austere science begets logically illegitimate children which it would be wrong to eliminate and which cannot be legitimated either. All we can do is to keep these modes of inferences under control, to which end we must study them rather than hold them in contempt. Beget first, then rear.

Just as in reality we find two kinds of pattern, namely laws and trends, so in argument, whether ordinary or scientific, we find two kinds of pattern: lawful or valid (deductive inference patterns) and trend-like or logically invalid but often fruitful—i.e. patterns of non-deductive or tentative inference. Valid, i.e. deductive inference is studied by formal logic. Among the most common modes of deductive inference we shall recall substitution (change of variables), detachment or *modus ponens,* rejection or *modus tollens,* instantiation, and generalization—in addition to a large number of specific (nonuniversal) inference schemata like those used in arithmetical computation. When a valid inference is made from premises held as true or accepted by convention, the inference is called a *proof.* Clearly, there are no proofs of factual formulas, for the simple reason that no initial assumption of factual science can be taken for granted: whereas in mathematics the premises validate (or justify or ground) the conclusions, in factual science it is the conclusions which afford a partial validation or justification of the premises. The "conclusion" from conclusion proper to premise is inconclusive. And it is of a number of kinds, usually lumped together under the ambiguous label of *induction.* We shall take a glance at some of the most oft-recurring patterns of tentative or plausible inference. Here they go:

1. Substantive analogy: *Similarity of components*

Pattern

a is $P_1, P_2, \ldots,$ and P_n
b is $P_1, P_2, \ldots,$ and P_{n-1}

It is likely that b is P_n

Example. Among animals cancer is often caused by viruses. Man is an animal. Hence it is likely that viruses are often responsible for human cancer.

2. Structural analogy: *Form similarity*

Pattern

Systems with a similar form (structure or law) often share a number of other properties
a and b have the same structure (or they "obey" formally similar laws)

It is likely that a and b share further properties

Example. The equation of motion of a quantum system is mathematically similar to a classical wave equation. Hence it is likely that the quantum equation refers to a wave motion.

3. First degree induction: *From instances to low level generalization*

Pattern

All A up to the n-th were found to be B

It is likely that all A are B

Example. All measured values of the electron charge are, to within experimental error, the same. Hence it is likely that all electrons have that same charge.

4. Second degree induction: *From low level generalizations to higher level generalization*

Pattern

Law L holds for every set S_i up to the n-th

It is likely that law L holds for every set S_i

Example. The basic laws of learning hold for all studied animal species. Hence it is likely that they hold for all existing animal species.

5. Statistical generalization: *Sample-population inference*

Pattern

S is a random sample of the population of *U*
The observed frequency of *P*'s in the random sample *S* of *U* is *p*

The expected frequency of *P*'s in *U* is near *p*

Example. In a random sample of college students amounting to the one-hundredth part of the student body, 10 per cent were found to master their mother tongue. Hence it is likely that the corresponding percentage in the whole student population is about 10 per cent.

6. Statistical specification: *Population-sample inference*

Pattern

The observed frequency of *P*'s in the population *U* is *p*
S is a random sample of *U*

The expected frequency of *P*'s in *S* is near *p*

Example. In the last election 80 percent of the nation voted Laberol, hence it is likely that about 80 percent of the population of Medianville, which is pretty representative, voted Laberol.

7. Weak modus ponens: *Weak detachment of consequent upon weak assertion of conditional and/or antecedent*

Pattern 7 a	*Pattern 7 b*	*Pattern 7 c*
If *p*, then *q* *p* is plausible	"If *p* then *q*" is plausible *p*	"If *p* then *q*" is plausible *p* is plausible
q is plausible	*q* is plausible	*q* is plausible

Example. It is likely that a mild climate has favored the birth of civilizations. Spain enjoys a mild climate. Therefore it is likely that a civilization was born in Spain.

8. Weak modus tollens: *Weak rejection of antecedent upon strong or weak assertion of conditional and weak or strong negation of consequent*

Pattern 8 a	*Pattern 8 b*	*Pattern 8 c*
If *p*, then *q* –*q* is plausible	"If *p*, then *q*" is plausible –*q*	"If *p*, then *q*" is plausible –*q* is plausible
–*p* is plausible	–*p* is plausible	–*p* is plausible

Example. It is likely that science flourishes only in freedom. Now, there is no freedom in Dogmaland. Hence it is unlikely that science flourishes in Dogmaland.

9. Strong reduction: *Weak assertion of antecedent upon assertion of consequent*

Pattern

If p, then q
q

It is possible that "p" is true

Example. If an electric current passes near a needle, the latter is deflected. A certain magnetic needle has suffered a deviation. Hence it is possible that an electric current exists near the needle.

10. Weak reduction: *Weak assertion of antecedent upon assertion of consequent and strong or weak assertion of conditional*

Pattern 10 a	*Pattern 10 b*	*Pattern 10 c*
If p, then q q is plausible	"If p, then q" is plausible q	"If p, then q" is plausible q is plausible
It is possible that p is true	It is possible that p is true	It is possible that p is true

Example. It is likely that, when a weather change approaches, fish jump out of water. Fish have been jumping out of water. Hence, it is possible that weather will change.

None of the above patterns is an inference *rule* or can be converted into such: they neither permit nor prohibit inferences. They are just among the various *heuristic* procedures in need of control. The conclusions of plausible inferences do not follow with certainty in every case: what "follows" is always a guess, and indeed in a psychologically seductive way rather than in a logically deductive manner—whence the corresponding "logic" might be called *seductive logic.* All attempts to sanctify such sins are doomed to failure because plausible inference depends on the nature of the case, i.e. on the content of the premises and not on their form: it is contingent and cannot be rendered necessary. This content dependence of tentative inference is what makes it logically weak and heuristically strong.

What is the place of the above-mentioned patterns of tentative inference in science? Substantive analogy, first degree induction, statistical generalization and statistical specification occur in the formulation of low level propositions both in ordinary knowledge and in science. Structural analogy leads occasionally to conjecturing high level hypotheses. The remaining patterns occur, alongside the deductive inference forms, first degree induction and statistical generalization, in the inferences we make when testing hypotheses and theories. It is true that in our example of a first degree induction a theoretical concept, namely "electron charge", occurred. But this only illustrates the point that fundamental laws are not obtainable by induction, let alone by direct jumps from individual propositions expressing empirical data: the mere measurement of the electron charge requires a number of theoretical formulas, including the high level hypothesis that there are units of electric charge. It is not just that substantive analogy, first degree induction and statistical generalization do not happen to occur in the formation of high level hypotheses: they are *de jure* inhibited to climb those heights because high level concepts must be available before they can occur in high level hypotheses. Besides, those among the latter which are universal statements do not usually emerge as laborious step-by-step generalizations from singulars, but are *born* general. The job of induction is not to generate such generalities but to take part in their empirical test, as we shall see in Sec. 15.4.

Induction, statistical generalization and instantiation, as well as weak *modus ponens,* weak *modus tollens* and reduction both weak and strong can be either wild or controlled. Thus, inductive generalization can be either a free jump from an arbitrary collection (a nonrandom sample) to a universe or it can be restrained by statistics. The *statistical control* of a tentative inference consists in deliberately constructing a representative random sample—for example of data—and weighing the evidence represented by the sample. Mathematical statistics supplies more or less grounded criteria for sample representativeness and for the reliability of the inferences. In particular, statistics is concerned with quantifying the risk assumed in drawing "conclusions" from empirical data in the absence of theory—i.e. in the protoscientific stage. In the early stages of a discipline most hypotheses are weighed almost exclusively on the basis of empirical data. Now, some such results can be due to chance: the randomness hypothesis cannot be excluded if little or nothing is known about the mechanism that brings such effects about. The conjecture that a set of data concerning the variables under control are due to uncontrolled variables is called the *null hypothesis* (recall Sec. 14.4); the rival conjecture that the data do reflect changes in the variables under control is one of a number of alternative specific hypotheses. If neither of the competitors has a theoretical backing, the experimenter has to rely on his data alone. Now, if he makes the decision to reject the null hypothesis, i.e. to accept any of the alternative hypotheses according to which the effect is real, he takes certain risks. If this decision (rejecting the null hypothesis) is wrong, he will be said to make an *error of the first kind.* The corresponding risk of making such an erroneous inference is called the *level of confidence*—a number α much smaller than unity, say 0.05. The wrong acceptance of the null hypothesis is called an *error of the second kind.* The corresponding risk is measured by another number, β, also much smaller than unity. The choice of α and β is up to the experimenter: the more severe a test the smaller will α and β be. (Unfortunately these numbers cannot be used to *define* the concept of test toughness because the converse proposition is false.) In any case, statistics supplies criteria for (tentatively) concluding that a set of empirical results is statistically significant and for determining the risk of drawing wrong inferences. (These errors in inference are not to be mistaken for measurement errors.) However, none of these criteria *secure* the truth of a generalization: they are just brakes preventing hasty jumps, not devices warranting happy leaps. Statistics, then, neither builds scientific theories nor establishes them but it takes part in their test: statistics belongs therefore to the methodology of experimental science and consequently any detailed treatment of plausible inference must match with statistics.

It must be realized, though, that the importance of statistics both in the design and in the interpretation of empirical operations is inverse to the definiteness and richness of the ideas that are being tested. A qualitative hypothesis such as that talent "depends on" early training will require a much bulkier body of data and a more complex statistical analysis of such information than, say, any quantitative hypothesis that can be checked by measurement. If the tested hypothesis itself is strong, we need not concentrate on looking for possible counterexamples or on indefinitely increasing the number of confirming instances: a few confirmations at strategical points may suffice to show whether the idea is worthwhile being further investigated (rather than accepted). And if the hypothesis, in addition to being definite and strong, is systematic, i.e. related to other scientific hypotheses rather than stray, then its testability will increase even further and accordingly the bulk of directly relevant data and the complexity of data processing may decrease. It pays more to conceive strong hypotheses, to work them out, or to plan their test, than to hope that statistical techniques or computers will do the thinking for us.

In conclusion, the science of nondeductive inference cannot attempt to validate or justify nonvalid inferences, i.e. to erect trend-like patterns into laws. Rather than aping deductive logic it should discern, analyze and criticize patterns of non-deductive inference, investigating with the assistance of mathematical statistics the conditions under which the chances of leaping to erroneous "conclusions" can be decreased rather than eliminated—i.e. the conditions under which such inferences though inconclusive, are plausible.

Problems

15.1.1. Examine the differences between an analogy and an argument from analogy or analogical inference. In particular, discuss whether a statement of homology—of the form "*A* is to *B* as *C* is to *D*", i.e. "*A*:*B*::*C*:*D*"—occurs in the former or in the latter. *Alternate Problem:* Examine the role played by the lock-and-key analogy in biochemistry—e.g., antigen-antibody reaction and molecule-odor receptor.

15.1.2. Comment on Aristotle's distinction between *apodeictic arguments,* which start from true premises, and *dialectical arguments,* which use premises the truth value of which is unknown. Are such modes of argument used in science?

15.1.3. To which discipline should the study of plausible inference be assigned: formal logic, statistics, or epistemology? Or does it call for a multidisciplinary approach?

15.1.4. Statistical specification (pattern 6 in the text) is plausible if a random sample is taken and if the sample size is not too small: otherwise violent fluctuations around the frequency will be observed. For example, from the generalization that most punks are uneducated we cannot conclude that a given punk is uneducated. That is, if the sample size is small the chances of error are large. Is the inference in such cases preposterous or can it be used as a preliminary indication? Recall Sec. 9.5.

15.1.5. Study the method of confidence intervals, applicable to statistical generalization (Pattern 5). *Alternate Problem:* Light waves, atoms, genes, viruses and the like are often believed to have been inferred from observations and are consequently called *inferred entities.* Is this name justified?

15.1.6. Is it possible to justify or validate plausible inference by casting it in some mathematical mold, such as probability theory? See Alternate Problem 5.1.3 and W. Kneale, *Probability and Induction* (Oxford: Clarendon Press, 1949). *Alternate Problem:* Physicists do not handle as sophisticated statistical tools as psychologists and sociologists do for testing their hypotheses and theories. It has been suggested that physicists should catch up with the behavioral scientists in this respect. Is this injunction correct?

15.1.7. The existing systems of inductive "logic" are philosophical games which are neither suggested by actual scientific research nor used in it. One reason for this is that they apply only to isolated low level, scientifically uninteresting generalizations or, at most, to protoscientific empirical generalizations. Does this failure to tackle the real problems of tentative inference in science proper condemn the whole enterprise?

15.1.8. Up to now either of the following two extreme attitudes toward plausible inference are usually taken. One is to regard it as a purely psychological process in no more need of logical investigation than the intuitive flash; the other is to build some a priori system of inductive logic aping deductive logic. Examine both stands and see whether there is room for a third one.

15.1.9. If a predicate *P* is true of an arbitrary individual *x*, then it can validly be generalized to every individual (in the same set): $P(x) \vdash (x)P(x)$. Is this inference rule applicable in factual science? If not, why? And if in the above formula we replace the phrase 'an arbitrary individual' by 'an individual picked at random', do we obtain an inference pattern suitable for factual science? What kind of pattern: valid or tentative (heuristic)? *Alternate Problem:* Must the reliability coefficients α and β (p. 295) be chosen arbitrarily? See A. Wald's work.

15.1.10. Consider the inference schema known as the (weak) *principle of mathematical induction* or pattern of proof by recursion. Let $p_0, p_1, \ldots, p_n, \ldots$ be a denumerable sequence of propositions. The pattern is: $\{p_0, p_k \rightarrow p_{k+1}\} \vdash (n)p_n$. Queries: Is it a primitive principle or can it be deduced? In particular: is it possible to deduce it from elementary (propositional) logic or is it an irreducibly mathematical principle? What has it got to do with induction? And could it be applied to establish unrestricted factual generalizations, e.g., law statements, on the basis of the examination of individual cases? *Alternate Problem:* Study the problem of the psychogenesis of inference patterns. See B. Inhelder and J. Piaget, *The Growth of Logical Thinking from Childhood to Adolescence* (New York: Basic Books, 1958).

15.2. Testing Observational Propositions

Both logically and epistemologically the simplest factual propositions are those which refer to single observable facts, such as "That thing is black". There is no particular methodological difficulty in testing this proposition as long as we remain content with the qualitative concept of blackness. Methodological problems will arise if we have in mind the physical rather than the commonsensical concept of blackness, because that will call for measurements of the radiation and absorption rates of the body concerned. The innocent-looking proposition "That thing is black" may not pose the methodological problem of ascertaining *how do we know* that it is true (or false), i.e. of picking the right conceptual and material tools enabling us to evaluate it. But it does pose the problem of what does it *mean* to say that the proposition is true or false, and this is a philosophical problem.

Truth criteria are elaborated by the special sciences depending on the nature of the case. But no set of truth criteria replaces the elucidation of the concept of truth however much they may clarify this concept—just as an acidity test does not solve the problem of the meaning of 'acid'. Sure enough, we shall say that an observational proposition is true if and only if it *agrees* with fact or if it *adequately* expresses the state of affairs it refers to, or if it *fits* actual observation. But what does it mean for a proposition to fit a nonconceptual entity such as a fact? This fitness, adequacy or agreement is metaphorical: a suit can be tailored to fit a body, or a proposition can be tailored to entail another proposition: here we are comparing objects of the same nature, both physical or both conceptual. But how can we compare two heterogeneous objects, such as an idea and a fact? We can compare *x* to *y* as long as *x* and *y* share some property, and precisely the one in whose respect the comparison is performed; thus we can say that *x* is longer than *y* if both *x* and *y* have a length. But we cannot say that an idea is blacker than its referent or that a thing is truer than its idea. We may, on the other hand, *confront* or contrast an idea with its referent—but again in a metaphorical rather than in a literal sense, because we cannot put the one vis à vis the other—for one thing, because ideas have no independent existence in physical space.

*The only possible comparisons, in a strict sense, would be between a fact and either (i) the sign (physical shape) representing the proposition expressing the fact, or (ii) the psychophysiological process going on in our brain when we conceive the proposition that "reflects" the fact. Either would be a relation between two physical objects. The first comparison is possible but useless since one and the same proposition is expressible in infinitely many ways. The second comparison is desirable and in principle possible in the case of perceptible facts. (Think of the physiological process induced first in the retina and then in the brain cortex by the passage of a friend across our visual field.) It might show the mechanism whereby true or false ideas are formed in response to physical situations and, conversely, the actions following the formation of certain ideas. Such an investigation might lead to a characterization of truth as a certain relation between two physical objects, a brain process and the corresponding external state of affairs. But it would be useless in the most interesting case, which is the one of nonobservational propositions, and it would fail to give us a characterization of truth as a property of *propositions* in certain conceptual contexts—unless we managed to characterize propositions in terms of psychophysiological processes alone. As long as such a program is neither carried out nor shown to be impracticable, we had better adopt a sort of methodological dualism, regarding ideas and facts as heterogeneous and accordingly truth as a basic property of propositions (in certain contexts), a property that is not further analyzable—for a further analysis of the truth concept would consist in filling in all the gaps between ideas and their referents, a task momentarily beyond our possibilities (perhaps because it has not even been seriously proposed).*

We cannot wait until the above difficulties are solved and we cannot dispense with the truth concept either, since without it the test of ideas makes no sense. Therefore we shall here adopt the practical and highly unphilosophical decision not to analyze the concept of truth as a property of singular observational propositions, taking it as a primitive concept in terms of which the truth of more complex propositions can be analyzed. Once we feign to know what it means to say that a certain proposition referring to a given observable fact is true, we may ask about the truth of empirical generalizations such as "All birds in this town are songbirds". The truth value of such a *limited generalization* might readily be established if we cared to. It would certainly involve technical problems but no fundamental problem, because such a generalization is just a conjunction of a finite number of singular observational propositions—and we have agreed to feign that we know what the truth of such propositions consists in.

More serious problems arise in relation with *unlimited generalizations* such as "All adult mockingbirds sing". But, again, these are technical problems. We would take random samples of mockingbirds, watch their singing performance, compute the frequencies of singing mockingbirds in the various samples, compute the over-all frequency, and jump to the tentative conclusion that the unlimited (past, present and future) world population of mockingbirds has the given property with the same frequency. That is, we can use well-known statistical techniques, such as random sampling, for the control of our inferences. Such techniques do not guarantee the truth of our "conclusions" (posits) from observational data but at least they enable us to maximize the degree of truth of our generalizations and, eventually, to correct them.

If we cared to go beyond empirical validation, i.e. if we wanted to obtain a law statement rather than an empirical generalization, we would have to embed it in a theory. Thereby, we might be able to explain the mechanism whereby mockingbirds cannot help to sing, and then if we managed to observe or to produce a mute mockingbird we would be able to test such a mechanism hypothesis. Should the hypothesis pass the test, we would end up with completing the initial empirical generalization in some such way as this: "All adult mockingbirds endowed with such and such biochemical, physiological and ecological characteristics sing". That is, we would explain and consequently understand the generalization. At any rate, the problem of the truth value of observational generalizations, whether limited or not, is in principle soluble—even though not finally soluble—with the help of statistics and substantive theory.

But even such qualitative empirical generalizations raise a philosophical problem: the one of *partial truth.* If we had initially assumed that all *A*'s are *B*'s and experience showed that ten per cent of *A*'s are not *B*'s, we would not conclude that "All *A*'s are *B*'s" is altogether false but, rather, that it is *partially* false or, if preferred, that the generalization is true to some extent or degree. Moreover, we might be tempted to say that its degree of truth is 9/10 since it comes out true 9 out of 10 times or instances. *If we did the latter we would be tacitly assuming that the truth value of the conjunction of N mutually compatible and independent propositions is the average of the truth value of the various conjuncts, i.e. $V(e_1 \,\&\, e_2 \,\&\, \ldots\, e_N)=(1/N)\Sigma_i V(e_i)$. Notice that this is very different from the probability of the conjunction, which equals the product of the probabilities of the conjuncts. Alternative formulas are possible. See M. Bunge, *Understanding the World,* pp. 272–275 (Dordrecht-Boston: Reidel, 1983).*

The concept of partial and, in particular, of approximate truth is explicitly employed in relation to quantitative statements and it underlies the various techniques of successive approximations. Thus, when we estimate the number of objects of a kind in a region as 100 and an actual count gives 110, we say our estimate was 10 per cent in error—or, equivalently, that it deviated from truth in a fraction 1/10. We may also say that the original assumption had a degree of truth equal to 9/10. In short, in everyday life and in science, both pure and applied, we employ the concept of partial truth and sometimes we go as far as quantifying it. It is only some logicians who still oppose the very idea of partial truth, as a result of which we continue using an intuitive, presystematic concept of partial truth.

The testing of observational propositions derived with the help of scientific theories poses special problems; these are simpler in one respect but more complex in another. The testing of a low level proposition, such as a prediction, is simpler than that of a nontheoretical proposition in the sense that the former boils down to the confrontation of two objects of the same kind: the proposition under test, t, and the corresponding empirical evidence, which is another proposition, e

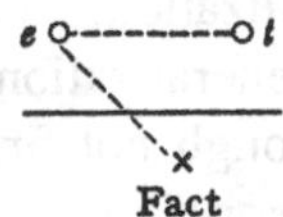

Fig. 15.1. Test: confrontation with a fact in the case of a nontheoretical statement *e*, and confrontation with another proposition in the case of a theoretical statement *t*.

(see Fig. 15.1). There would be no difficulties if *t* and *e* were or could become identical—but they never are and this is a source of complexity. In fact, as we have seen time and again, the referent of a theory is not the same as the evidence relevant to it.

Let us, for the sake of definiteness, take the case of a quantitative prediction concerning a directly observable property *P*, which prediction we check by means of a run of measurements. Suppose we predict a certain numerical value *n* for the property *P* of a system *c* in a certain state, and upon measurement we obtain a somewhat different value *n′* with a standard error σ:

$$t=[P(c, s)=n] \qquad [15.1]$$

$$e=[\overline{m}P(\dot{c}, s) =n' \pm\sigma] \qquad [15.2]$$

where '$\overline{m}P(\dot{c}, s)$' stands for the average measured value of the property *P* of the real object $\dot{c}$ on the scale-cum-unit system *s*. These two propositions are different even in the infrequent case when the numerical values coincide, i.e. when $n'=n$ and $\sigma=0$. In fact, (i) the theoretical proposition *t* refers in an immediate way to a more or less idealized system *c*, whereas the empirical proposition *e* refers to a real system $\dot{c}$ which *c* is supposed to represent (recall Sec. 8.4); (ii) the empirical proposition contains the concept of measured value, which is absent from the theoretical one. (A further difference would arise if the theoretical proposition included a theoretical error introduced by simplifying assumptions.) If we stopped at this point we would have to conclude that no empirical operation, however precise, could ever either confirm or refute a theoretical prediction. But if we did this we would have to condemn theorizing or experience or both.

The above difficulty is resolved by abandoning the bizarre doctrine that scientific theories entail observational propositions, recognizing the gap between theoretical propositions and empirical ones, and bridging it by means of the following *Convention:* A theoretical quantitative proposition has the same truth value as the corresponding empirical proposition if and only if both coincide to within the experimental error—i.e. if the difference between the corresponding numerical values is not larger than the standard error. This convention, of an epistemological rather than a logical nature, may be regarded as elucidating the concept of *epistemic equivalence* between propositions that are formally and semantically inequivalent. In fact, for the case of the quantitative propositions [15.1] and [15.2], the preceding convention may be reworded in the form of a definition, namely

$$t\ Eq\ e =_{df} |n - n'| < \sigma. \qquad [15.3]$$

Otherwise, i.e. if $|n - n'| \geqq \sigma$, we "conclude" that e is not epistemically equivalent to t.

The discrepancies between the measured and the calculated values become significant if they exceed the experimental error. This excess may be regarded as measuring the differences in truth value between the empirical and the theoretical findings, but there is so far no accepted theory of partial truth. Accordingly, what is usually done is just to indicate such differences and to "draw" either of the following qualitative "conclusions":

1. *The available evidence is favorable to the prediction (e supports t).*
2. *The available evidence is unfavorable to the prediction (e undermines t).*
3. *The available evidence is inconclusive with respect to the prediction (e is neutral to t).*

Either of the first two "conclusions" is in turn qualified in some such way as "*e* supplies a strong (weak) confirmation of *t*" or "*e* supplies a strong (weak) disconfirmation of *t*".

At any rate, what is called *scientific inference*—in contrast to deductive inference—is precisely the evaluation of the weight of empirical evidence relevant to certain propositions. One usually speaks in this regard of "drawing conclusions" from empirical data, but this is an incorrect expression: only strict deduction can "draw" or "extract" conclusions from premises. It would be equally incorrect to say that either of the three "conclusions" above is a *decision:* they are *valuations* and as such they can provide grounds for the decision to (provisionally) accept *t*, reject *t*, or leave *t* in suspense. What is at stake in scientific inference is (i) the *confrontation* of two sets of propositions,

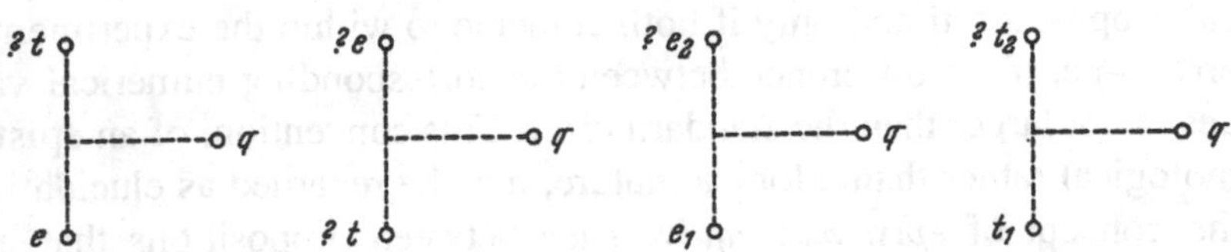

Fig. 15.2. "Conclusions" *q* are "drawn" from the confrontation of two sets of propositions, one of which is assumed to be true. None of the broken lines symbolize logical (deductive) relations or operations.

at least one of which is empirical, and (ii) the formulation of a *third* set of propositions (actually metastatements) concerning the truth value of one of the given sets on the strength of the other. The valuational statements (third set) are built with the help of vaguely stated nonlogical conventions like the one regarding epistemically equivalent propositions, and are seldom regarded as being more conclusive than the statements they "conclude" about.

Accordingly there can be four classes of problem regarding the valuation of observational propositions: (i) theoretical propositions are gauged against empirical propositions; (ii) empirical propositions (regarding, e.g., a new apparatus) are tested on the basis of the predictions derived with the help of a theory; (iii) two sets of empirical data, one of which is somehow better backed up than the other, are confronted; (iv) two theoretical formulas, one of which is for some reason taken as a base line, are confronted. These four possibilities are diagrammatically shown in Fig. 15.2.

The inference pattern becomes even more complex in the case of transobservational statements (hypotheses); in this case the *t* will be the testable consequences of the hypotheses, so that the new pattern will absorb the previous one, which remains fundamental. Let us glance at the logic of testing hypotheses.

Problems

15.2.1. Examine R. Carnap's "Truth and Confirmation" (1936) repr. in H. Feigl and W. Sellars, Eds., *Readings in Philosophical Analysis* (New York: Appleton-Century-Crofts, 1949).

15.2.2. Suppose you drive under the rain and the rain appears to fall obliquely. You know this appearance could be the resultant of two mutually perpendicular motions relative to the ground. How do you proceed to infer reality from appearance? Recall that astronomers correct their observations for light aberration, a similar effect.

15.2.3. List and exemplify possible sources of disagreement between theoretical predictions and empirical findings relevant to them. See, e.g. M. Bunge, *Causality,* 2nd ed. (New York: Meridian Books, 1963), sec, 12.4.3.

15.2.4. Compare any theoretical statement with a piece of empirical evidence relevant to it and disclose the differences between the two propositions. Thereupon examine the popular view of scientific inference according to which the pattern of scientific inference (in turn equated with induction) would be this: (i) hypothesize "$h \rightarrow e_i$"; (ii) test for a number of e_i; (iii) assign h a truth value.

15.2.5. What, if any, are the differences among the following statements? (i) The individual c has the property P. (ii) Assume that the individual c has the property P. (iii) The proposition "The individual c has the property P" is true. (iv) The proposition "The individual c has the property P" has a high degree of confirmation. (v) The proposition "The individual c has the property P" is true if and only if, when c is subjected to the test condition T, it exhibits the behavior B.

15.2.6. Do we accept the statements about the reasonable agreement (or disagreement) among theoretical predictions and empirical data on the sole authority of those who utter them? See T. S. Kuhn, "The Function of Measurement in Modern Physical Science", *Isis*, **52**, Pt. 2, 161 (1961), pp. 165–66.

15.2.7. Is there any difference between truth and credibility? And can "credence" be an *Ersatz* for "truth"?

15.2.8. Examine Tarski's elucidation of the classical concept of truth as adequation to fact. Does it apply within formalized theories alone? Does it apply to scientific hypotheses, i.e. statements that are not directly confronted with matters of fact although they refer to them? And does it apply to systems of hypotheses, i.e. theories? See A. Tarski, "The Semantic Conception of Truth", *Philosophy and Phe-*

nomenological Research, **4**, 341 (1944), repr. in the volume quoted in Problem 15.2.1.

15.2.9. See whether the method of successive approximations involves the concepts of partial truth and of sequence of statements approaching truth. Recall Sec. 1.3, see W. S. Jevons, *The Principles of Science,* 2nd ed. (1877; New York: Dover, 1958), Ch. XXI, and any treatise on numerical analysis.

15.2.10. Sketch a theory of partial truth. Hint: Consider the problem of assigning a truth value to a binary propositional compound, such as $p \& q$, given the truth values of the components. Take into account that the resulting formulas must include the special case of total truth and total falsity, so that, in particular, it must be $V(p \vee -p)=1$.

15.3. Testing Hypotheses

So far we have been concerned with *directly* testable propositions, that is, with statements the test of which does not depend on the results obtained from testing other propositions. All singular and existential propositions in which observational predicates alone occur are directly testable. All other factual statements are *indirectly* testable, i.e. their test involves the empirical checking of further propositions. For example, while "The pointer of this ammeter has suffered a deflection" is directly testable by visual inspection, the test of "This ammeter indicates that a 10 ampere current is flowing through that circuit" requires not only the inspection of the pointer but also the acceptance of a law relating the deflection to the current intensity—a law built into the instrument design and into the code for its use. All hypotheses are indirectly testable since, by definition, they contain nonobservational concepts. Furthermore, they are *incompletely* testable by experience because, also by definition, they go beyond it (recall Sec. 5.1).

Suppose we want to test the surmise that a given object is made of gold. At first sight this looks as an observational proposition, but if we recall how easily we are deceived by good gold imitations we realize it is a hypothesis proper: while "yellow", "shiny" and "weighty" are observational concepts, "golden" is not—at least nowadays. Our proposition "This is golden" is, then, a singular hypothesis. The simplest test to which we can subject it is to spill some aqua regia on the thing we suspect is made of gold and watch the effect. The hypothesis underlying this test technique is that golden things are dissolved by aqua regia. The inference will then be the following instance of weak reduction (recall pattern 9 in Sec. 15.1): If a thing is golden then it is dissolved by aqua regia; this thing was dissolved by aqua regia; therefore it is possible that this thing is golden. Now, aqua regia can dissolve many things other than gold: it is sufficient but not necessary for a thing to be golden in order that it be dissolved by aqua regia. If it were necessary the major premise would be "If a thing is dissolved by aqua regia then it is golden" and the argument would fit the valid *modus ponens* form. As it stands, the original argument is the most common logical fallacy. Yet, although it is logically worthless, it does give some hint and is therefore cognitively valuable—as long as we realize its inconclusiveness. Once we have applied the aqua regia technique and shown the possibility that the thing be golden, we may care to employ more complex and powerful techniques, such as analyzing the reaction products, finding out the crystal structure with the help of *X*-rays, or even determining the atomic number. If, on the other hand, the aqua regia test gave a negative result, our inference would have been a conclusion proper: we would plainly assert that the given thing was not golden, and this would be logically valid. In fact, our inference would fit the *modus tollens* pattern: If golden, then dissolving; now, not dissolving; therefore, not golden. In short, the empirical test of a hypothesis involves (i) assuming or trying *further* hypotheses and (ii) using patterns of conclusive and/or tentative inference.

Some hypotheses are less directly testable than others, i.e. their test involves a longer chain of propositions. Take, for instance, Galilei's law of inertia, which is often (wrongly) supposed to be an inductive generalization from observation. No experiment can test it directly because no region of the universe satisfies exactly the antecedent of the law, namely that no forces whatsoever be present. What experiments and even ordinary observation show is a definite decrease of accelerations as external influences weaken: from this we jump to the "conclusion" that if the force is zero the acceleration is zero as well, a hypothesis amenable to not direct test. But even if no such indirect experimental evidence were at hand we would accept the law of inertia in the domain of classical mechanics because it is indirectly corroborated by every successful application of Newtonian mechanics, of

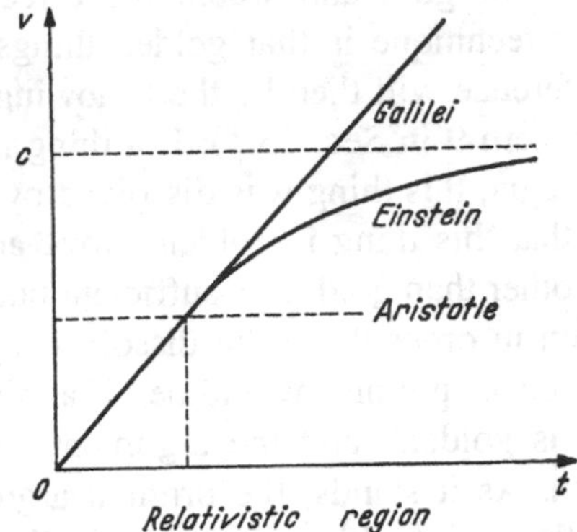

Fig. 15.3. Falling body. Classical law: $v=gt$. Relativistic law:

$$v = gt\Big/\left(1+\frac{g^2t^2}{c^2}\right)^{\frac{1}{2}}.$$

Einstein's law reduces to Galilei's for $gt \ll c$, i.e. for small durations and/or weak fields. In the extreme relativistic limit (unknown to Aristotle) Einstein's law happens to coincide with Aristotle's.

which it is a part.

*Or, again, take the relativistic law concerning the free fall of a body in a constant gravitational field: see Fig. 15.3. At the date of writing, this law does not seem to have been subjected to empirical tests, whereas the corresponding classical law has satisfactorily been confirmed myriads of times. Yet we believe the relativistic law to be truer than the corresponding classical law because it is one of a large number of logical consequences of principles that have been tested by examining alternative logical consequences. That is, the relativistic law has the indirect support of all the hypotheses to which it is logically related (including classical electrodynamics) and which have successfully passed the test of experience. Moreover, when estimating the degree of truth of Galilei's law we take Einstein's as a base line (fourth case in Fig. 15.2). In other words, we compute the relative error of Galilei's law (directly tested) relative to Einstein's law (untested in a direct way): $\varepsilon(G|E)=|E-G|/E=\left|1-\left(1+\frac{g^2t^2}{c^2}\right)^{\frac{1}{2}}\right|$, whence the truth value of Galilei's law relative to Einstein's can be taken as $V(G|E)=1-\varepsilon=1-\left|\left(1+\frac{g^2t^2}{c^2}\right)^{\frac{1}{2}}-1\right|\cong 1-\frac{1}{2}\left(\frac{g^2t^2}{c^2}\right)$. We conclude then that our limited experience with falling bodies is in error by an amount of the

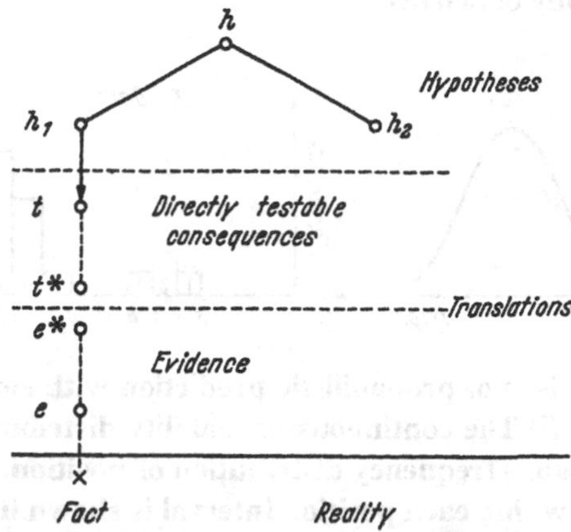

Fig. 15.4. Empirical testability: h_1 is more directly testable than h_2, which enjoys the indirect support of the evidence e relevant to h_1. But e is relevant to h_1 via its refinement e^*, comparable to the translation t^* of the consequence t of h_1.

order of v^2/c^2, which is negligible as far as ordinary experience goes. In short, sometimes we correct some experiences on the basis of theories tested in other areas. Direct testability has no special merit for the assignment of truth values in science.*

A scientific hypothesis, then, is *empirically testable* if it either implies directly testable statements or is implied by higher-level hypotheses having independently testable consequences: recall Fig. 5.6. In the first case the hypothesis is said to be more directly testable than in the second case. But in neither case is the test reduced to a direct confrontation of the hypothesis with a set of data. That is, notwithstanding the official doctrine we never have $h \rightarrow e$ and e, but rather $h \rightarrow t$ and e, where 't' stands for a directly testable yet nonobservational statement, that is, for a proposition that, subject to certain modifications, can be contrasted with an empirical report. This confrontation requires the previous translation of both t and e into a common language, semitheoretical and semiempirical, as shown in Fig. 8.15 of Sec. 8.4. (Recall also the discussion at the end of Sec. 8.1.) This intricate process is summarized in Fig. 15.4, where 't^*' and 'e^*' stand for the translations of t and e respectively into a half-theoretical and half-empirical language—a translation that renders their comparison possible. In short, the test of hypotheses is very indirect. Let us insist on this point.

We can test the hypothesis that a given individual is absent-minded

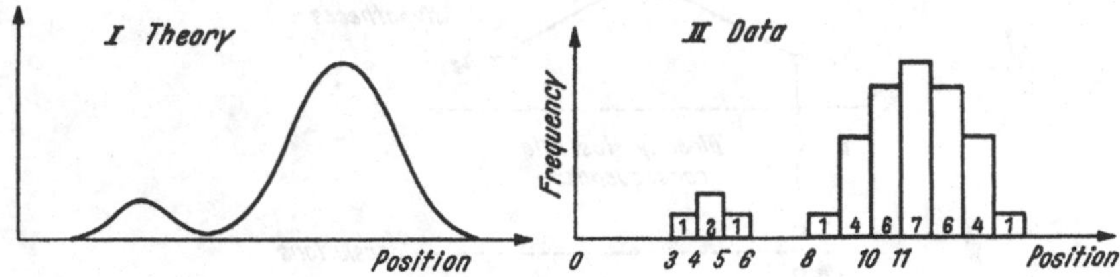

Fig. 15.5. Comparison of probabilistic prediction with empirical findings (imaginary case). (i) The continuous probability distribution of position. (ii) The discontinuous frequency distribution of position. The number of corpuscles found within each position interval is shown in the base of the corresponding rectangle; thus between 10 and 11 length units off the origin 6 particles were found.

because an absent-minded person does not react to small environmental changes that are normally perceived by alert people—or this at least we assume to be true. If our subject does respond to such changes we conclude (by *modus tollens)* that he was not absent-minded. Whereas if he seems to be unaware of them we may presume (by reduction) that he is in fact absent-minded—but in fact he might be just deaf, or blind or even simulating or dead. No hypothesis is testable by itself: in order to be subjected to an empirical test it must have observational consequences, which are derived with the help of some body of knowledge that is not questioned during the test: recall our discussion in Sec. 5.2. More precisely, in order to test a hypothesis we must make the *further* assumption that, if the hypothesis is true, certain facts will be perceivable, which facts will be regarded as manifestations, objectivations or indices of the state of affairs referred to by the hypothesis. (Recall the unobservable-index relation $U=F(0)$ discussed in Sec. 12.3.) This further hypothesis is of course built with the help of the body of antecedent knowledge and, in particular, with the help of some theory. This is a first sense in which the empirical test of scientific hypotheses is indirect.

A second sense in which the test of scientific hypotheses and theories is indirect is, that such constructs do not refer to real experience but to aspects of idealized models: they contain theoretical concepts with no exact counterparts in reality, much less in human experience. (That this further indirectness too should escape to the usual treatment of the problem is not surprising, since it is confined to nonscientific empirical generalizations, whence the accent on their generality with neglect of their referent.) A simple yet glaring illustration of this point is provided by any theory predicting the probability that a point particle [construct] will lie somewhere on a straight line [construct]. Since there are infinitely many points in any segment, there will be infinitely many probability values, the set of which will constitute a probability distribution such as, e.g., the one on Fig. 15.5 (i). Our hypothesis (the theory's projection) will involve then the following constructs: "point particle", "infinite set of possible positions of a point particle", and "probability distribution". On the other hand, in experiment we shall deal with corpuscles (or with wave packets) rather than with unextended entities; we shall be concerned with a finite set of position measurements and with a histogram rather than with a smooth distribution curve (see Fig. 15.5 (ii). The empirically found frequency distribution and the theoretically calculated probability distribution are separated by an unbridgeable gap: no addition of data can perform the miracle of converting corpuscles into point particles, finite sets into infinite ones, and histograms into continuous functions. We must reckon with such gaps between theory and experience and learn to establish the right correspondences between them rather than ignoring such differences or trying to reduce constructs to data.

The over-all pattern of hypotheses testing is, in a nutshell, the following:

1. *Antecedent knowledge* A unquestioned in the present research and relevant to a hypothesis *h* to be tested. A
2. *Problem*: Is *h* true? *h*?
3. *Additional* (test) *hypothesis:* From A and *h* derive "If *h*, then *t*". Unless *t* derives from A alone, $h \to t$ will have to be subjected to a further, independent test. $h \to t$
4. *Translate t into t**—for example, refer planetary positions, calculated in the sun's reference frame, to the earth. t^*
5. *Experience:* test for t^*, i.e. find a set of data *e* comparable to t^*. *e*
6. *Translate e into e**—for example, express planetary position readings in geocentric coordinates. e^*
7. *Compare t* to e** and decide whether they are epistemically equivalent or not; if they are, assert t^*, if not negate t^*.
8. *Inference concerning h:* if t^* was confirmed, declare *h* supported by *t*; otherwise conclude that *h* is undermined by *t*.

Step 3, i.e. the introduction of the auxiliary hypothesis "$h \to t$" enabling us to make contact with experience, may be far from obvious: it may be false that *t* is relevant to *h* or at least that *h* is sufficient for *t*. At any rate the new hypothesis, "$h \rightarrow t$", is not questioned when *h* is put to the test but, quite on the contrary, the test is designed and executed on the strength of "$h \rightarrow t$". Yet, since this auxiliary hypothesis is supposed to be scientific, an independent support must be found for it: not only should it be founded on the antecedent knowledge (step 1) but it should be subjected to independent empirical checking. That is "$h \rightarrow t$" should imply some different proposition t' such that it can be tested with a different technique; otherwise it will not be much better off than the projective tests. This independent test of the bridging hypothesis "$h \rightarrow t$" will pose fresh methodological problems but no new logical problem. Therefore we are free to return to our original question of "drawing conclusions" from data designed to test scientific hypotheses.

Let us begin by recalling what deductive logic has to say in this regard. Logic contains three elementary rules of inference involving $h \rightarrow t$ and *t*. First: *t, hence* $h \rightarrow t$, where *h* is an arbitrary statement. This inference pattern leads nowhere precisely because *h* is arbitrary. It just serves as a reminder that, given *t*, logic will allow for infinitely many hypotheses implying *t*—i.e. there is no safely inferring back from consequents *t* alone to antecedents *h*. Second: *h and* $h \rightarrow t$, *hence t* (*modus ponens*). This, the central inference pattern, is of paramount importance in formal science, where anything free from contradiction can be asserted *ex hypothesi* as long as it promises to be fruitful. It also occurs in the indirect test of factual hypotheses: thus, in the case of Fig. 15.4, if for some reason we dare assert that the highest-level hypothesis *h* is true, then we can detach h_2 because we had previously assumed or proved that *h* implies h_2. But the *modus ponens* is useless for the left-hand branch of Fig. 15.4, i.e. for the empirical test of hypotheses, because our problem is now to find out the truth value of our posit *h* on no ground other than the truth value of a statement implied by it—which, as shown by the previous inference rule, is a poor indication. Third: $-t$ *and* $h \rightarrow t$, *hence* $-h$ (*modus tollens*). This argument form (a variant of the former) is useful in all cases of scientific inference because it serves to weed out false hypotheses, at least to the extent to which we can definitely negate the consequent on the strength of empirical information.

The conclusion is clear: deductive logic, powerful as it is in the working out of initial assumptions to extract whatever testable consequences they may entail, is powerless to validate those initial assumptions; it only helps to show which of them are false. In short, to the extent that there are definitely assertible empirical propositions, they can *refute* hypotheses but not verify them conclusively. The only positive indication empirical information can supply is that the hypothesis concerned *may* not be false. (Recall pattern 9 in Sec. 15.1.) This should suffice to dispose of the belief that the progress of knowledge consists in piling up "positive facts" (data) without paying attention to falsities, and that science can be built on empirical data. But what can be done apart from criticizing? There are several ways out. One is to declare that the search for truth in science is pointless. Another is to propose that, since falsity alone is conclusively ascertainable on empirical grounds, these are insufficient and must be supplemented with nonempirical tests. A third possibility is to say that, since deductive logic is insufficient, we need a new kind of logic, inductive logic, to establish patterns of nondeductive inference. Let us glance at the three proposals.

That the search for factual truth is pointless has been claimed on very different grounds, such as (i) man's unsurpassable limitations, (ii) the contingency (lack of logical necessity) of experiential propositions, and (iii) the worthlessness of truth as compared to practical efficiency or even credence. That human knowledge is limited could not be denied; nor could it be established, however, that the limits of human knowledge are all *fixed:* after all, among our limitations we must reckon the inability to demarcate such limits once and for all. That experiential propositions are not logically necessary and are consequently fallible, is equally undeniable; yet it does not follow that they are all *equally* worthless: to be sure some are falser than others and we often attain approximate truths, i.e. propositions that are "practically true". Finally, whether ideas work or not and whether they are unanimously believed or at least received by some experts, is quite irrelevant to their being true. False ideas may work in practice and true ones may fail in practice (recall Sec. 11.1); and false hypotheses are very often entertained because they are believed to be true. In conclusion, the above are not valid arguments against the search for objective truth

The second proposal was to regard empirical tests as insufficient—which suggests trying to supplement them with nonempirical criteria

This was, in fact, the standpoint adopted in Sec. 5.5 when scientific hypotheses were required to be not only testable but also grounded on part of the available knowledge. That some such nonempirical tests must be given to scientific hypotheses in addition to empirical tests, is suggested by (i) the multiplicity of hypotheses consistent with any given set of empirical information, and (ii) the dependence of empirical tests upon alternative hypotheses and theories. That a whole battery of nonempirical tests is in fact employed in weighing the truth claims of scientific ideas will be argued in the last section. But these nonempirical tests cannot *establish* scientific hypotheses any more than the empirical tests can conclusively validate them: a large number of successful tests, both empirical and nonempirical, renders our surmises more verisimilar and cogent, not infallible. The quest for final certainty characteristic of nonscience, is replaced in science by the quest for approximate but perfectible objective truth.

The third proposal we mentioned was to forget about deductive logic and look elsewhere for certainty—in inductive logic. But this is a chimera: there are no infallible hypotheses and, strictly speaking, all nondeductive inferences are fallacious: they allow us to conclude nothing although they are as many ways of suspecting the truth or the falsity of our surmises. The proper job of nondeductive "logic" should not be to consecrate fallacies but to study the actual patterns of nondeductive scientific inference occurring in the test stage, as well as the conditions for their optimal performance. For example, in our illustration of the absent-minded subject, if we find him nonresponsive to noise, we "conclude" that it is plausible to assume that he is absent-minded (a case of strong reduction). But our subject might be simulating, so that alternative tests would have to be tried—such as whispering that an important letter has just arrived for him; and, if possible, some objective (e.g., physiological) indicator of absent-mindedness should be sought and theoretically justified. Or the absent-mindedness hypothesis might be plausible on account of previous experience with the subject, suggesting that he is absent-minded most of the time: in this case we may argue on the assumption that he will probably not respond to small environmental changes: $A \rightarrow -R$, and A is plausible, therefore so is $-R$ (a case of weakened *modus ponens*). Or finally, we may be fairly sure, but not quite, that the subject is responding to noise, in which case we will conclude that the subject is most likely alert: $A \rightarrow -R$, and R is plausible, hence $-A$ is plausible (a case of weakened *modus tollens:* see Sec. 15.1).

But *how plausible* can a plausible assumption be and how plausible the tentative inference involving it? None of the existing theories of inductive "logic" give acceptable answers to these questions: they are still open problems. Moreover, the very legitimacy of the attempt to build an inductive "logic" has been doubted on the ground that nondeductive inference is illogical or lawless; and the worth of such a theory, in case it were possible, has been questioned because hypotheses and theories are never built inductively. The first argument seems beside the point: nondeductive inference is logically worthless but cognitively indispensable, and we had better face it than pretending that it does not occur in science. The second argument is not much better because induction, impotent as it is to *build* hypotheses and theories, is employed in the comparison of their testable consequences with empirical data: thus, if all the tested *t*'s agree closely with the corresponding *e*'s, we make the inductive (and fallible) jump to the "conclusion" that the hypothesis entailing the *t*'s has been satisfactorily confirmed—which, strictly speaking, is a half-truth bordering on a lie. In short, nondeductive "logic" seems to be a legitimate enterprise relevant to the *empirical test* of hypotheses and theories although it is irrelevant to their *construction.* And the sad circumstance that the discipline hardly exists is a challenge rather than a proof of its impossibility—a challenge to build an a posteriori or factual theory of tentative inference rather than an a priori and therefore inapplicable one.

Let us approach some of the problems facing any nondeductive "logic".

Problems

15.3.1. In the Middle Ages people accused of crimes would be thrown in deep water in order to test the guilt hypothesis. If they survived they were declared either innocent or in connivance with the devil; and guilty—or innocent, as the case may be—if they drowned. This water test is now regarded as altogether irrelevant to the hypothesis concerned. Was it irrelevant in the context of medieval beliefs? Draw some moral concerning the contextual character of the relevance of empirical information to hypotheses, and on the independent test of every hypothesis-indicator hypothesis.

15.3.2. Until recent times certain illnesses (particularly psychoses) were regarded as effects, hence as proofs, of demoniacal possession. Accordingly people affected by those illnesses were handled by exorcists (demonextractors). If the subjects cured, faith in the technique was reinforced. If they failed to cure, the particular exorcisms practiced on them were declared to be insufficient but not inefficient. Exhibit the inferences involved in the diagnosis and in the prognosis. *Alternate Problem:* Examine the Chinese practice of acupuncture, also grounded on the belief that sickness consists in or is produced by evil spirits inhabiting the body, and kept because it is alleged to work.

15.3.3. Study the symptom-illness relation. Show what the ontological and the logical relations are—i.e. exhibit the cause-effect relation and the antecedent-consequent relation. And examine the inference pattern physicians employ in guessing causes: is it a conclusive inference? And is it entirely wild when backed by physiology or by biochemistry?

15.3.4. How might we test the hypothesis that a given biological character, e.g. a plant having succulent leaves, is adaptive? Take into account that, although we may be able to perceive the character, its adaptive value must be inferred.

15.3.5. Consider the following statements: (i) "Drinking decreases willpower." (ii) "Weakness of character may lead to drinking". (iii) "Weakness of character and drinking reinforce one another mutually". Do they prove or do they exemplify the theses of (i) epiphenomenalism, (ii) priority of the soul over the body, and (iii) mind-body interaction?

15.3.6. Analyze the following sentence by E. D. Adrian, *The Basis of Sensation* (London: Christophers, 1928), p. 15: "In every case so far investigated whenever there has been reason to suppose that a nerve is in action, the usual potential changes have been detected, provided that the conditions for recording them were favorable". What if no changes in the potential are (indirectly) observed: shall we reject the hypothesis or shall we conclude that the conditions of observation were unfavorable?

15.3.7. Pick any scientific hypothesis and examine in succession the following problems raised by its test: (i) the logical problem of finding the relations between the hypothesis and its possible evidences; (ii) the methodological problem of designing empirical procedures for testing the hypothesis; (iii) the problem of weighing the evidence, i.e. of evaluating the hypothesis.

15.3.8. According to apriorism, data are deducible from a priori assumptions. Empiricism, on the other hand, claims that hypotheses, when necessary at all, are inducible from data. Does any of these doctrines fit actual scientific practice—i.e. are either hypotheses or data self-sufficient?

15.3.9. Examine the widespread opinion that hypotheses can be gradually corrected until entirely certain and immutable principles are reached. See C. Wolff, *Preliminary Discourse on Philosophy in General* (1728; Indianapolis and New York: Bobbs-Merrill, 1963), 126 and 127.

15.3.10. According to R. Carnap, in his influential "Testability and Meaning", *Philosophy of Science*, **3**, 419 (1936) and **4**, 1 (1937), the logical form of a statement expressing the test- and truth-condition of a hypothesis attributing an unobservable property P to an object x is: If x is subjected to the test condition T, then: if x exhibits the behavior B, then x has the property P. In short: $Tx \rightarrow (Bx \rightarrow Px)$. By the principle of exportation, this formula is equivalent to: $(Tx \,\&\, Bx) \rightarrow Px$. If this were the implication actually involved in scientific inference, we would be able to conclusively establish all our hypotheses on the basis of their empirical test, since the antecedent of the previous conditional involves only observational concepts. Is this possible? If not, what is wrong with the above logical analysis of empirical tests? How about the following schema: $Px \rightarrow (Tx \rightarrow Bx)$? Does this enable us to draw conclusive inferences? And finally: are any such conditionals relating unobservables and observables freely posited, or are they controlled both theoretically and empirically? *Alternate Problem:* The stronger a hypothesis the more it will explain and the more occasions it will have to come to grips with experience. But experience authorizes jumping to the weakest "conclusions". Which of the horns shall we choose? Or do we have to make a decision at all?

15.4. Confirmation and Refutation

If we meet a blonde, shall we conclude that we have confirmed the hypothesis "All ravens are black"? Intuitively we would deny it: our finding the blonde is *irrelevant* to the given hypothesis, hence it cannot count as either a confirmer or a refuter of it. Yet the logician might argue as follows. "All ravens are black" is logically (syntactically) equivalent to "Anything is either a nonraven or it is black". (Indeed, the equivalence "$(x)[Rx \rightarrow Bx] \leftrightarrow (x)[-Rx \vee Bx]$" is logically true.) Since a blonde is not a raven, meeting her confirms the blackness of all ravens. This exemplifies one of the paradoxes of confirmation (C. G. Hempel).

If the logician is right, instantiation will count for nothing, just because almost anything can be made to count as evidence. This leads the refutabilist (e.g., K. R. Popper) to insist that looking for confirmation (for inductive support) is pointless. If, on the other hand, only refutations are sought, there is no paradox of confirmation, for we shall try to find *nonblack* crows, paying no attention to the known black ones. That is, if the logician is right then the refutabilist is right in claiming that only attempted yet unsuccessful disconfirmations matter. But failing in an attempt to falsify a hypothesis is succeeding in confirming it—if *relevant* evidence alone is gathered. Consequently, if the logician is right, both the confirmationist (inductivist) and the refutabilist (deductivist) are at a loss, and we must conclude that empirical testing is useless. And if this is so, then any hypothesis is as good or as bad as any other; in particular, pseudoscience is as good as serious science.

Since we clearly reject this conclusion, some of premises used in our argument must be false. The culprit is, of course, the tacit assumption that any data whatsoever are relevant to any given hypothesis—or, equivalently, that data can be gathered from an arbitrary universe of discourse. This is clearly false: when we set out to test "All ravens are black" we tacitly fix the universe of discourse restricting it to ravens: we do not investigate anything that is not a raven. In other words, we do not blank-mindedly care for anything that may come up: we focus our attention, we select the kind of data, we restrict the range of our individual variable x to the set of ravens. This limitation of the domain of individuals dissolves the paradox and acknowledges the usual practice, both in everyday life and in science, of distinguishing relevant from irrelevant evidence. Incidentally, the empirical generalization in question has been falsified by the discovery of albino ravens.

Such a restriction of the domain of individuals or universe of discourse is clearly indicated in the following symbolization of the general conditional: "$(x)_R Bx$", read 'every x in R has the property B'. No such clear indication of the domain of individuals occurs in the usual symbolization of the general conditional, namely "$(x)[Rx \rightarrow Bx]$", which gives rise to the paradox of confirmation. Since the latter formula is equivalent to $(x)[-Bx \rightarrow Rx]$, a naive amateur who knew nothing of ravens—and nothing of science either—might carefully inspect every nonblack object, particularly blondes, instead of examining ravens alone. In order to stop him chasing blondes we must remind him that his context is ornithology. That is, instead of tampering with stray statements we must begin by placing them in their proper context, since this alone will tell us what the domain of individuals is.

In short, then, in order for a datum to be an evidence for or against a hypothesis it must, to begin with, be relevant to it. In the case of a low level empirical generalization g, we may say that the empirical report e is *relevant* to g if and only if g and e have the same referents; e will be irrelevant to g if and only if it is not relevant to it. If the individual c happens to be both a raven and black, i.e. if Rc & Bc, then c will confirm the infamous generalization—though not instantiate it, because the previous conjunction is an instance of the existential generalization "$(\exists x)[Rx$ & $Bx]$, which is not implied by the corresponding universal generalization. If c is a nonblack raven, i.e. if Rc & $-Bc$, it will disconfirm the generalization. And if c is not a raven, whether it is black, i.e. $-Rc$ & Bc, or not, i.e. $-Rc$ & $-Bc$, it will make no difference to the generalization: in either case c will be irrelevant to it. In short, the datum e will be of either of the following kinds:

Rc & Bc confirming	*relevant,*	*–Rc & Bc*	*irrelevant*
Rc & –Bc refuting	*(evidence),*	*–Rc & –Bc*	*(no evidence)*

In short, by pointing out the context and therefore the referent of the generalization under test we dissolve the paradox and retain the usual meaning of 'confirming case'. If, on the other hand, we tear off the generalization from its context and play with purely syntactical transformations, we run into trouble.

The above criterion of relevance is inapplicable to scientific hypotheses proper, the referents of which are unobservable and therefore

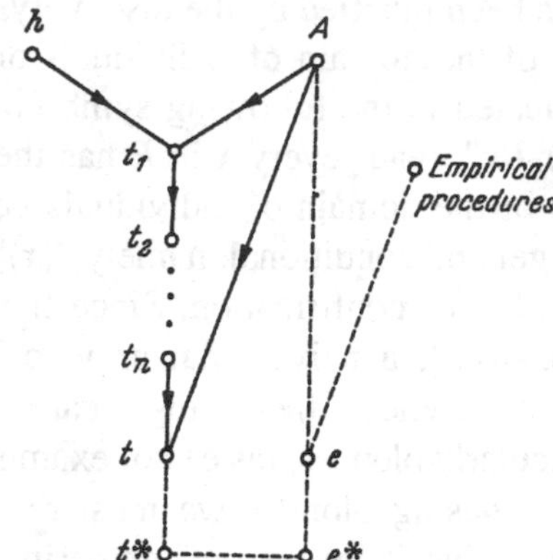

Fig. 15.6. The devious test of scientific hypotheses: only the last links t_n of a chain of implications are empirically tested. This requires the introduction of further conditionals $t_n \rightarrow t$, on the strength of the antecedent knowledge A, and such that *t* may be confronted with *e* after appropriate translation.

different from the referents of the evidences for or against the hypotheses. Yet such statements, if testable at all, must be logically related to propositions capable of being somehow confronted with empirical evidence. That is, the hypothesis itself will have no observational referent and no observational instance but, if testable, it will be a link in a longuish chain the last member of which should somehow be compared with some empirical report or other. The last link refers just as the first to an idealized object, such as a perfect gas or a purely competitive economy: deduction need not bring us any closer to experience. In order to bridge the gap between the last link and data we must introduce an additional hypothesis linking referent to evidence—step 3 in the process sketched in the last Section.

Now the hypothesis itself, in conjunction with the antecedent knowledge, will suggest the kind of evidence relevant to it—as pointed out in Secs. 5.2, 10.4 and 15.3. For example, the hypothesis that thought is a brain process will direct us to study the brain processes accompanying abstract thinking, in search for evidence of the latter. But this sounds circular: if a hypothesis is allowed to appoint its own confirmers then the test cannot be more impartial than the verdict pronounced by the defendant. (This is what in Problem 8.4.10 was called the paradox of empirical validation.) If *e* were all the support *h* could ever get, then empirical validation would indeed be circular and therefore useless.

But we have outgrown the view that testing h consists just in finding a set of e's and placing them vis à vis h on the scientist's word that e is, in fact, a piece of evidence relevant to h. We have got used to the idea of multiple controls, both empirical and theoretical: (i) the hypothesis under test is required to obtain the additional support of *further hypotheses* logically related to them—as was stipulated in Sec. 5.5 and will be argued again in Sec. 15.6; (ii) the test hypotheses involved in the testing of a high level hypothesis are themselves required to be *independently checked* both empirically and theoretically (recall Secs. 14.5 and 15.3). In short, theorification and independent cross-checking make up for the weakness of every single empirical test, which by itself is or can be made circular. An empirical test is effective to the extent to which it is neither single nor purely empirical.

In any case, every single process of looking for relevant data (evidence) pro or con a scientific hypothesis can be summed up in a diagram like Fig. 15.6. In it, A represents the body of antecedent knowledge employed in working out the consequences of h and in designing the test hypothesis and the test procedure. The test hypothesis or cause-symptom conjecture is an implication of the form "$t_n \rightarrow t$", where t_n is one of the low level consequences of h and A, and t stands comparison with e via t^* and e^* (recall the last Section). We shall accept e as *relevant* to h not if they have the same referent, which they cannot share by definition of "hypothesis", but if h and A entail t_n which, in conjunction with A, leads to a $t_n \rightarrow t$ such that e (an outcome of empirical procedures performed with the help of A), once translated into e^*, is relevant to t^*, which is in turn the translation of t into the common half-empirical and half-theoretical language (recall Secs. 8.1 and 8.4). In this way, the concept of relevance of a datum to a hypothesis subsumes the concept of direct relevance introduced in relation with empirical generalizations. If, in addition, the semiempirical translation t^* is epistemically equivalent to the semiempirical translation e^* (recall Sec. 15.2), we say that *e confirms h*. And if e^* and t^* are epistemically inequivalent, we say that *e disconfirms h*. In short, we have the following definitions:

e confirms $h =_{df} [\{A, h\} \vdash t_n]$ & $[\{t_n, A\} \vdash t_n \rightarrow t]$ & (t^* translates t & e^* translates e) & (e^* is relevant to t^*) & (t^* *Eq* e^*)

e disconfirms $h =_{df} [\{A, h\} \vdash t_n]$ & $[\{t_n, A\} \vdash t_n \rightarrow t]$ & t^* translates t & e^* translates e) & (e^* is relevant to t^*) & $\neg$(t^* *Eq* e^*).

What is the *weight of evidence,* i.e. how much do confirmations and disconfirmations count? Clearly, this depends on the degree of generality and systemicity of the hypothesis. A singular hypothesis, such as "There is oil 1,000 m below the surface of this place", can be tested once and for all with a single empirical information: a drilling will either refute or confirm it in a conclusive way. In other words, confirming and disconfirming evidence are equally valuable in the empirical test of singular qualitative and nonsystemic (isolated) hypotheses. But the most interesting hypotheses in science are both *general* and *systemic* to some extent: they refer to sets of facts of a kind and they hang together with other hypotheses (recall Sec. 5.7). Consequently they must be tested by confronting them with (i) whole sets of data of a kind and (ii) entire bodies of related hypotheses. The latter is a test of external consistency (see Sec. 7.6)—a purely conceptual operation as far as it goes. But, of course, the hypotheses acting as controllers of the one under test must in turn be checked both theoretically and empirically, so that in any case we must get down to empirical tests: there are no eternal truths of fact serving as absolute standards. Let us therefore focus on the question of the *weight of empirical evidence* pro or con a scientific hypothesis. This concept can be quantified in various ways. We shall recall the simplest.

Let $\{t_i|1 \leq i \leq m\}$ be the set of testable consequences of h so far derived and tested, and let $E=\{e_i|1 \leq i \leq m\}$ be the set of empirical reports relevant to h, i.e. comparable, either directly or not, with the t_i. Suppose N tests are performed, $n \leqq N$ of which are favorable to h, the rest, $N-n$, being unfavorable to h. Renumbering the empirical reports e_i in such a way that the first n of them are favorable and the rest unfavorable, we have: t_i *Eq* e_i for $i=1, 2, \ldots, n \leqq N$, in the case of empirical generalizations. In the case of hypotheses proper we replace t_i by t_i^* and e_i by e_i^*. The *degree of empirical confirmation* of h with respect to the empirical information $E=\{e_i\}$ may then be defined as

$$C(h|E)=\frac{n}{N}. \qquad [15.4]$$

If $n=0$, i.e. if all the empirical data gathered so far disagree with the t_i entailed by h, then $C(h|E)=0$. If all the tests performed so far agree with the corresponding t_i (or else with their translations t_i^*), then $C(h|E)=1$. In all other cases $C(h|E)$ will lie between 0 and 1. Needless to say, in computing $C(h|E)$ we shall take the total available evidence into account; in particular, we shall not suppress unfavorable evidence unless it is disqualified by antecedent knowledge—as is the case of experimental data lying too far away from the most probable value.

The next step would be to build a *calculus* of degrees of confirmation, i.e. a theory allowing us to compute the degree of confirmation of complex hypotheses, such as $h_1 \vee h_2$ or h_1 & h_2, given the degrees of confirmation of the individual hypotheses h_1 and h_2. No such theory exists at the time of writing. True, according to a widespread opinion worked out by R. Carnap, the probability theory is such a calculus; in other words, the logic of confirmation would be a model of the mathematical theory of probability. But this is mistaken: the concept of degree of confirmation fails to satisfy the central axiom of the elementary calculus of probability, i.e. formula [7.25] of Sec. 7.5. In fact, the probability of any tautology is maximal (=1), whereas the degree of confirmation of it should be nil since experience is altogether irrelevant to logical truth: far from confirming logical identities by empirical tests, we design empirical tests with the help of logic. For example, whereas $P(h \vee -h)=1$ (as is easily seen from [7.25] and [7.26]), we should have $C(h \vee -h|E)= 0$. Similarly, while $P(h \,\&\, -h)=0$ (as can be seen using the former result, [7.26] and De Morgan's theorem), we should have $C(h \,\&\, -h\, |E)\neq 0$. A theory of empirical confirmation that assigns degrees of confirmation identical with probabilities is certainly not a theory of *empirical* confirmation. (On the other hand probability theory enters legitimately the theory of errors of measurement, which is not concerned with the probability of *hypotheses*—a fuzzy concept anyway—but with the probability distributions of numerical *data*, as mentioned in Sec. 13.6. But this theory has failed to attract the attention of empiricist philosophers.)

Leaving aside these difficulties, what do the numbers [15.4] tell us? An overwhelming majority of favorable test outcomes is quantified as a high degree of confirmation whereas a very small fraction of favorable results is interpreted as a low degree of confirmation. But a high degree of confirmation is valuable only if the set of empirical data relevant to the hypothesis is both *random* (picked without bias) and *numerous*. The randomness condition can be replaced by the much more exacting condition of deliberately picking the least likely or riskier cases (e.g., intervals of a magnitude), as directed by Rule 1 in Sec. 14.4; if this is done, a lower bound of the degree of confirmation is obtained. What about the numerosity condition? How large should the sample of data be? Will 100 cases be enough, or shall we need 100 times as many? Obviously no number of cases, however numerous, will be *sufficient* to declare that the hypothesis is true even if its degree of confirmation turns out to equal unity; but a certain minimum will be *necessary* to conclude that the hypothesis is verisimilar, i.e. to attach some value to the degree of confirmation.

What is this minimum and who stipulates it? There is no fixed number, there is no way of establishing it save by convention, and there is no point in trying to lay down a universal rule both because it would furnish no guarantee and because the minimum is determined by the nature of the case. In fact, the state of the science determines a lower bound for the number of cases deemed necessary to infer anything about the truth value of a hypothesis on the strength of those cases. Thus, in the early days of atomic and nuclear physics half a dozen data would usually suffice to either encourage or discourage further research on a hypothesis, and every one such data might have been affected by an error as high as 30%, whereas nowadays one hundred data with a 0.1% error are often insufficient in the same field. This dependence of N, hence of $C(h|E)$, on the nature of the case—on the state of scientific knowledge—is one more reason not to regard the theory of scientific inference as a formal enterprise on a par with deductive logic.

At any rate, when a hypothesis is assigned a high degree of confirmation on the basis of a "large" number of tests, subject to either the randomness or to the extreme toughness condition, it can be said to enjoy *pro tempore a strong inductive support*; otherwise it may be said to have a *weak inductive support* or even none. Exact numerical values of the degree of confirmation are very seldom computed by scientists and, when given at all, they are not assigned a permanent value. There is little point in computing such values, and this for the following reasons.

First, the ratio n/N of favorable cases to the total number of cases approaches no limit in the mathematical sense: it is an empirically found frequency which fluctuates subject to *no law*. (H. Reichenbach's theory of induction applies to the case where such a limit exists. If a sequence of relative frequencies has a limit, then the last empirically found member of the sequence may rightly be taken as the best estimate of the unknown but existing limit. But since the assumption of the existence of this limit is groundless, the whole theory collapses.) Second, the formula [15.42 does not account for the degree of refinement of the empirical technique leading to the data E, much less for the logical strength and the ground of the hypothesis, all of which are important indicators of verisimilitude. Third, scientists do not spend their lives retesting the same hypotheses over and again; they often replace them by alternative hypotheses entailing partially or even totally different consequences and calling for, as a consequence, different empirical evidence. (This is one reason why it does not pay to accumulate empirical evidence not digestible by some respectable theory: it may become obsolete too soon. Certain objective facts, and particularly objective patterns, are more permanent than any of our theories about them; but empirical data are as mortal as the ideas to which they are relevant and by means of which they are produced and interpreted.) Fourth, a high degree of empirical confirmation, i.e. a high inductive support, is but one among a number of virtues we require from a scientific hypothesis before courting it—others being the variety of the evidence and the fitting of both evidence and hypothesis in fundamental theories.

We have been speaking of the *inductive support* of hypotheses without, however, subscribing to *inductivism*, the doctrine that hypotheses are both built and justified by first kind induction alone. We are assigning induction a role in evaluating the *empirical* support of hypotheses, none in framing them or in justifying them. We do not build hypotheses, much less theories, by generalization from observational data, if only because hypotheses proper do not refer to anything observable. And we do not justify or ground a hypothesis just by saying that it has a strong inductive support: there must be some other ground for a hypothesis to be relied on: some theory or other must suggest at least that it is possible. Induction appears when we *sum up* the results of the empirical testing of a hypothesis in a statement such as "The available evidence gives strong (weak, no) support to our hypothesis". This is the main role of induction in science: it is not a method but an act, and one performed in the empirical phase of the test of hypotheses. To say that a hypothesis enjoys a strong inductive support does not mean that the hypothesis *entails*, covers or subsumes a large body of evidence, but rather that the set of its testable consequences agrees reasonably well with the set of relevant empirical data. (Recall that a universal statement does not imply, let alone entail, the corresponding existential generalization, and that this implies no individual statement. For example $(x)[Rx \rightarrow Bx] \nrightarrow (\exists x)[Rx\ Bx]$, and

$R(c)$ & $B(c) \rightarrow (\exists x)[Rx$ & $Bx]$ but not conversely.) Nor even empirical generalizations entail data—much less do data entail empirical generalizations.

According to inductivism—as represented by Carnap and Reichenbach—we should try to *maximize* the degree of confirmation of our hypotheses; moreover, this is all we can do and it is quite a lot. To this it must be objected that a high degree of confirmation can be attained by tampering either with the hypothesis or with experience. In fact, any of the following tricks would enhance a degree of confirmation to practically any desired extent: (i) reinterpreting unfavorable evidence as favorable by introducing *ad hoc* hypotheses; (ii) reformulating the hypothesis in a loose (semantically weak) way so that it will span almost any evidence; (iii) disregarding (hiding) unfavorable evidence, i.e. selecting the data; (iv) subjecting the hypothesis to slack tests. If all we can do is to enhance the degree of confirmation of our hypothesis, then we can do nothing to establish it, as a high value of $C(h|E)$ can be obtained with any of the aforementioned unscrupulous tricks. In other words, a high value of the degree of confirmation as defined by [15.4] is by itself of little value since the function $C(h|E)$ does not depend on either the testability of the hypotheses or the toughness (accuracy) of the test techniques. In view of these objections one may ask why a high degree of confirmation is so often regarded as an absolute desideratum. A possible explanation is that inductivists never have hypotheses proper in mind but rather information-packaging statements of the type of "All ravens are black". These generalizations may be interpreted as conjunctions of singular observational propositions—although it may also be argued that this interpretation is unduly restrictive. Anyway, if such an interpretation is accepted, then the only way in which we could empirically test the conjunction is by subjecting every conjunct to direct empirical scrutiny, or by taking large random samples of the total population. If all scientific statements were of that ravenous kind, knowledge would advance by just accumulating empirical evidence, which is always "positive" and directly verifiable. Since scientific hypotheses are not of the same kind as the statements handled by inductive "logic", the latter is irrelevant to science.

To the foregoing objections deductivists—notably Popper—would add the following. First, a single unfavorable evidence, if established beyond practical doubt, is sufficient to ruin a general hypothesis that has received 99 confirmations—that is, a load of confirming cases is weightless relative to a single refuting case; in short, a value $C(h|E)=0.99$, far from being weighty, goes to show that h is false. Second, from the confirmation of a number of directly testable consequences of a hypothesis it does not follow logically that the hypothesis itself is true: to assert this is to indulge in a logical fallacy (reduction). Third, the favorable cases found in an empirical investigation may happen to constitute a nonrepresentative sample: any number of falsifiers may still be in store. In effect, the directly testable consequences that have so far been derived are after all a finite sample of a potentially infinite set (in the case of unrestricted universal hypotheses), and there is no reason for being sure that it is a fair (representative) sample: what is a moment of pleasure compared to a possible eternity of hell? Deductivists would therefore conclude that a high degree of confirmation is not only of little value but altogether worthless, and that all we may aspire to is that our hypotheses be not falsified by tests aiming at refuting them. Their aim is to weed out falsity rather than to find truths.

The first argument is valid on the presupposition that *partial truth* is worthless. But the argument has no appeal for those who believe partial truth is *all* we can get: to them, a hypothesis with a high degree of confirmation—one just below unity—suggests that it *may* have a grain of truth; moreover, it will stimulate the search for a truer hypothesis, perhaps one not too distant from the previous conjecture. The second argument is impeccable and so is the third. Yet all they prove is that a high degree of confirmation does not *warrant* our concluding to a high degree of truth, a high probability, or a high credence. What the deductivist's arguments show is that the inductivist is wrong in feeling the more secure the higher the degree of confirmation of his hypothesis becomes: the next run of observations may be unfavorable to the point of altogether ruining the hypothesis—as has so often been the case in the history of science. (In fact, remember that the variation of $C(h|E)$ with n and N is *lawless*.

Still, these arguments do not establish the complete worthlessness of a high degree of confirmation: after all, a well-confirmed assumption is one which, being testable at all, has passed tough tests. And a test can be regarded as tough, hence valuable, (i) if it bears on a hypothesis sensitive to empirically controllable changes, (ii) if it is designed on previously corroborated (yet fallible) theories and tested techniques, and (iii) if the data and the inferences are screened. It is

Table 15.1. Value of data in science

Value of data for the inception of h	Value of data for the test of h
Confirmationism	
Since all scientific hypotheses are inductive generalizations, empirical data are decisive in their framing.	Since empirical tests are the sole possible ones, a great majority of favorable instances will establish a hypothesis .
Falsificationism	
Since all scientific hypotheses are transempirical and universal, data play no logical role in their origin, however psychologically effective they may be.	Since empirical tests are the sole possible ones and deductive inference alone is valid, data are decisive if unfavorable, useless if favorable.
Actual scientific procedure	
Data are (i) occasions for the discovery of problems and (ii) instrumental in the framing of empirical generalizations, all-important in protoscience (early stages). They cannot even suggest hypotheses proper.	Since empirical tests are part of a complex test battery, they are not decisive although they are significant: (i) if favorable, they are suggestive though inconclusive; (ii) if unfavorable, they are even weightier yet inconclusive, because the test may be wrong .

not enough for a hypothesis not to have been refuted: the failure to empirically disprove a factual statement may be due to slackness in the test procedure, or to technical difficulties, or even to lack of interest in it on the part of experimenters. Positive evidence, not just lack of negative evidence, is needed as well as some ground or matching with fundamental theories when available.

In the light of the foregoing, inductivists are seen to exaggerate the value of empirical confirmation whereas deductivists are seen to underrate it, both being misled by the wrong presupposition that empirical tests, which are indeed necessary, are also the only possible tests and moreover definitive. Table 15.1 summarizes the three positions at stake.

Problems

15.4.1. Does the confirmation of a hypothesis consist in the possibility of interpreting (explaining) *ex post facto* a set of facts with its help? *Alternate Problem:* Do scientists normally compute degrees of confirmation? Do they attempt to enhance them by relentlessly piling up evidence? And does every progress of knowledge consist in increasing the degree of confirmation of a set of hypotheses?

15.4.2. H. Bondi, in *Cosmology,* 2nd ed. (Cambridge: University Press, 1960), p. 143, states that a merit of the hypothesis of continual creation of matter out of nothing is that no observational evidence conflicts with it. Is this alleged lack of unfavorable evidence sufficient to entertain that conjecture?

15.4.3. No hypothesis is ever *verified,* since all that can be empirically tested is some logical consequence of it, and the verification of the latter does not authorize the assertion of the corresponding premises. On the other hand, hypotheses can be refuted by *modus tollens*—on condition that the empirical evidences be asserted without further ado. Query 1: Does this render confirmation worthless? Query 2: Since no hypothesis can be conclusively verified, should we replace the quest for truth by the quest for probabilification (increasing the probability of hypotheses)? (On the probability of hypotheses see Sec. 10.4.) Query 3: Does every negation of a set of refuted hypotheses (such as "It is not so that there are electrons with worms inside") constitute scientific knowledge, and is consequently refutation a method for obtaining positive truth rather than for eliminating falsity?

15.4.4. Are there conclusive empirical refutations? In particular, is a single unfavorable evidence always enough to disown a hypothesis, or is it legitimate in some cases to invoke an *ad hoc* hypothesis to save it—such as that the observation was wrongly conducted, or wrongly interpreted, or that an unknown external disturbance was responsible for the anomaly?

15.4.5. Does a tough test have to be conducted with the aim of *refuting* a hypothesis? Take into account that, as a matter of historical record, one of the very few known cases in which a scientist is known to have tried to refute rather than just *test* a hypothesis was R. A. Millikan's, who spent ten years (1906—1916) trying to debunk Einstein's photon hypothesis of 1905 (a "bold, not to say reckless, hypothesis"). Make a case for *impartial* tests which attempt neither to confirm nor to refute hypotheses, however partial the individual scientist may be. Try to specify the conditions for a tough test in objective, nonpsychological terms.

15.4.6. Philosophers have often asked for a rigorous justification of induction, on the assumption that induction is the method whereby science extrapolates to the unknown from a sample of observed "instances". Others have claimed that induction, precisely for being such a method for the expansion of knowledge, can justify every single projection but need not be justified itself—nay it cannot be justified because a recourse to the method's high percentage of successes would be inductive, whence the attempt would be circular. Examine both stands. Hint: Begin by finding out whether induction *is*, in fact, such a method. *Alternate Problem:* R. Carnap, in *The Continuum of Inductive Methods* (Chicago: University of Chicago Press, 1952), holds that there are infinitely many "inductive methods" by which the degree of confirmation of a given hypothesis can be determined on the basis of a given body of empirical evidence. (That is, data are not only powerless to suggest hypotheses: they do not even yield an unambiguous value of the inductive support of hypotheses.) These "inductive methods" are mutually incompatible and none is definitely the best—whence the choice among them is arbitrary. Furthermore, all of them apply only to properties of individuals belonging to finite universes; finally, hypotheses involving continuous magnitudes, such as time, must be excluded from Carnap's system of inductive "logic"—so far the most highly developed. Point out the use of this theory for the effective assaying of scientific hypotheses. See I. Lakatos, Ed., *The Problem of Inductive Logic* (Amsterdam: North-Holland, 1968).

15.4.7. A high degree of confirmation is valuable only if it refers to a highly testable hypothesis that has been subjected to exacting tests. Are these two conditions contained in the formula [15.4] or in any known formula for the probability of h given E, such as Bayes' formula?

15.4.8. Study the paradox of confirmation. See C. G. Hempel, "Studies in the Logic of Confirmation", *Mind*, **54**, 1 and 97 (1945), "Empirical Statements and Falsifiability", *Philosophy*, **33**, 342 (1958); R. Carnap, *Logical Foundations of Probability* (Chicago: University of Chicago Press, 1950), Secs. 87–88; G. H. von Wright, *A Treatise on*

Induction and Probability (London: Routledge & Kegan Paul, 1951), Ch. 9, sec. 5; J. W. N. Watkins, "Between Analytic and Empirical", *Philosophy*, **33**, 112 (1957) and "Confirmation, the Paradoxes, and Positivism", in M. Bunge, Ed., *The Critical Approach* (New York: Free Press, 1964); H. G. Alexander, The Paradoxes of Confirmation, *British Journal for the Philosophy of Science*, **9**, 227 (1958); J. Agassi, "Corroboration versus Induction", *ibid.*, **9**, 311 (1959); L. E. Palmieri, "Confirmation, Intention, and Language", *Methodos*, **11**, 33 (1959); D. Stove, "Popperian Confirmation and the Paradox of the Ravens", *Australasian Journal of Philosophy*, **37**, 149 (1959); J. L. Mackie, "The Paradox of Confirmation", *British Journal for the Philosophy of Science*, **13**, 265 (1963); M. Bunge, *Sense and Reference*, pp. 79–80 (Dordrecht-Boston, 1974).

15.4.9. The logic of confirmation disregards the relation of hypotheses and theories to the prevailing world view and the prevailing theory desiderata. Is this abstraction realistic, i.e. does it conform to actual practice of theory weighing? For the matching of scientific ideas with model patterns or paradigms, see T. S. Kuhn, *The Structure of Scientific Revolutions* (Chicago: University of Chicago Press, 1962). *Alternate Problem:* Formula [15.4] for the degree of confirmation of a hypothesis refers to a finite set of *actual* data, n of which confirm whereas $N–n$ refute h. What happens to $C(h|E)$ if N is interpreted as the total number of *possible* evidence relevant to h and n as the number of actual confirmers of h? And if $C(h|E)$ is interpreted—as in Carnap's theory—as the probability of h given E, what is the probability of a universal law statement and what does it mean to say that such a probability is zero: that the hypothesis is false, that the law is impossible, that probability is not a correct measure of degree of confirmation, or what?

15.4.10. Elucidate the concepts of *convergent evidence* (i.e. of evidence that "points to" the likelihood of a hypothesis or theory) and of *circumstantial* or presumptive evidence (i.e. of evidence that is not directly relevant to the construct under consideration but rather to some of those logically related to it). For the circumstantial and converging character of the evidence in favor of Darwin's theory of evolution at the time it was forwarded, see W. Le Gros Clark, "The Crucial Evidence for Human Evolution", *Proceedings of the American*

Philosophical Society, **103**, 159 (1959) and *The Antecedents of Man* (1959; New York: Harper Torchbooks, 1963). *Alternate Problem:* Galilei and his disciple Torricelli, both champions of rational criticism, did not seem to care much for contrary evidence. Believing their science to be purely mathematical, they claimed that, if a theory did not fit certain facts, then the theory failed to refer to them but was not therefore false. For example, if certain bodies did not move in accordance with their predictions, Torricelli would conclude: "*suo danno: noi diremmo che non parliamo di esse*." Was this an acceptable conclusion? If not, can we say that the present methodology of science is completely faithful to Galilei's?

15.5. A Case History: Torricelli

In 1643 Torricelli faced the problem of explaining why lift pumps could not lift water to more than 18 cubits—around 10 m. This problem was of interest not only to mining engineers and plumbers but had also a theoretical and even an ideological interest. If, as the Aristotelian physicists held, water mounted a pump of this kind because it was being sucked by the vacuum (this being why we still call them suction pumps), then there should be no limit to this manifestation of the aversion of Nature to vacuum *(horror vacui)*. Actually there were additional, nonscientific reasons, for doubting the Aristotelian explanation in addition to its reduced explanatory power: the *horror vacui* was an anthropomorphic and inscrutable force rather than a manifest, mechanical and measurable force; orthodox science was crumbling down in other fields and therefore becoming suspect altogether; and last, but not least, fighting traditional science was part of the war against the feudal social and cultural order.

The schoolmen tried to rescue the orthodox hypothesis by introducing a number of *ad hoc* hypotheses. Even Galilei tried this strategy, assuming that the water column broke down under its own weight in analogy with a long metal rod. His pupil Torricelli did not try to patch up the orthodox hypothesis but invented an alternative conjecture which fitted the hydrostatics of Archimedes and Stevinus as well as the new, mechanistic world view that was being built on the ruins of the traditional organismic and theological one. Moreover, Torricelli designed experiments to test his idea.

Torricelli's solution was not tailored to fit just the lift pump phe-

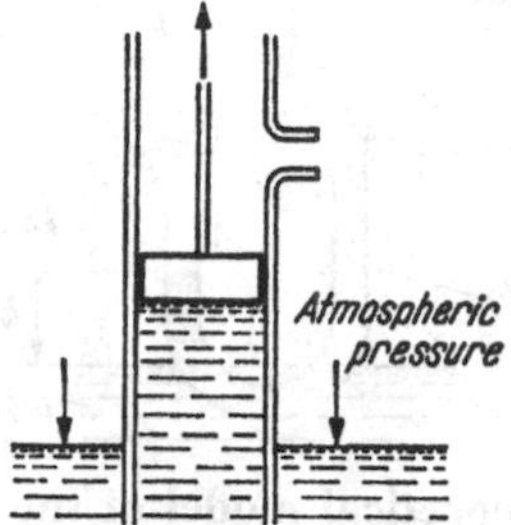

Fig. 15.7. Interpretive explanation of the working of the lift pump: the atmospheric pressure is transmitted to the liquid body and the piston only frees the liquid column from that pressure.

nomena but had a far wider coverage: it was a small theory and, for being a general theory, it could be expected to be applicable to various domains of technology. Torricelli's two initial assumptions were:

*A*1. Air is a fluid that obeys the laws of the statics of liquids.

*A*2. The earth is surrounded by an air layer—or, as he put it himself, "We live immersed in a sea of elemental air".

From *A*1 in conjunction with hydrostatics he deduced

t_1. Air exerts a pressure on all bodies in contact with it.

From *A*2 and t_1 follows

t_2. Atmospheric air exerts a pressure on the terrestrial surface (the modern "atmospheric pressure").

Notice that *A*2 can be reformulated thus: "The atmosphere exists". *A*1 does not state the existence of anything but the basic identity of gases and liquids: their inclusion in the genus fluid. Its consequence, t_1, extends to air a property discovered by Stevinus for liquids. As for t_2, another law statement, it has a more restricted scope and will be employed in explaining the working of pumps. In fact, Torricelli explains the ascension of a liquid in a pump as an effect of the pressure of the atmospheric air on the free surface of the liquid: the function of the piston is not to suck the water but to dislodge the air thereby freeing the ascending water from the atmospheric pressure (see Fig. 15.7). The logic of this explanation is the following. From t_2, by specification one derives

t_3. Atmospheric air exerts a pressure on the free surface of liquids. From *A*1 it follows that the laws of communicating vases hold for air

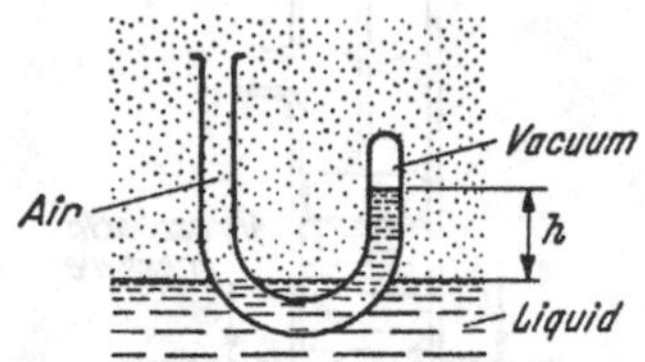

Fig. 15.8. Torricelli's theoretical model of the pump-air-liquid complex, as expressed by axiom *A*3.

liquid systems. In particular, *A*1 in conjunction with t_3 and with hydrostatics entails

t_4. If no force acts on a portion of a liquid surface while the rest of the surface is acted on by atmosphere, the former mounts until its weight is balanced by the weight of the air column.

We now introduce the concept of hydrostatic equilibrium in a U-shaped tube:

Df. Two fluids are in equilibrium in a U-tube if and only if the pressures they exert mutually on the intersurface are the same.

Df. and t_4 entail

t_5. The maximum height attained by a liquid mounting under the action of the atmospheric pressure is the one corresponding to equilibrium. (In quantitative terms: the atmospheric pressure P equals the pressure of the liquid, i.e. its specific gravity times its height: $P=\rho h$, from which $h=P/\rho$.)

A small step is still needed: the relation among the lift pump, the liquid reservoir, and the atmosphere must be stated. This crucial statement is not a pure intuition because it has been preceded by an analysis; it is not an induction either because it does not generalize a number of observations; nor is it a deduction, because it is not derived from previously stated premises. It is a new axiom suggested by the comparison between two things, and stating their similarity: it is an analogy. The new hypothesis is

*A*3. A lift pump, the liquid in which it is partially immersed, and the atmosphere, constitute a U-tube the peculiarity of which consists in that the atmospheric pressure does not act on one of its branches (see Fig. 15.8).

From *A*3 and t_5, the unheard-of possibility of actually producing a vacuum is deducible:

t_6. If, once the liquid column has reached the equilibrium height, the pump continues to operate, a vacuum appears between the surface of the liquid column and the bottom of the piston.

This conclusion, which contradicts an axiom of the plenist cosmologies of both Aristotle and Descartes, is the theoretical basis of the vacuum techniques. Anyway, we have now all we need to explain why liquids ascend in lift pumps and why this ascension is limited: *A*3 and t_4 entail the former explanandum whereas *A*3 and t_5 entail the latter. Moreover, we have gained a way of objectifying and measuring the atmospheric pressure by measuring the height of a liquid column in equilibrium with the atmosphere: the unobservable-observable, or cause-symptom hypothesis, is part of the theory.

In a nutshell, the main stages of Torricelli's reasoning up to this point are the following:

1. *Problem statement:*

   Limited lifting power of "suction" pumps and inadequacy of the *horror vacui* hypothesis to explain this empirical generalization.

2. *Initial hypotheses:*

   *A*1: Aerostatics and hydrostatics basically one. — *Analogy*
   *A*2: Existence of the atmosphere ("sea of elemental air"). — *Analogy*

3. *Working out of initial hypotheses:*

   $\{A1, \text{Hydrostatic theory}\} \vdash t_1$ — *Deduction*
   $\{A2, t_1\} \vdash t_2$ — *Deduction*
   $t_2 \vdash t_3$ — *Deduction*
   $\{A1, t_3, \text{Hydrostatic theory}\} \vdash t_4$ — *Deduction*

4. *Expansion of the initial theory:*

   Df. — *Stipulation*
   $\{Df., t_4\} \vdash t_5$ — *Deduction*
   *A*3: Pump part of U tube. — *Analogy*
   $\{A3, t_5\} \vdash t_6$ — *Deduction*

5. *Solution of the initial problem:*

   $\{t_4, A_3\} \vdash$ Mounting of liquids in pumps. — *Deduction*
   $\{t_5, A_3\} \vdash$ Limited height of liquid column in pump. — *Deduction*

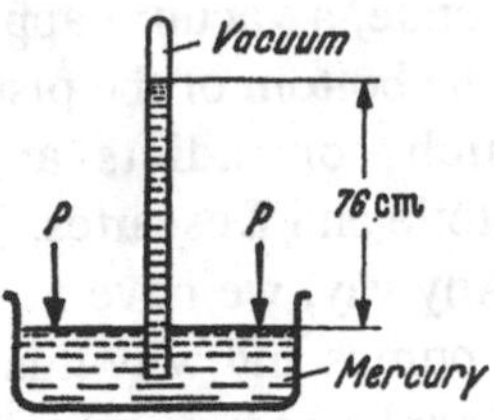

Fig. 15.9. Torricelli's first experiment: the "appearance" of a vacuum.

Torricelli, a modern scientist, did not rest content with inventing hypotheses and working out their consequences, i.e. with building a theory: he tested it. Since his basic hypotheses were not directly testable, he had to work them out: the building of the theory and its test went hand in hand. Consider, in effect, any of the hypotheses $A1$ to $A3$ and t_1 through t_3: they cannot be contrasted directly with empirical evidence because they refer to nonperceptible physical objects—the atmosphere and atmospheric pressure. In other words, the concepts "atmosphere" and "atmospheric pressure" are transempirical and, moreover, theoretical concepts: if this sounds surprising it is only because we have grown accustomed to them. Torricelli's initial assumptions could only be tested indirectly, by checking the adequacy (truth) of those of their consequences which referred to perceptible manifestations of the atmosphere and its pressure. These lowest-level propositions are the theorems t_4 through t_6, which refer to the ascension of liquids and their height. Actually t_4 is only partially testable by confrontation with observable things and events: the mounting of the liquid is observed but its cause, atmospheric pressure, is not—and this is what makes t_4 valuable as a test hypothesis. Anyhow statements t_4 through t_6 can be tested by handling liquids and tubes. And, being universal statements (referring to all liquids, all tubes and all lift pumps, in every place and time on the Earth and under every circumstance), they can be refuted or supported by choosing a number of specific cases.

Torricelli selected the air/mercury pair for technical reasons: given the great specific gravity of mercury, the height of the liquid column would be moderate and the glasswork simple. Viviani performed the first experiment, shown in Fig. 15.9. The result is well-known: contrary to common sense, mercury did not drain off completely but, as predicted, stayed to a little more than one and a half cubits—76 cm.

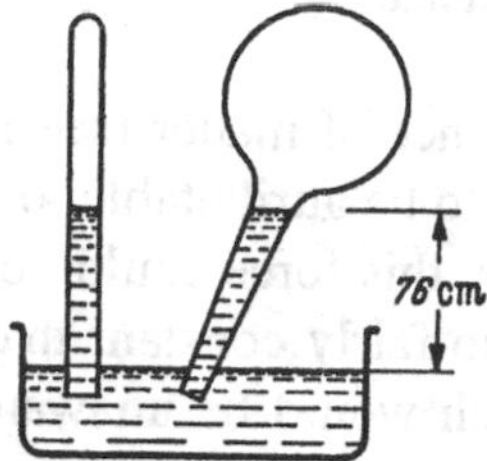

Fig. 15.10. Torricelli's second experiment: refutation of the hypothesis that the vacuum exerts a sucking force.

And in the upper part of the tube "we saw that a void space was formed and that nothing happened in the container in which this space was formed" (Torricelli). Why should anything have happened? Because Descartes and other partisans of the *plenum* hypothesis had predicted that the tube walls would collapse since space is a substance, "distance being a property of extension, which could not subsist without something extended" (Descartes). The experiment refuted the *plenum* hypothesis and confirmed Torricelli's theory. In particular, it confirmed the vacuum hypothesis, unthinkable on the rival views. As a bonus "an instrument that will show the changes of the air"—the barometer—was invented. One more unobservable had ceased to be inscrutable.

The first experiment might, however, be interpreted in alternative ways. It was possible to argue that the relevant variable was not the atmospheric pressure but an extremely rarefied substance—an aether—exerting an attractive force on the mercury column, thereby keeping it suspended. The Aristotelian F. Linus claimed, in fact, that there was no void in Torricelli's inverted tube, which was on the other hand occupied by an invisible elastic band, which he called *funiculus,* from which the mercury column hung. This was, of course, an *ad hoc* hypothesis incapable of being independently tested: the *funiculus* had no other function, hence no other manifestation, than preventing the mercury from draining off—or, rather, it discharged the function of protecting the plenist hypothesis from adverse experience. Additional empirical evidence was however desirable; this was obtained from two further experiments.

If the void sucks the liquid, then the greater the void at a top of a tube the higher must be the liquid column. Torricelli observed that this was not the case: the vacuum was not an active physical agent but just the old Democritean absence of matter (see Fig. 15.10). Hence, if the motion of the liquid was to be attributable to a force (and on this there was universal consensus), this force could not be internal: it had to be an external force—one so fairly constant that it had been overlooked much as the deep sea fish would be unaware of the existence of the sea.

A third experiment was designed by Pascal. If nature abhors vacuum, the abhorrence must be the same at sea level as in the mountain; and if the mercury column is balanced by the atmosphere, the liquid column must be the shorter the less atmosphere is left. In other words, if Aristotle was right the barometer must show the same pressure at different altitudes; whereas if Torricelli is right, the barometer must show a decreasing atmospheric pressure as the altitude increases. The experiment confirmed Torricelli's theory although it was a qualitative experiment: no barometer and thermometer readings were made by the mountain climber. Anyhow the so-called barometer law of the exponential decrease of pressure with altitude could not have been established at that time because the exponential function, a necessary formal ingredient of the law statement, had not been invented.

This third experiment, performed by an amateur—Périer—at the suggestion of another amateur—Pascal—was hailed by all progressive scientists as a conclusive *proof* of Torricelli's theory. Actually it only refuted the rival views and lent a great *support* to the new theory. Luckily, no philosopher remarked—rightly, alas—that there was nothing conclusive in the three experiments for, after all, a number of variables—temperature, time of year, clouds, winds, kind of glass, origin of mercury, and so forth—had been neglected, and all those experiments had been executed just once or twice each. Nobody demanded an accumulation of favorable cases in order to probability (increase the probability) of the theory: no inductive logicians were around to pester Torricelli. What happened was rather the opposite: Torricelli's theory gave birth to a number of new fertile hypotheses, experiments, and instruments: Guericke's spectacular experiments with the Magdeburg hemispheres, Boyle's epoch-making experiments and hypotheses on gas compressibility, the hypothesis that sound was a wave motion of the air, meteorology as the science of atmospheric changes (rather than as the natural history of "meteors"), vacuum pumps and pressure gauges, artifacts using vacuum, and so on. The new theory did not come *post festum* to summarize data but led the way and went so far as to stimulate the development and adoption of the Democritean cosmology and had in turn the support of it. But no theory is eternal if it is factual: Torricelli's conclusion that a complete or "absolute" void was possible was corrected two centuries later when the universe was repopulated with a "subtle matter", the gravitational and electromagnetic fields: even the vacuum of Torricelli's barometer is filled with them. But this time a comparatively small correction was sufficient: "corpuscular vacuum" replaced Torricelli's "absolute vacuum".

In his work, Torricelli and his collaborators made ample use of analogies or homologies (recall Problem 15.1.1) and of deduction. Induction, on the other hand, took no part in either building the theory or in designing the experiments aiming at testing it: it could play no role in these activities since neither hypotheses nor experimental designs are observable singulars. Induction did intervene in the *final evaluation of the empirical tests* of Torricelli's theory. In effect, the qualitative "conclusion" (summary statement) that the theory had been confirmed by a large number of variegated data was inductive. (But so far nobody seems to have computed the probability or the degree of confirmation of Torricelli's theory, or even of its central hypothesis, that we dwell at the bottom of an airy sea.) Empirical confirmation was not piled up by repeating and perfecting observations and experiments: the theory kept pointing to new possible evidence, namely, the facts it could explain even if they had not yet been observed. And empirical support was far from being the sole criterion of evaluation: the joy of refuting official science was at least as important—but this aspect of theory assaying will concern us later on.

Note finally that, in testing his central hypothesis about the hydrostatic effect of the atmosphere, Torricelli was forced to introduce additional hypotheses; as a matter of fact he ended by unwittingly building a small theory. This is characteristic of scientific hypotheses in contrast with commonsensical conjectures: the former face experience in bunches, and even theories require the assistance of further theories in order to make contact with facts (recall Sec. 8.4). It could not be otherwise given the interrelation of properties and events. In any case, the important thing in science is not the test of stray conjectures—the subject of the existing systems of inductive "logic"—but the test of theories. To this subject we now turn.

Problems

15.5.1. Torricelli claimed he *saw* the formation of a vacum. Was he right?

15.5.2. Draw the tree exhibiting the structure of Torricelli's argument.

15.5.3. What role did the *modus tollens* and reduction play in Torricelli's work?

15.5.4. Did Torricelli employ arguments from analogy?

15.5.5. Torricelli's theory was resisted for some years both by the Aristotelians and the Cartesians. Was this a case of sheer obstinacy, of refusing to believe what everyone could see?

15.5.6. Did barometer readings start the study of the atmosphere or, conversely, was the instrument an outcome of Torricelli's theory? Caution: this might be a question of the hen-and-egg type.

15.5.7. Suppose Torricelli had mastered mathematical statistics and had access to an automatic computer. What would he have gained? Conversely: put a statistician in the place of Torricelli: what would he have discovered?

15.5.8. Make a study of a case history chosen from the recent or the remote history of science, disclosing the inference patterns actually employed. This may require a bit of logical reconstruction, i.e. the formulation of some hypotheses concerning the actual but now effaced path of research. Recall that only the scientist's paperbasket knows what the original zig-zagging path was.

15.5.9. Some false hypotheses have suggested fruitful research projects. Why is this so? And what bearing does this have on the pragmatist equation of truth with usefulness?

15.5.10. Study the problem of the probability of hypotheses. In particular, determine (i) the meanings of the various probabilities, (ii) the domain of application of Bayes' formula, and (iii) its use in scientific inference and in metascientific elucidation. See M. Bunge, *Interpretation and Truth* (Dordrecht-Boston: Reidel, 1974), M. S. Bartlett, *Probability, Statistics and Time*, pp. 98–110 (London: Chapman and Hall, 1975).

15.6. Testing Theories

In order to regard a hypothesis as verisimilar it is necessary, but not sufficient, that it gets high grades (a strong inductive support) upon passing tough empirical tests. Favorable empirical evidence is inconclusive because one-sided. In effect, empirical tests may show that a given hypothesis covers or fails to cover a set of data, but the most exacting empirical tests are silent concerning other desirable properties, especially the way the hypothesis fits or fails to fit the rest of the relevant knowledge—which is no less entitled to pass judgment on the worth of the hypothesis than newly arrived empirical data, if only because in producing these very data we have used part of our background knowledge. The more such contacts with available knowledge the more varied the tests can be and accordingly the more varied the (indirect) empirical support can become. In short, the *variety* of the supporters (or opponents) or a hypothesis counts at least as much as their sheer *number*.

Now, variety is added to numerosity of confirmers (or refuters) if the hypothesis is logically related to other statements, i.e. if it is or becomes part of a *theory*. (For the concept of theory see Sec. 7.1.) In fact, if two hypotheses, h_1 and h_2, join forces to constitute a theory, then more numerous and varied data will weigh for or against them than if they remained isolated from one another: the set $T1$ of directly testable consequences of h_1 will now become an indirect test for h_2 and, conversely, $T2$ will indirectly support or undermine h_1. In addition, a whole new set $T3$ of consequences may be derived from the conjunction of h_1 and h_2, so that in the end every one of the assumptions of the theory will have gained in testability: the total number of empirically controllable consequences will now be the union of the sets $T1$, $T2$ and $T3$ (see Fig. 15.11). The testability, and therefore the empirical support or the undermining of the initial assumptions has increased because the hypotheses themselves have become richer upon systematization. Consequently the building of theories, far from being required by economy, is necessary both for enriching the stock of

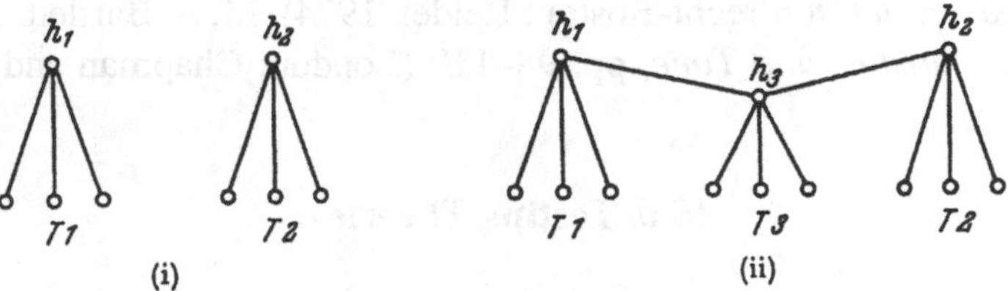

Fig. 15.11. The testability of two hypotheses, h_1 and h_2, before (i) and after (ii) they are fused in a single theory; h_3 is deduced jointly from h_1 and h_2 and it entails *T*3, a new set of consequences.

hypotheses with new deductive consequences of their conjunction and for improving their empirical checking.

Suppose now we are given a theory and are asked to subject it to empirical tests. Shall we rush to the laboratory or to the field, as the case may be? The answer is in the negative: a number of preliminary conceptual operations come before the empirical ones. First: before planning the test of a theory we must make sure it is a *theory* proper, i.e. a hypothetico-deductive system and not just a heap of guesses: not knowing whether a given proposition of the theory follows from another, we might not know what are the assumptions our tests render likely or unlikely. If such a logical organization is wanting and yet the given set of conjectures seem worth testing, we should try to build a theory out of some of the given hypotheses, or test them one after another. Something of the sort has been proposed, but never been carried out, in the case of psychoanalysis. Second: we should make sure the theory is empirically *testable* in principle. In case it is not, we should sharpen the vague conjectures and eliminate the *ad hoc* assumptions that render the theory immune to refutation or nearly so. Otherwise every conceivable datum will confirm the theory. A high degree of confirmation is valuable only if it is genuinely a posteriori, i.e. if the theory is not built so as to match every possible circumstance. Third: we should realize what are the *presuppositions* of the theory, for it might be that these, rather than the theory itself, are in need of test or even that they have already been discarded by previous tests. However, care must be exercised in this regard because no factual statements, whether they occur as presuppositions or as hypotheses to be tested, need be wholly true. Fourth: we should collect the testable consequences we wish to check against empirical data, i.e. we should effectively *derive* some directly testable yet theoretical (nonobservational) logical consequences t_i of our theory's initial assumptions. This will require not only deduction but also the adoption of certain empirical data obtained from previous observations, measurements, or experiments. It may also require the introduction of simplifying assumptions making the deductions and/or the application to special cases possible. *If the theory is mathematically complex, the simplifications (e.g., linearizations) can be such that a *new theory*, weaker but testable, is produced—which goes to show that testability cannot be equated with logical strength (see Sec. 5.7). Hence the empirical tests will be relevant to the weaker theory. If they are unfavorable to it we are at liberty to blame either the original "exact" theory or some of the brutal simplifying assumptions. Only if, upon gradually relaxing the latter, a better agreement with experience is obtained, shall we suspect that the "exact" theory is partially true.* Fifth and last: we should *translate* the testable consequences t_i into semiempirical statements t_i^* capable of being confronted with similarly translated data e_j into pieces of evidence e_j^* (see Sec. 8.4). In fact, let us recall that even the lowest level consequences of a theory, being theoretical statements, refer to the theory's ideal model and need certain semantical changes before they can be confronted with empirical data: e.g., theoretical probabilities will have to be estimated by countable frequencies.

After the above preliminary stage has been gone through we can go over to empirical testing proper. The testing process will start with planning and will end up by producing a body E of filtered evidence relevant to the theory's testable consequences T (actually via translations E^* and T^* into a common semiempirical language). The final act will consist in comparing these two sets of propositions with a view to passing judgment on T and so, indirectly, on the whole theory T of which T is only a part—or with a view to passing judgment on E on the strength of a previously corroborated T. The process of theory testing can then be analyzed into three main stages: theoretical processing, planning and execution of tests, and inference. In more detail, we have the scheme below, based on the diagram 8.15 in Vol. I, p. 502 and on the discussion of Sec. 15.4.

1. Theoretical stage: Finding T

***1a. Presuppositions** P* of the theory. (Not questioned by the test but criticizable before and after.)

1b. Specific assumptions A of the theory: axioms and auxiliary premises referring to the theory's model. (The indirect subject of test.)

1c. Data e_i from former and/or present experience, stated in the language of the theory. (Not questioned by test.)

1d. Deduction of particular testable consequences t_i:

$$\{P, A, e_i{}^*\} \vdash t_i, \qquad i=1, 2, \ldots, m<\infty \qquad [15.5]$$

The effectively derived consequences t_i are the direct subject of test; they are in finite number even though the total possible consequences constitute an infinite set.

1e. Interpretation of t_i in semiempirical terms, i.e. as potential evidence:

$$I(t_i)= t_i{}^*, \{t_i\}=T^* \qquad [15.6]$$

These translations require often an auxiliary theory and they may be questioned by test.

2. Empirical stage: Finding E

2a. Theoretical presuppositions Q of the empirical procedures, i.e. fragments of theories built into the design, execution and interpretation of the test.

2b. Observational statements (protocols) e_j.

2c. Interpretation of observational propositions in terms of theory Q: deduction of evidence $e_j{}^*$ from the protocols or data e_j in conjunction with theory Q and rules of interpretation I:

$$\{Q, I, e_j\} \vdash e_j{}^*, \quad j=1, 2, \ldots, n<\infty, \quad \{e_j{}^*\}=E^* \qquad [15.7]$$

Recall that, whenever magnitudes are at play, $e_j{}^*$ is not identical with $t_j{}^*$, if only because the former includes a reference to the experimental error, which is absent from t_j^*.

3. Inference: Confronting T^* with E^* and deciding about T

If the collected evidence E^* is relevant to T^* at all, two cases are possible: either the two agree with each other or they do not.

3a. Case 1: T^* agrees with E^*, i.e. the evidence is consistent with the consequences of the theory (to within the accepted experimental error). If t^* is empirically confirmed we regard t_i as equally confirmed even though the latter has not, strictly speaking, been subjected to test since it refers to an ideal model. We now look at the source of t_i, namely formula [15.5], and realize that the confirmation of t_i is of no great value because (i) from a true conclusion we cannot unambiguously refer back to a true premise but at most to a possible one, and (ii) a true consequence may derive from a set containing false propositions since, for any proposition p, whether true or false, $\{p, q\} \vdash q$. Accordingly, what can be concluded from the agreement of E^* with T^* is just this: (i) the evidence E^* is *relevant* to the theory and (ii) the theory is *consistent* with that evidence. Since this is the most we can expect from an empirical test, we regard confirmation as a support of the theory but not as its proof.

3b. Case 2. T^* disagrees with E^*, i.e. the discrepancy between the projections and the data exceeds the tolerance agreed on beforehand. (The discrepancy may be total or partial: a subset $E^{*\prime} \subseteq E^*$ may disagree with the corresponding subset $T^{*\prime} \subseteq T^*$.) Two courses of action are open: either accepting the nonmatching evidence as true or rejecting it as suspect.

Subcase 2a: The nonconfirming evidence $E^{*\prime}$ is accepted on the ground that it is well backed up by another piece of knowledge: $E^{*\prime}$ may have been checked independently, by a different technique, or it may be consistent with another, already accepted theory. Consequently we doubt the theorems t_i incompatible with the corresponding evidence, which leads us to doubt the whole set $\{P, A, e_i^*\}$ of premises entailing t_i (see formula [15.5]). Still, nothing can be concluded about the specific assumptions A of the theory, which are the ones we wish to test: the error may lie in the presuppositions P or in the empirical data e_i, or in their translations e_i^* employed in the derivation of t_i. In other words, the refutation of a number of consequences of a theory need not point to the latter's total falsity: the fault may lie with either the presuppositions or with the empirical information fed into the theory. How can we proceed to identify the culprit? We shall return to this point in a while.

Subcase 2b: The nonconfirming evidence $E^{*\prime}$ is rejected on the strength of the strategic hypothesis that it results from a mistake in either the design, the execution or the interpretation of the empirical procedures. The remaining data of the same run of observations (i.e., the set $E-E'$) are accepted; or, if anomalous and likewise subject to suspicion (i.e., if $E'=E$), they are all rejected and the whole observation, measurement, or experiment, is redone, eventually on a new design. This course—rejecting "anomalous" data—is not infrequent and is chosen whenever the theory under test has previously passed tests in the same range whereas the experimental set-up is being tried for the first time. In short, there are times when theory checks experiment in addition to intervening in its planning and interpretation. On such occasions nothing can be concluded about the theory under test, since it has hardly been questioned: to test a theory is to doubt it.

Subcases 2a and 2b obviously exhaust the class of disagreements of theoretical projections with data. And, save for borderline cases—which call for the improvement of the empirical procedures—agreement and disagreement with experience is all there is. In short, we have considered all possible cases. But then we face the distressing consequence that neither the confirmation of the testable consequences of a theory nor their refutation can force us to either *conclusively* uphold or *conclusively* reject a theory. With experience alone we may at most *accumulate evidence* pro or con a theory. In the former case we get empirical confirmation or reinforcement (Case 1). Empirical disconfirmation will be regarded as definite so far as the present test goes if the unfavorable evidence is in turn backed up by a well tested body of knowledge. (New theories are often built to account for data that conflicted with previous theories.) Even so, negative evidence will not point unambiguously to the defective specific assumptions of the theory (Subcase 2*a*). If the evidence itself is suspect (Subcase 2*b*), the matter will remain even less settled.

Favorable evidence is inconclusive and unfavorable evidence is not much more conclusive in the case of theories—unlike the case of single hypotheses, which do not face experience backed up by a more or less bulky body of theoretical assumptions. Empirical confirmation entitles the theory to make a step forward, whereas empirical refutation pushes it some steps backward—but this is all. We have apparently reached a stalemate: either we suspend judgment on the truth value of scientific theories or we adopt a conventionalist criterion of theory acceptance, such as simplicity of some sort, or a pragmatist one, such as practical usefulness or aesthetic appeal. The first course amounts to stop theorizing since it deprives this activity of its motivation, namely, attaining the truest possible conceptual representation of reality. The second course bypasses the burning question 'How true are our theories?' and gives no hint as to how to improve their adequacy to fact. Fortunately the situation is complex but not as grave as this.

The skeptical conclusion about the impossibility of assigning a truth value to any scientific theory rests on two presuppositions that are far from established: (i) that all we are interested in is a *conclusive*, hence definitive proof of the *complete* truth or the *complete* falsity of our theories; (ii) that experience is an *absolute* standard and, particularly, one independent of every theory. The first assumption is wrong: in fact, we know theories are probed rather than proved, and all we hope is to be able to temporarily show that any given theory is partially true. That this is the *summum bonum* in the field of theory evaluation is suggested not only by the history of science but also by an analysis of theory construction (recall Sec. 8.1). As to the assumption that experience is an absolute standard against which theories are gauged, it occurred in the above argument in the following tacit methodological hypotheses: (*a*) that the presuppositions *P* of the theory under test were questioned by the test, and (*b*) that empirical tests are the sole kind of test of factual theories. We shall show that these two assumptions are false and accordingly, that we need not accept an all-round skeptical attitude toward the possibility of estimating the truth value of scientific theories.

Whenever a hypothesis, say t_i, is questioned a certain set of hypotheses and data must be taken for granted, i.e. assumed to be true, at least provisionally: otherwise there would be no standard against which the hypothesis concerned might be tested. In other words, any given proposition can be doubted and criticized only on the strength of a certain set of propositions which are for the purpose of criticism treated *as if* they were wholly true even though they may be questioned later on. Accordingly we can never question the whole body of knowledge at a given time: scientific criticism is always piecemeal and progressive. Not even scientific revolutions are total.

This has the following consequence for the evaluation of the disagreement of a theoretical projection with the corresponding empirical evidence (Subcase 2*a* above). In order to derive t_i from *P*, *A* and e_i^* (recall formula [15.5]) we have independently asserted *P* and e_i^*; consequently, the empirical test does not test the whole body of premises but just the *specific assumptions A* of the theory. In particular, the empirical test does not bear on the logical or mathematical formulas presupposed by the theory but which are not introduced by the theory: if they were questioned they would not be called presuppositions. Consequently, if t_i is confirmed, the assumptions *A* can be declared supported though not true; and if t_i is refuted on good evidence,

A falls as well without necessarily dragging *P* and $e_i{}^*$ in its fall. In short, in Subcase 2a (unfavorable sound evidence) we have the following pattern of inference:

$$\frac{\vdash P,\ \vdash e_i{}^* \qquad \{P, A, e_i{}^*\} \vdash t_i \qquad \dfrac{-t_i{}^*}{-t_i}}{-A}$$

that is, a conclusive refutation of the set of specific premises. We can therefore weed out false theories on condition that the body of premises (presuppositions and data) used to derive the falsity is not questioned in the experience concerned although it may well be questioned in the course of a different experience. Still, we cannot in this way *individualize* in the conjunction *A* of assumptions, the falsest members. The discovery of the culprit(s) is however possible by means of further *independent* tests—that is, tests involving all the specific assumptions save the suspect one—and also, if possible, tests involving just the suspect assumption but none of its companions. The same procedure can in principle be applied in succession to every member of the set of specific premises until every false assumption has been identified. So much for refutation.

In the case of confirmation no such clear-cut conclusion can be reached. If the theory is confirmed—or, if preferred, if the theory fails to be refuted by tough tests—we do not accept it right away as true but subject it to nonempirical tests such as continuity with the bulk of knowledge and capacity to forecast unexpected facts (serendipity). That theory will usually be regarded as the truest—*pro tempore*—which, having being empirically confirmed, meets best a battery of nonempirical tests that will concern us in the next section.

In conclusion, it is usually possible to reach *definite* though *not definitive* conclusions concerning the truth value of any scientific theory, or of part of it, on condition that certain other theories or parts of them are not questioned, and that empirical tests are supplemented by certain nonempirical tests. (Very often, some parts of the theory are used in the design of experiments aiming at probing other parts of the theory: in such cases one cannot try to test the whole theory at a time.) Furthermore, although theories—in contrast to stray conjectures—do face experience as wholes, it is possible to analyze such wholes and identify their grossly false components. This disposes of the view that scientific theories are altogether conventional (neither true nor false but just convenient) and therefore adamant to experience.

This completes our account of the *empirical* testing of theories. Lest the impression may remain that empirical tests are just means for estimating the truth values of theories we shall note that, whether the aim of an empirical test has been attained or not, a number of side-results may be gained. First, the theory may have been made more precise either because its formulation has been rendered more accurate or because it has been illustrated. Second, the theory may have been worked out in greater detail precisely in order to test its consequences. Third, the gap between the theoretical model and its real referent may have become better delineated; i.e., the domain of empirical adequacy of the theory may have been demarcated. Fourth, test techniques may have been invented or refined in the course of testing. Fifth, new facts—side effects—may have emerged and been discovered in the course of the test. Some of these *boni* may turn out to be more important than either the theory or the empirical results. No theory testing, if thorough, is likely to be altogether sterile even if it yields no other truth than that the theory under test is false.

Problems

15.6.1. Examine the philosophy of science of C. Huygens as exposed in the Preface to his *Treatise on Light* (1690; London: Macmillan, 1912), pp. vi–vii: . . . “whereas the Geometers prove their Propositions by fixed and incontestable Principles, here [in physics] the Principles are verified by the conclusions to be drawn from them, the nature of these things not allowing of this being done otherwise. It is always possible to attain thereby to a degree of probability which very often is scarcely less than complete proof. To wit, when things which have been demonstrated by the Principles that have been assumed correspond perfectly to the phenomena which experiment has brought under observation; especially when there are a great number of them, and further, principally, when one can imagine and foresee new phenomena which ought to follow from the hypotheses which one employs, and when one finds that therein the fact corresponds to our prevision”.

15.6.2. Examine W. H. Sewell’s “Some Observations on Theory

Testing", *Rural Sociology*, **21**, 1 (1956), on the need to reformulate the diffuse psychoanalytic guesses (therein called theories) before subjecting them to test. *Alternate Problem:* A theory of perfect or quasiperfect fluids is all that can be built for the time being. But then it cannot be tested precisely because there are no perfect or quasiperfect fluids. What is to be done?

15.6.3. The central axiom of Newton's mechanics, namely the law "*f=ma*", was not tested directly and independently until, one century later, G. Atwood designed and operated his machine. How come Newton's mechanics was regarded as true long before, and even as true *a priori? Alternate Problem:* Could every unfavorable evidence be reinterpreted as an experimental error?

15.6.4. Newton's corpuscular theory of light can be compressed into the following axioms

*A*1. Light consists of corpuscles obeying the laws of mechanics.

*A*2. Light corpuscles are attracted by the particles constituting transparent media.

From *A*1 the true theorem of the rectilinear propagation of light in a vacuum follows. From *A*1 and *A*2 it follows that light is deviated when passing very near stars (true) and that the speed of light is larger in transparent media that in void (false). Draw the logical tree and indicate what crucial experiments might refute the theory despite its having some true consequences. *Alternate Problem.* Examine the situation in contemporary learning theory, in which alternative models (theories) account for roughly the same data. In particular, show how the various assumptions are tested and find out whether they are checked with other pieces of knowledge (e.g., physiology). See S. Sternberg, "Stochastic Learning Theory", in R. D. Luce, R. R. Bush and E. Galanter, *Handbook of Mathematical Psychology* (New York: Wiley, 1963), vol. II, pp. 72 and 102–116.

15.6.5. A *crucial* test is one enabling us to draw a conclusive inference regarding the truth claims of pairs of rival theories. Are experiments crucial for the acceptance or rather for the rejection of constructs? *Alternate Problem:* How long do scientific theories remain on probation (as biologist P. Medawar would say)?

15.6.6. Examine P. Duhem's following contentions; (i) there are no crucial experiments because whole theories rather than isolated hypotheses are subjected to empirical tests; (ii) in the face of unfavorable evidence any constituent of the theoretical body might be arbitrarily modified, and consequently (iii) theories are conventions that can be tampered with *ad libitum.* See his classic *The Aim and Structure of Physical Theory* (1914; New York: Atheneum, 1962), Part II, Ch. vi, W. V. Quine, *From a Logical Point of View* (Cambridge, Mass.: Harvard University Press, 1953), Ch. 2, and A. Grunbaum, "The Duhemian Argument", *Philosophy of Science*, **27**, 75 (1960).

15.6.7. Suppose two rival theories are tested which, as is so often the case, happen to share some assumptions. Suppose further one of the theories is definitely refuted by experience whereas the other is confirmed. How can we separate the wheat from the chaff in this case? Hint: Simplify the problem by handling the case $T1=\{A1, A\ 2\}$, $T2=\{A2, A3\}$.

15.6.8. Suppose in the foregoing case *T*2 turns out to be confirmed. Should we confidently assert *A*2 and *A*3 or should we keep trying to find fault with them—or else continue working with them in a critical way? *Alternate Problem:* Is it possible to test a scientific theory in an *exhaustive* way, i.e. to subject every one of its formulas or at least every one of its lowest level consequences to actual empirical tests? What if not?

15.6.9. The laboratory or field test of any factual theory presupposes (utilizes) at the very least fragments of physical geometry, optics, mechanics, and thermodynamics. Draw some morals, for in this case the Duchess would be right. *Alternate Problem:* According to our account of theory testing, one or more additional (auxiliary) theories take part in the empirical test of every factual theory. (See text and recall Sec. 8.4.) Compare this view with the popular account according to which theory testing requires the replacement of theoretical concepts by observational ones—i.e. a detheorization rather than the enrichment of the theory under test.

15.6.10. Discuss the claim that the empirical evidence in favor of extrasensory perception is overwhelming, the only thing that remains to be done being to systematize such data into a theory. Hint: Distinguish reports e from evidences e^*. *Alternate Problem:* Every concrete system has a number of traits, just a few of which are covered by the corresponding theories. Consequently any factual theory will yield consequences that are either untestable or false under real circumstances unless they are supplemented by fragments of alternative noncompeting theories. Does this support the thesis of the absolute character of experience and its right to pass judgment on theory without the assistance of it?

15.7. Theory Assaying

Several models of the universe are compatible with the available astronomical data, which are both scanty and imprecise: for all we know the universe might be spatially finite or infinite, space might be uniformly curved or not, and so on. In atomic and nuclear physics we find the partisans of a theory yielding just as much information as experiment can supply, and those who favor the introduction of further hypothetical constructs referring to unobservable properties. In meteorology we meet the view that weather is a result of the interplay of a large number of factors, and accordingly that small variations in any of these variables may be neglected, and the rival view that the atmosphere is an unstable system, so that any small single factor might, say, trigger rain. In biology we find two kinds of theories of mortality: those assuming that the life span of the individual is genetically determined and those which assume that death is the outcome of a long cumulative sequence of small hits—and so forth. There is hardly a scientific field where an important theory reigns undisputed or, at least, where no alternative theories are conceivable. The more important a theory the more rivals it is likely to have. And such rivalry will be a source of progress unless some of the theories are institutionalized into dogmatic schools in imitation of school philosophies.

Two scientific theories can be regarded as *rival* if they deal differently with roughly the same subject matter or problem system. If, in addition, they yield much the same testable consequences or projections, the theories are often called *empirically equivalent,* however conceptually different they may be. Empirically equivalent theories need not be mutually compatible as are the various cartographic representations of the globe: if they were they would be just different formulations of the same theory, i.e. they would be *conceptually equivalent*. Theories that are empirically equivalent but conceptually inequivalent can be generated almost at will. Thus, any physical theory in which the length concept occurs presupposes some metrical geometry, usually Euclid's. The flat space portrayed by the latter can be approximated by infinitely many curved spaces handled by as many Riemannian geometries; for small curvatures or for small space volumes there will be no empirically noticeable differences between the corresponding physical theories, i.e. they will be empirically equivalent under those conditions. Anyhow, disputes over empirically equivalent theories are usually the bloodier, just as fights among sects of the same religion: in either case there is a single set of subjects to be saved.

Experience will be weighty, perhaps decisive at one point, in the evaluation of empirically *inequivalent* theories. But how does one proceed in the presence of two or more empirically equivalent theories? One possible move is to wait and see further empirical evidence: if the theories are really different, if they are not just two equivalent ways of saying the same thing, a situation may eventually arise which at least one of them will not be able to cope with. But this will not be enough: the partisan of the empirically defective theory may resort to patching it up either by retouching some of its initial assumptions or by introducing *ad hoc* hypotheses to save them. (Recall Sec. 5.8). Since this process might be continued forever, some battery of *nonempirical tests* seems to be called for. That is, tests for properties other than agreement with observed fact are needed when two or more theories agree equally well, or almost so, with empirical information.

To ascertain what such desirable properties of scientific theories are is not a matter of purely logical work: we must use the history of science in order to discover the criteria actually at work in the evaluation of scientific theories. Yet this is insufficient, for some such criteria might be wrong. In order to sift out which among them are desirable for the progress of knowledge we must subject them to logical and methodological scrutiny. Let us then begin by taking a look at some important scientific controversies.

Our first case history will be the Ptolemy-Copernicus—or geostatic-heliostatic—controversy. According to eminent philosophers this dispute has not yet been settled nor will ever be settled, because the two "systems of the world" are equivalent: the sole difference between them would be one of complexity and convenience. This alleged equivalence is taken by inductivists as a point in favor of simplicity as the decisive criterion of choice among alternative theories, and is regarded by conventionalists as supporting the view that the search for the simplest theory consistent with experience should replace the search for the truest theory. The heliostatic picture of planetary motions is said to be preferable to the geostatic account merely because it simplifies astronomy—which is not true, as the former view conflicts with common sense (by containing transempirical concepts such as "planet orbit") and uses refined hypotheses of dynamics and subtle tools of calculus, such as the perturbation theory. As a matter of fact the equivalence between the two accounts is very limited: it covers only the geometry of planetary motion, in the sense that either theory can be adopted to account for the apparent positions of the sun and the planets. In every other respect the two theories are inequivalent, as suggested by the following synoptic table:

Ptolemaic "System"	*Copernican System as embedded in Newtonian mechanics*
Conventional combination of cycles epicycles, excentrics, and other intervening variables	Equations of real motion
	↓
	Solutions (referred to the Sun)
↓	↓
Observable orbits	Observable orbits (referred to the Earth)

Let us list the more remarkable differences between the two "world systems" since each may suggest a criterion for assaying scientific theories. First: *systemicity.* The heliostatic system does not deal with each planet separately, as the geostatic does, but introduces the physical concept of solar system and attains thereby a conceptual and methodological unity; on the other hand Ptolemy's was actually a set of theories, one for each planet, and if one went wrong it could be adjusted without modifying the others. Second: *external consistency.* Whereas the Ptolemaic "system" is isolated from the rest of science, the heliostatic system is contiguous to (not just compatible with) dynamics, gravitation theory, and the theory of stellar evolution, in the sense that its main features are not taken for granted but can be explained, some in detail and others in outline, by those other theories. In short, the heliostatic system is embedded in science whereas the

Ptolemaic "system" was not even consistent with Aristotelian physics. Third: *explanatory and predictive powers.* The heliostatic system accounts for facts which the geostatic view ignores, such as the Earth's seasons, the planets' and satellites' phases, the aberration of light, and so on. Fourth: *representativeness.* Far from being a merely symbolic device for linking observations—as the conventionalist and pragmatist view of scientific theory would like—the heliostatic system is a visualizable model of a piece of reality and was meant by Copernicus, Bruno, Galilei and others to be a true image of the solar system. Fifth: *heuristic power.* The heliostatic system stimulated new discoveries in mechanics, optics, and cosmology. For example, it made possible the search for Kepler's laws, the explanation of which was in turn a major motivation of Newton's dynamics, the shortcomings of which prompted some of the ulterior developments. Sixth: *testability.* By not admitting groundless repairs through the addition of *ad hoc* hypotheses devised to save the theory from discrepancies with observation, the heliostatic system lends itself to refutation by empirical data. (A data-fitting model can be adjusted and readjusted without end; a fact-fitting model is not so adjustable.) Moreover, the heliostatic system has been corrected a number of times: first by Kepler, then by the discovery that, owing to the mutual perturbations of the planets, the real orbits are more complex than Kepler's ellipses and even Ptolemy's curves. Seventh: *world-view consistency.* The new theory, which was definitely inconsistent with the official, Christian cosmology, was on the other hand compatible with the new physics, the new anthropology and the new, naturalistic world-view. In fact, the Copernican system was the bridge that led astronomy to physics, it transformed the heavens into a natural object—a piece of machinery—and it showed that Terra was not the basest spot in the cosmos—which in turn enhanced the dignity of man. In short, although in the beginning the two theories were empirically equivalent, empirical differences were worked out in time and, from the start, the theories were conceptually inequivalent and were backed up by conflicting philosophies.

Our second example will be the strife over Darwin's theory of the origin of species, which triumphed over its two main rivals, creationism and Lamarckism, after a long and bitter struggle which in some quarters is hardly over. Darwinism was far from perfect and it did not fit the data better than would a theory of supernaturally guided evolution, for which every datum would count as a supporter, none as a falsifier. It contained several unproved, disputable, or even false assumptions, such as "The units of heredity result from the fusion or blending of the parental contributions", "Natural populations are fairly constant", and "Acquired characters, if favorable, are transmitted to the offspring". The theory had not been tested by direct observation, much less by experiment on living species under controlled conditions; the development of antibiotic-resistant strains of bacteria and *DD7*-resistant insects, industrial melanism in certain butterflies, and competition among the individuals of dense vegetal populations, were observed only decades after *The Origin of Species* (1859) appeared. Darwin's theory was suspiciously metaphorical, having been suggested in part by Malthus' work on population pressure and the competition for food in human societies. Furthermore, Darwin's theory was not inductive but, as every other genuine theory, it contained transempirical concepts. And, as if these sins were not enough to condemn the theory on the strength of the metascientific canons prevailing in the 19th century, Darwin's system was far more complex than any of its rivals: compare the single postulate about the special creation of every species, or Lamarck's three axioms (the immanent tendency to perfection, the law of use and disuse, and the inheritance of acquired characters), with Darwin's theory. The latter was by far less commonsensical and more complex, as it contained among others the following independent assumptions: "The high rate of population increase leads to population pressure", "Population pressure results in the struggle for life", "In the struggle for life the more fertile prevail", "Favorable characters are inheritable and cumulative", and "Unfavorable characters lead to extinction".

The characters which ensured the survival of Darwin's theory despite both its imaginary and real shortcomings seem to have been the following. First: *external consistency*. The theory, though inconsistent with traditional biology and cosmology, was consistent with the new evolutionary geology and the evolutionary theory of the solar system. Second: *extensibility and heuristic power*. The theory was quickly, boldly, and fruitfully exported to physical anthropology, psychology, history, and linguistics—and it was unwarrantedly extrapolated to sociology (social Darwinism) and ontology (Spencerian progressivism). Thirdly: *originality*. Although the idea of evolution was old, the mechanism proposed by Darwin was new and suggested daring fresh starts in all related fields, as well as the very relation among hitherto disconnected fields. Fourth: *testability*. Darwin's theory involved unobservables (e.g., "population pressure" and "evolution") but not inscrutable concepts, such as "creation", "purpose", "immanent perfection", and the like. And it did not involve unscientific modes of cognition such as revelation or metaphysical intuition. Fifthly: *level-parsimony*. No spiritual entity was invoked by Darwin to account for lower-level facts, and no purely physico-chemical mechanism was resorted to either: evolution was treated on its own, biological level—which did not prevent it from eventually taking roots in the cellular and the molecular levels. Sixthly: *metaphysical soundness*. Darwinism was consistent with the genetic postulate—"Nothing comes out of nothing or goes into nothing"—and with the principle of lawfulness, which was on the other hand violated by the creation dogma. But it was inconsistent with the inductivist methodology then dominant and sounded suspect to many on this count. Seventhly: *world-view consistency*. The theory was definitely consistent with the naturalistic, agnostic, dynamicist, progressive and individualistic outlook of the liberal *intelligentsia* on which the recent social and cultural changes (1789, Chartism, and 1848) had made a deep impression. These various virtues of Darwinism overcompensated for its shortcomings and made it worth correcting on several important points until it merged, in the 1930's, with modern genetics, to constitute the synthetic (or neo-Darwinian) theory.

Our third and last example will be the theory of the origin of organisms as the last links of long chains of chemical reactions starting with comparatively simple compounds. This theory is generally accepted although it does not systematize a large body of empirical data: moreover, it has so far not been tested although a start has been made. It is generally felt that it will eventually be confirmed in the laboratory but this may take generations of researchers willing to risk their time in long-term investigations. Anyhow the theory is accepted not because it has a strong inductive support but because it solves an important problem by using well-tested laws of chemistry, biochemistry and biology and because it is expected to guide the synthesis of living beings in the laboratory. The virtues that make up for the lack of direct empirical support are in this case the same which inclined people to favor Darwin's theory: external consistency, fertility, originality, testability, ontological soundness, and compatibility with the naturalistic *Weltanschauung* prevailing in contemporary scientific circles.

As may be gathered from the foregoing, a number of criteria are at play in the evaluation of scientific theories and, in particular, in the discussion of the merits and shortcomings of rival systems. To be sure not all of these criteria are explicitly mentioned, much for the same reason that not all relevant respects are mentioned in the evaluation of men: namely in part for ignorance and in part for hypocrisy. This is particularly so in the case of the philosophical features of scientific theories. Yet if the criteria of theory assaying were always explicitly stated, disputes over the relative merits and demerits of rival scientific theories and rival research programmes would resemble less the religious and political disputes than they do in many cases. Since it is a job of metascientists to uncover the presuppositions, criteria and rules of scientific research, we shall do well to present the chief nonempirical criteria of theory evaluation in an explicit and orderly fashion. These criteria may be grouped into formal, semantical, epistemological, methodological, and ontological. They are:

Formal criteria

1. *Well-formedness:* the formulas of the theory should be well-formed—no gibberish.

2. *Internal consistency:* the formulas of the theory should be mutually consistent (recall Sec. 7.6).

3. *Validity:* the derivations in the theory should follow as closely as possible the patterns set by ordinary (two-valued) logic and/or mathematics.

4. *Independence:* both the primitive concepts and the primitive assumptions of the theory should be mutually independent (see Sec. 7.6).

5. *Strength:* the initial assumptions of the theory should be as strong as truth permits (recall Sec. 5.7).

The first three conditions are mandatory while the remaining ones are desiderata to be satisfied as closely as possible.

Semantic criteria

6. *Linguistic exactness:* minimal ambiguity and vagueness (see Sec. 3.1).

7. *Conceptual unity:* the theory should refer to a definite universe of discourse and its predicates should be semantically homogeneous, connected, and closed (recall Sec. 7.2).

8. *Empirical interpretability:* some of the lowest level theorems of the theory should be interpretable in empirical terms, eventually with the help of further theories (see Sec. 8.4).

9. *Representativeness:* the more representational or "mechanistic" (the less phenomenological) a theory, the deeper it will go beyond appearances, the more effectively it will guide new research and the better testable it will be (see Sec. 8.5).

The first three conditions are mandatory. Mathematization may be regarded as included in (6) but it may also be rated separately. As to representativeness, it is a desideratum that should not preclude the construction of black box theories, indispensable in the early stages and in technology.

Epistemological criteria

10. *External consistency:* compatibility with the bulk (not the whole) of reasonably well-tested knowledge; if possible, continuity with it (see Sec. 7.6). This requirement extends to theories the requirement of foundation stipulated for hypotheses. Example: psychological theories should be consistent with physiology and, if possible, they should employ some of the latter's results either explicitly or as presuppositions.

11. *Inclusiveness:* the theory should solve to a good approximation a substantial part of the problems that stimulated its construction (recall Sec. 9.6). Also, the better theory will be the one which can answer the more ambitious questions. But it should not attempt to solve every possible problem: this must be left to pseudoscience.

12. *Depth:* other things being equal, deep theories, involving fundamentals and basic mechanisms are preferable to shallow systems that do not commit themselves to any unobservable mechanism (see Sec. 8.5). But we need both bones and skin to make a growing organism; thus, statistical mechanics does not enable us to dispense with thermodynamics.

13. *Originality:* bold theories, containing shocking (but not wild) high-level constructs, yielding unheard-of projections and unifying apparently unrelated fields, are more valuable than safe, down-to-earth systems (recall Sec. 10.4). Theories systematizing what is known are certainly necessary, but the inspiring revolutions in knowledge have been those consisting in the introduction of theories which, far from packaging what was known, have forced us to think in new ways, to

formulate new problems and to seek new kinds of knowledge: in short, all-round original theories.

14. *Unifying power:* capacity to encompass hitherto isolated domains. Examples: Newtonian mechanics (unification of terrestrial and celestial mechanics), Maxwellian electromagnetic theory (unification of the theories of electricity, magnetism, and light), and utility theory (applicable to psychology, economics, and management).

15. *Heuristic power:* a new theory should suggest and even guide new research in the same or in allied fields. Fertility is often a bonus of representativeness and depth, but it is not necessarily linked to truth: true theories may be barren for being shallow or uninteresting and, conversely, false theories may be fruitful for allowing the statement of challenging problems and the design of enlightening experiments: think of the fertility of the mechanistic models of life and mind as contrasted to the vitalist and spiritualist doctrines.

16. *Stability:* the theory should not tumble down in the face of the first new datum but should be capable of growing, up to a point, along the same line: it should be capable to "learn" from the new experience it was not able to predict. Rigid theories, on the other hand, are apt to succumb at the first unfavorable evidence because they were designed to slavishly account *ex post facto* for a handful of data. But the elasticity or stability of theories has a limit: they should be neither insensitive to new experience nor too accomodating to it and, in particular, they should not be consistent with contrary pieces of evidence. Good theories, like good cars, are not those that cannot collide but rather those which can stand some repairs.

The first two requirements—external consistency and coverage—are mandatory. The remaining are desiderata which only great theories fulfil. Projective power is not an independent property: it results from coverage and originality (recall Sec. 10.4).

Methodological criteria

17. *Testability:* the theory, its presuppositions, and even the techniques employed in its test must be capable of test: they must all be open to scrutiny, control and criticism. The theory as a whole must be both confirmable and refutable, although it may contain a few hypotheses which, individually, are only confirmable (recall Sec. 5.6).

18. *Methodological simplicity:* if the tests proposed for the theory are so complicated that there is no fear of refutation within the fore-

seeable future, it will not be possible to pass judgment on the inclusiveness and the stability of the theory. Yet methodological simplicity must not be understood in an absolute sense: it means just technological feasibility, in principle, of empirical test.

The first condition is mandatory; notice that, by contrast with the case of single hypotheses, we require that theories as a whole be both confirmable (applicable) and refutable (conceivably at variance with data, even if no refuter has so far appeared). On the other hand, the condition of methodological simplicity should be required to a moderate extent, as unexpected technological innovation may reduce the waiting time; moreover, the theory itself may make such important claims as to stimulate new technological developments primarily aimed to test or apply the theory itself.

Ontological criteria

19. *Level-parsimony:* the theory should be parsimonious in its reference to levels other than those directly concerned. Particularly, the higher levels of reality should not be appealed to if the lower are enough, and distant levels should not be introduced, whenever possible, without the intermediate ones (see Sec. 5.9).

20. *World-view compatibility:* consistency with the prevailing outlook or, at least, with the common core of the world views held by the most competent scientists at the moment.

Condition 19th is desirable, but parsimony in the multiplication of levels does not mean forced reduction to a single level: a number of levels may have to be invoked in order to explain a single fact, such as the explosion of an atomic bomb or the writing of a good novel. World-view compatibility, on the other hand, is not a desideratum as long as the world-view itself is not scientific. If applied rigidly, the requirement of world-view compatibility may kill theories (and even theoreticians) at variance with a given outlook, particularly if this is centered in a school backed up by political power. Yet this criterion filters some crackpot theories and, for better or for worse, it does intervene in the evaluation of scientific theories. The remedy does not reside in demanding the neutrality of science vis à vis world-views and philosophies. Firstly, because it is impossible to strip a scientist off his general outlook. Secondly, because world-views and philosophies are among the stimuli to theory construction—and often also among its inhibitors. There seems to be but one way to prevent world-views and philosophies from distorting science: namely, to consistently embed scientific theories in definite world-views and philosophies, to control the latter by the former, and to resist internal crystallization and external (chiefly political) guidance.

The above criteria are not all mutually independent. Thus, depth depends on strength and heuristic power depends on depth, representativeness, originality, and unifying power. Nor are they all mutually compatible. Thus inclusiveness is greater for phenomenological and traditionalist theories than for representational and revolutionary ones. In the third place, none of these requirements can probably be satisfied entirely. For example, syntactical correctness and linguistic exactness may not be perfect in the initial stages; the important thing is that they can be improved. In the fourth place, some of the above requirements are double-edged, particularly so methodological simplicity and world-view compatibility. In all these respects the evaluation of scientific theories is very much like moral evaluation: in either case some desiderata are mutually dependent, others are mutually incompatible, still others unattainable ideals, and finally some are double-edged. This is one reason for doubting that mechanically applicable *decision procedures* (true-false tests) will ever be invented to effect clear-cut choices among competing theories or alternative courses of action—except in trivial cases. In this field, decisions are made in the absence of decision *rules.* In many cases, perhaps in most cases of rivalry among scientific theories, the decision may require both scientific and philosophic "sound judgment", not by one man or even by a committee of sages, but by generations of experts and competent critics. It would be silly to deplore this situation for, after all, science is a social enterprise and we are not chasing the definitive perfect theory. Not only the process of theory construction with its necessary simplifications and its bold leaps beyond experience, but also the complexity of the battery of theory assaying, show that there *can* be no perfect factual theory. Stated positively: there will always be room for new, better theories. Let no *ism* keep us from trying them and no pride from caning them.

We have discussed a set of metascientific principles fulfilling two functions: (i) they restrict the number of theories worth considering at any stage, and (ii) they constitute a battery of nonempirical tests. What these tests test for is factual *truth,* i.e. conformity with fact. There is not a single test—let alone an empirical one—of factual truth: every one of the above constitutes a *nonempirical* criterion of factual truth. In order to estimate the degree of truth of factual theories we have to use them all, in conjunction with multiple (numerous and varied) and tough empirical tests. A theory found to be approximately true (semantical aspect) is normally adopted or accepted (pragmatic aspect), hence believed (psychological aspect) to some extent. Nearly false theories, on the other hand, may still be adopted for restricted domains or for technological purposes—although they will not be believed, at least if known to be nearly false. Theories are accepted *faute de mieux.* This explains why use, or acceptance—a pragmatic category—does not occur among the above tests for truth.

A subtler analysis should disclose further evaluation criteria, and the advancement of both science and metascience is likely to correct some of the current criteria and introduce new ones. The large number of evaluation criteria may seem bewildering to the student used to read about simple truth conditions for isolated propositions. There are no such clear-cut text-book "truth conditions" for the systems of hypotheses occurring in real science: there are consequently no decision criteria, and no hint that there can be any. The most that can be secured is a set of numerous and nearly independent *controls,* which are singly insufficient to guarantee complete truth but which can jointly detect partial truth. What is important about these tests is not that they supply decision rules concerning our acceptance or rejection of scientific theories as if these were good or bad eggs: this they cannot do. What they do is to show the extent to which any factual theory succeeds and the extent to which it fails, and by so doing they can occasionally show new research lines likely to be rewarding.

This brings the book to its end. The preceding views on theory testing conflict with the popular idea that theories are tested more or less like fertilizers, whence the whole methodology of science would be reducible to statistics. They also conflict with the various philosophical schools, if only because every school commits itself to a set of fixed tenets rather than to an increasingly ambitious though attainable goal and a self-correcting method. Yet certain key contributions of a number of mutually conflicting philosophical views have been employed in building the image of scientific research presented in this treatise. From realism we have taken the theses that there is an external world, that scientific research aims at depicting it, and that this is an endless task, as every picture of the world misses some of its traits and adds others which are fictitious. From rationalism, that such pictures are symbolic rather than iconic, hence original creations and not photographs; that logic and mathematics are a priori, and that signs are meaningless unless they stand for ideas. And from empiricism, that experience is a touchstone of factual theories and procedures, that there can be no certainty in them, and that philosophy ought to adopt the method and the standards of science.

The position reached, scientific realism, is therefore a sort of synthesis of rationalism and empiricism—without, it is hoped, the extremism and rigidity characterizing every philosophic school. Philosophic *isms* are the grave of inquiry, for they have got all the answers, whereas research, either scientific or philosophic, consists in wrestling with problems rejecting dogmatic strictures. Hence let us welcome any alternative view giving a more accurate account of scientific research *in vivo* or promoting it more effectively—for these are the ultimate tests of every philosophy of science.

Problems

15.7.1. Show how a decision was eventually reached concerning a pair of rival scientific theories, such as, e.g., the following: caloric vs. kinetic theory of heat, phlogiston theory vs. oxidation theory, action at a distance vs. near action (in electrodynamics or in gravitation theory), continuum theory vs. atomic theory, creationism vs. evolutionism, spontaneous generation vs. seminal theory. *Alternate Problem:* Whereas biologists, psychologists and sociologists are aware of the multiplicity of theoretical viewpoints or tendencies, hence of the importance of discussion and criticism, physicists usually disclaim the existence of such a variety in physics. Is it true that there are not different tendencies in physics? If so, why? If not, how can the resistance to acknowledge such a diversity be accounted for? Because of ignorance, naiveté, shame? Because of a Baconian philosophy of science according to which it is fact, not theory, what matters in science? Because the deviants from the main trend are always in a small minority and nobody wants to swim against the current?

15.7.2. Averpes, and later on Copernicus, maintained that astronomical theories should meet two requirements: (i) they should account for appearances and (ii) they should accord with physics. Which "system of the world, Ptolemy's or Copernicus, matched better with Aristotelian physics and which with Newtonian physics? And what role, if any, did considerations of external consistency play in their evaluation? *Alternate Problem:* Newton's corpuscular theory of light was criticized at the beginning of the 19th century from two opposite angles: by *Naturphilosophen* like Goethe and by physicists like Young and Fresnel. What were these criticisms?

15.7.3. Comment on B. Barber's "Resistance by Scientists to Scientific Discovery", repr. in B. Barber and W. Hirsch, Eds., *The Sociology of Science* (New York: Free Press, 1962), where the following internal factors of resistance to scientific change are pointed out: (i) prestige of an old theory; (ii) prestige of a methodological tenet; (iii) religious ideas of scientists; (iv) low professional standing of innovators; (v) tribalism of specialists; (vi) schoolism. *Alternate Problem:* Examine some of the recent debates over the foundations of quantum mechanics, cosmology, evolutionary biology, economics, political science, or history.

15.7.4. Comment on the following idea of G. Spencer Brown, in *Probability and Scientific Inference* (London: Longmans, Green & Co., 1957), p. 23: One and the same set of observations can be accounted for in at least two ways: either in scientific terms or in terms of miracles, magic, spirits, etc. "Some people would like to ask which of these kinds of description was the *true* one; but there is no sense in this. Both descriptions are ways of saying what we observe, and can therefore be true". Would this statement stand if one added the requirements that theories must be testable, that even the tests must be testable, and that no theory contradicting basic scientific principles is acceptable—on pain of falling into contradiction? And is it true that scientific theories describe observations? *Alternate Problem:* Einstein's hypothesis of light quanta was received coolly because, although it explained the photoelectric effect, it conflicted with the wave theory of light. Einstein himself called it "a heuristic viewpoint". Study this conflict and its eventual resolution via the building of quantum electrodynamics.

15.7.5. Should theories be evaluated apart from the bulk and depth of the problem system that originated them? In particular, should a modest but accurate theory be always preferred to an ambitious but comparatively inaccurate theory? Recall the discussion on the volume of theories in Sec. 9.6. *Alternate Problem:* Examine the claim that the next computer generation will be able to form new concepts and hypotheses. Is it reasonable to prophesy that they will invent algorithms, and thus program themselves? What grounds are there to prophesy that computers will go beyond inductive generalizations of the type of curve fitting? What, short of a brain soaked with a multiple cultural heritage, could invent the constructs typical of theories proper? And what kind of theory assaying could be performed by a system lacking any philosophical bias?

15.7.6. Work out and illustrate the following scheme concerning the varieties of discrepancies within the scientific community. (i) Linguistic: due to ambiguity, vagueness, inadvertent shift of meaning or of context, etc. (ii) Concerning the record: resulting from either different means of observation or different interpretations of data. (iii) Regarding the truth value of hypotheses and theories. (iv) Valuational: different emphases on the value of problems, lines of research, techniques, results, etc. (v) Philosophical.

15.7.7. In 1911 E. Rutherford conceived a planetary model of the atom with the sole purpose of explaining the scattering of alpha and beta rays by thin metal foils. Immediately after performing that feat he turned to other problems and scarcely anybody grasped the size of his conjecture. Only N. Bohr suspected that this model, if conjoined with M. Planck's quantum hypothesis, might explain the optical and chemical properties of atoms, and undertook to build the corresponding theory. Discuss this story in terms of extensibility, predictive power, and boldness. *Alternate Problem:* Discuss N. Bohr's *correspondence principle,* which prescribes that the results of quantum theory for macroscopic systems must approach the corresponding results of classical theory. Is this principle extensible to other fields? How is it related to the principle of external consistency? Does it function as a test for truth? And is it scientific or metascientific?

15.7.8. Is the economy of presuppositions (background simplicity) an absolute desideratum for scientific theories? In particular, consider the following questions. Is the economy of presuppositions favorably or unfavorably relevant to empirical testability? Is background simplicity compatible with external consistency? Should background simplicity be demanded from theories consisting in the application of other theories? *Alternate Problem:* Evaluate the claim of the constructivist-relativist philosophers and sociologists of science that there are no objective criteria for the choice among rival theories. See their main periodical, *Social Studies of Science*, and M. Bunge, "A critical examination of the new sociology of science", *Philosophy of the Social Sciences* 21, 524 (1991); 22, 46 (1992).

15.7.9. The number of problems a theory can handle (formulate and solve) should be a criterion of theory evaluation. A theory may be good in many respects but it may handle a narrow range of problems. The domain of facts a theory covers may be restricted either because the theory cannot cope with them in principle or in practice (owing, say, to computational difficulties). Is this an independent trait of theories or does it follow from the traits dealt with in the text? *Alternate Problem:* Is it possible to rank competing scientific theories, i.e. to order them linearly, as regards overall merit? And does the proposal to compute their probability (or their improbability, as the case may be) satisfy the desideratum of ranking?

15.7.10. Work out and illustrate the following scheme concerning the kinds of novelty a theory may exhibit. (i) The theory is a reformulation of a known theory. (ii) The theory generalizes a known theory. (iii) The theory is formally analogous to a known theory. (iv) The theory consists in the merging of various fragments of known theories. (v) The theory is altogether new either because it contains radically new concepts or because it relates known concepts in an unheard-of way. *Alternate Problem:* Discuss and illustrate the distinction drawn by T. S. Kuhn, in *The Structure of Scientific Revolutions* (Chicago: University of Chicago Press, 1962), between *normal* scientific research—which fits some pattern or paradigm—and *extraordinary* research, which brings about a scientific revolution, consisting in the replacement of prevailing paradigms by entirely new ones, enabling one to pose problems of a fresh kind. Elucidate the concept of paradigm and relate it to the concepts of theory, background knowledge, and "spirit of the times".

Bibliography

Blalock jr., H. M.: Causal inferences in nonexperimental research. Chapel Hill: University of North Carolina Press 1964.

Bogdan, R. J., Ed.: Local induction. Dordrecht-Boston: Reidel 1976.

Bunge, M.: The myth of simplicity, Chs. 7 and 9. Englewood Cliffs, N. J.: Prentice Hall 1963.

——— (Ed.): The critical approach, Part I. New York: Free Press 1964; especially the articles by P. Bernays, J. Giedymin, and J. W. N. Watkins.

Carnap, R.: Logical foundations of probability. Chicago: University of Chicago Press 1950.

Czerwinski, Z.: Statistical inference, induction and deduction. Studia logica **7**, 243 (1958).

Duhem, P.: The aim and structure of physical theory (1914), Part II, Ch. VI. New York: Atheneum 1962.

Fisher, R. A.: The design of experiments, 6th ed. Edinburgh and London: Oliver and Boyd 1951.

——— Statistical methods and scientific inference. Edinburgh and London, Oliver and Boyd 1956.

Goodman, N.: Fact, fiction, and forecast. London: Athlone Press 1954.

Harris, M. : The rise of anthropological theory. New York Thomas Y. Crowell 1968.

Hume, D.: A treatise of human nature (1739, various editions), Book 1, Part iii, Sec. vi.

Jeffreys, H.: Scientific inference, 2nd ed. Cambridge: Cambridge University Press 1957.

Kempthorne, O. and J. L. Folks: Probability, statistics and data analysis. Ames: Iowa State University Press 1971.

Keynes, J. M.: A treatise on probability. London: Macmillan 1929.

Kneale, W.: Probability and induction. Oxford: Clarendon Press 1949.

Kuhn, T. S.: The structure of scientific revolutions. Chicago: University of Chicago Press 1962.

——— The essential tension. Chicago: University of Chicago Press 1977.

Kyburg jr., H. E., and E. Nagel (Eds.): Induction: Some current issues. Middletown, Conn.: Wesleyan University Press 1963.

Lakatos, I.: The problem of inductive logic. Amsterdam: North-Holland 1968.

Lakatos, I. and A. Musgrave, Eds.: *Criticism and the growth of knowledge*. Cambridge: Cambridge University Press 1970.

Merton, R. K.: Social theory and social structure, rev. ed., Part I. Glencoe, Ill.: Free Press 1959.

Pólya, G.: Mathematics and plausible reasoning, 2 vols. Princeton: Princeton University Press 1954.

Popper, K. R.: The logic of scientific discovery (1935), London: Hutchinson 1959.

——— Conjectures and refutations. New York: Basic Books 1963.

Reichenbach, H.: The theory of probability. Berkeley and Los Angeles: University of California Press 1949.

Rescher, N. : Objectivity. Notre Dame: University of Notre Dame Press 1997.

Russell, B.: Human knowledge: Its scope and limits. New York: Simon and Schuster 1948.

Wright, G. H. von: The logical problem of induction, 2nd ed. Oxford: Blackwell 1957.

Afterword: The Goals and Means of the Philosophy of Science

I submit that any authentic philosopher of science has two aims, one epistemic, the other pragmatic. The former is to understand scientific research and some of its findings. The other goal is to help scientists hone some concepts, refine some of theories, scrutinize some methods, unearth philosophical presuppositions, resolve controversies, and plant doubts about seemingly uncontroversial points. The two goals complement one another.

To achieve these aims, philosophers have traditionally adopted either of two methods: the direct of looking at science, and the indirect of looking at the philosophy of science. The approach adopted in this work is direct rather than the usual one of discussing the views of other philosophers of science. No doubt, many of these opinions are interesting and worth examining. But I suggest that such examination falls to the historian of the philosophy of science rather than to the philosopher of science.

To understand how science works one must look at it in the face—to indulge in metaphor. Otherwise one will not contribute anything true, let alone original, to the knowledge of scientific knowledge. Moreover, one will be unable to help scientists tackle or even recognize the philosophical problems that raise their head at critical junctions.

The direct look is rewarding because it can reveal the way scientists actually work, and what drives them in addition to peer recognition. It confirms certain elementary truths that often get lost in the rarefied heights of philosophy, as well as in the "post-modern" or punk sociology of knowledge. These truisms are that scientists explore the real world to find out how it really works; that to conduct such exploration they invent and discuss ideas, instruments and procedures, and perform complicated calculations and experiments; that they check and recheck whatever they create or use; that they always start from a body of background knowledge, some of which they question, and which they try to enrich; that they normally abide by rigorous rules of conduct; and that, although occasionally basic research produces findings that may be misused in other fields, the aim of scientific research is not profit or power, but objective truth—which, however, is only partial in most cases.

Such a view of basic research as a disinterested and disciplined quest for truth is particularly important at a time when many students suspect that it is only a tool of domination and destruction. This suspicion derives from two confusions: those of science with technology, and of the latter with business or politics. To be sure, technology draws on science, and both business and government use technology. But it is technologists, not scientists, who design useful or destructive artifacts, from pharmaceuticals and dish-washers to weapons and mass-persuasion techniques. And it is business or government, not technology, that manufactures or uses such artifacts. So, if anyone is to blame for the evil uses of technology, it is people in power, not scientists: these are powerless and seek only knowledge for its own sake. When they don't, their peers do not regard them as bona fide scientists. For example, if a scientist cheats, he is banned from his scientific comunity; and if he craves power more than truth, he is kicked up to the administration.

Yet, it would be unrealistic to suppose that scientific research is conducted in a social vacuum, so that its history can be regarded as a string of disembodied ideas. In fact, serious sociologists and historians of science have shown conclusively that science interacts with technology and ideology and, through these, with the economy and the polity. For example, further research on the relation between mood and neurotransmitters such as serotonin is likely to enrich not only neurobiology but also pharmacology and psychiatry. In turn, useful findings in the latter are bound to attract the interest to the pharmaceutical industry, which is subject to government regulations. By contrast, other types of research are likely to be starved or even suffocated by certain governments, for being thought to be useless or even to question the dominant ideology. Thus investigations on the origin of life, the neurophysiology of the mind, the connection betwen religion and politics, the virtues and flaws of the market, or the history of the ruling party, have been discouraged or even banned altogether by the powers that were.

In short, science, technology, the economy, the polity, and ideology may be pictured as forming a pentagon whose sides symbolize flows of information or influence. And the philosophers, sociologists or historians of science or technology wishing to be useful as well as truthful will place themselves at the center of that pentagon. By so doing they will avoid missing or confusing any of the five sides. And only from the center may they succeed in crafting a realistic view of science or technology.

The philosophical stand adopted in this work is scientific realism. This is a combination of two rival classical philosophical traditions: rationalism and empiricism. Scientific realism encourages the construction of theories capable of accounting for empirical data, as well as the search for data in the light of theories. Hence it discourages unbridled speculation as well as mindless data-gathering. Scientific theories should be subjected to reality checks, and some data should be relevant to high-level theories. For instance, ambitious social scientists do not remain satisfied with establishing the occurrence of social changes of some kind. They imagine mechanisms explaining such processes, and they check theories with the help of social indicators that bridge theory to data.

Scientific realism emphasizes the constructive nature of ideas, in particular concepts, problems, rules, and experimental designs. It makes room for conventions, such as definitions, but stresses that scientific theories are expected to represent their referents as accurately as possible. It demands the justification—conceptual, empirical, or both—of all claims to knowledge. However, it is fallibilist as well as meliorist. That is, it recognizes no final truths except in logic and mathematics; but it also trusts that it is possible to correct errors. Scientific realism is also scientistic, in that it regards the scientific approach as the most likely to lead to general and deep truths, in particular scientific laws. Finally, scientific realism is systemic: It places every question and every answer in some system, such as a description, classification, theory, or a well-designed empirical research project.

Scientific realism is an alternative to all of the major schools in the philosophy of knowledge. In particular, it conflicts with radical rationalism as well as with irrationalism; with intuitionism and pragmatism; with radical empiricism and logical positivism; and with Popper's critical rationalism as well as with constructivism-relativism. Let us take a quick look at the salient differences between these schools and scientific realism.

At first sight it would seem that every problem of knowledge can be approached with the exclusive use of either of two "sources": reason and experience. For example, whereas mathematical problems rely on pure reason, problems in anthropology call for observation. However, this does not imply that observation suffices in anthropology or any other branch of empirical science. Indeed, empirical work is worthless unless guided by some hypotheses. After all, to find something we must know beforehand what to look for, where we may find it, and how we may recognize it if stumbling upon it, thus putting an end to that particular search. And the meshing in of data with generalizations, as well as the weighing of the worth of either, requires a modicum of rational argument.

The preceding suffices to indict all kinds of irrationalism, from Neo-Platonism to existentialism and "post-modernist" relativism-constructivism, passing through intuitionism and pragmatism. However, this is not to deny either the existence of the non-rational aspects of scientific research (unlike its depurated results), or the value of insight. I am merely stating that irrationalism is worse than useless to get to know things: It is the greatest obstacle to the exploration of the unknown—in particular, the scientific exploration of intuition.

Note that radical empiricism too is a kind of irrationalism, even when combined with the fashionable cult of data procesing. Scientific experience—observation, measurement, and experiment—is reasoned. Indeed, it is designed, planned and interpreted in the light of hypotheses or even entire theories. Moreover, the data resulting from such empirical procedures are no more sacrosanct than conjectures. Indeed, they are usually affected by errors of some kind or other, which can be partially corrected with the help of further doses of theory, statistical analysis, or improved experimental design. Experimental scientists know this and act accordingly.

The upshot is that scientific research combines reason with experience. This being so, we need a theory of knowledge combining the valuable components of rationalism and empiricism. Such synthesis has been tried several times before. Thus, Kant attempted to join Leibniz's rationalism with Hume's empiricism. But I submit that he chose the wrong halves of each: Leibniz's apriorism and Hume's phenomenalism. Worse yet, he glued them with intuitionism, on top of which he decreed that neither psychology nor the study of society could ever become scientific. Not surprisingly, Kant's philosophy turned out to be useless in mathematics and natural science, and obnoxious in social studies.

Another attempt to synthesize rationalism with empiricism was logical empiricism (or neo-positivism). This school admitted the formal nature of mathematics, but it offered no viable philosophy of mathematics. It recognized the need for scientific theory, but attempted to eliminate theoretical concepts in favor of empirical ones. Besides, it was afraid of metaphysics—as if scientists could dispense with the metaphysical postulates of the autonomous existence and lawfulness of the external world. Not surprisingly, there are no logical positivists left in the philosophical community. However, when discussing metatheoretical problems, most scientists adopt a positivist stance. For example, they claim that every scientific concept ought to be given an "operational definition". Upon examination this claim is nothing but the correct demand to anchor constructs to data via indicators.

Popper's was a third attempt in the same direction. He corrected some empiricist mistakes, in particular the myths that one always starts by observing and then makes inferences with the help of inductive logic. But Popper's view of science is untenable, for holding that scientific research is just a trial-and-error process identical with a lowly animal's; that one should never ask questions of the "What is?" kind; that positive evidence is worthless; that empirical operations are only good to refute theories; that all hypotheses, even the well-tried ones—such as that all atoms contain protons—may turn out to be false; that all questions of meaning ought to be dropped; that there is such thing as knowledge without a knower; and that science is disjoint from metaphysics. Moreover, Popper's adherence to psycho-neural dualism is inconsistent with neuropsychology. And his defense of methodological individualism in social studies, while useful to discredit holism, has discouraged research into social systems—which is what the social scienists do.

Finally, there is constructivism-relativism, or the view that every "scientific fact" is constructed by some scientific community, whence there are no objective and cross-cultural truths, but only local conventions. I submit that this view, currently fashionable among literary critics, has a grain of truth, namely the thesis that scientific research must be placed in its social context. But the rest is utterly false. It derives from a conflation between objective facts and our ideas about them. It offers no evidence for its thesis that all science, even mathematics, has a social content. It fails to account for empirical tests, without which science would not be such. It ignores that all scientists act on the assumption that scientific truths and methods can cross all sexual, ethnic and political borders: that there is no such thing as masculine (or feminine), Aryan (or Jewish), or bourgeois (or proletarian) science. Finally, for all its talk of social context, constructivism-relativism fails to explain its own success among humanities students concerned about environmental and social issues.

So much for some of the best known philosophies of science. Because of their failure to account for scientific research, I have advanced the version of scientific realism argued for in the present work. Readers interested in special topics in the philosophy of science, in the perspective of scientific realism, are urged to consult some of the references below.

References

Bunge, Mario. 1959. *Causality.* Cambridge MA: Harvard University Press. Reissued by Dover in 1976.

———. 1959. *Metascientific Queries.* Springfield IL: Charles C. Thomas.

———. 1962. *Intuition and Science.* Englewood Cliffs NJ: Prentice-Hall. Reissued by Greenwood Press in 1975.

———. 1963. *The Myth of Simplicity.* Englewood Cliffs NJ: Prentice-Hall.

———. 1973. *Philosophy of Physics.* Dordrecht-Boston: Reidel.

———. 1973. *Method, Model, and Matter.* Dordrecht-Boston: Reidel.

———. 1974. *Sense and Reference.* Dordrecht-Boston: Reidel.

———. 1974. *Interpretation and Truth.* Dordrecht-Boston: Reidel .

———. 1977. *The Furniture of the World.* Dordrecht-Boston: Reidel.

———. 1979. *A World of Systems.* Dordrecht-Boston: Reidel.

———. 1980. *The Mind-Body Problem.* Oxford-New York: Pergamon Press.

———. 1981. *Scientific Materialism.* Dordrecht-Boston: Reidel.

———. 1987. *Philosophy of Psychology* (with Ruben Ardila). New York: Springer.

———. 1989. *Ethics.* Dordrecht-Boston: Reidel.

———. 1996. *Finding Philosophy in Social Science.* New Haven CT: Yale University Press.

———. 1998. *Social Science Under Debate.* Toronto: University of Toronto Press.

Mahner, Martin and Mario Bunge. 1997. *Foundations of Biophilosophy.* Berlin-Heidelberg-New York: Springer.

Author Index

Subject Index

Science and Technology Studies

This series includes monographs in the philosophy, sociology, and history of science and technology. "Science" and "technology" are taken in the broad sense: the former as including mathematics and the natural and social sciences, and the latter the social technologies or policy sciences. It is hoped that this series will help raise the level of the current debate on science and technology, by sticking to the standards of rationality and the concern for empirical tests that are being challenged by the current Counter-Enlightenment wave.

Printed in the United States
by Baker & Taylor Publisher Services

Printed in the United States
by Baker & Taylor Publisher Services